THE INTERNATIONAL SERIES OF
MONOGRAPHS ON CHEMISTRY

THE INTERNATIONAL SERIES OF
MONOGRAPHS ON CHEMISTRY

BIOSYNTHESIS OF INDOLE ALKALOIDS

by

ATTA-UR-RAHMAN

and

ANWER BASHA

CLARENDON PRESS · OXFORD · 1983

CHEMISTRY

6659-7213

Oxford University Press, Walton Street, Oxford OX2 6DP

London   Glasgow   New York   Toronto
Delhi   Bombay   Calcutta   Madras   Karachi
Kuala Lumpur   Singapore   Hong Kong   Tokyo
Nairobi   Dar es Salaam   Cape Town
Melbourne   Auckland

and associate companies in
Beirut   Berlin   Ibadan   Mexico City   Nicosia

Published in the United States by
Oxford University Press, New York

British Library Cataloguing in Publication Data

Atta-ur-Rahman
   Biosynthesis of indole alkaloids.—(The
International series of monographs on chemistry;7)
1. Indole
I. Title   II. Basha, Anwer   III. Series
547'.593   QD401

ISBN 0-19-855610-1

Library of Congress Cataloguing in Publication Data

Rahman, Atta-ur-, 1942-
   Biosynthesis of indole alkaloids.

   (The International series of monographs on
chemistry ; 7)
   Includes bibliographical references and index.
   1. Indole alkloids—Synthesis.   I. Basha,
Anwer, 1947-   .   II. Title.   III. Series.
QP801.I45R256   547.7'2   81-22591
ISBN 0 19 855610-1   AACR2

Printed in Great Britain by
Butler & Tanner Ltd, Frome and London

Foreword

The indole alkaloids are a compact, if complex, group of natural products having, in many cases, a long history. This past interest derived initially from pharmacology and toxicology, though as time went on structural elucidation presented a growing challenge to the ingenuity of the organic chemist. Indeed it is fair to say that a large part of classical heterocyclic chemistry owes its origin to alkaloid degradative studies. In recent times we have, of course, seen X-ray diffraction becoming the preferred tool for structure determination — perhaps to the regret of the organic chemist.

As more structures became established many ingenious ideas were put forward concerning structural relationships between the various families of alkaloids and other natural products; the names of Sir Robert Robinson and R. B. Woodward are familiar in this context. But these notions could be no more than speculative, and it is only in the past twenty years or so that studies of the precise pathways of biosynthesis have been made. These studies have proved a triumph for the employment of isotopic tracers in plant-feeding experiments and it is with these studies that the authors of this work are mainly concerned.

There are, of course, some problems still remaining; but this book is most timely in presenting a comprehensive, up-to-date, and detailed picture of a field in which a satisfying overall view is now possible.

Cambridge, 1982   *John Harley-Mason*

PREFACE

The indole alkaloids, with their highly complex structures, have presented organic chemists with a challenging problem regarding their mode of biosynthesis. With the advent of isotopic labelling techniques, the major biosynthetic pathways to the various classes of indole alkaloids have been painstakingly unravelled during the last two decades. We therefore felt that the subject had reached a sufficient state of maturity to deserve a full-length monograph. While some gaps still exist and the biosynthetic pathways to many indole alkaloids remain to be properly elucidated it is hoped that this monograph will focus attention on these unresolved problems and serve as an incentive for further efforts in this field. A number of plausible routes to several alkaloidal systems have been proposed by the authors, and no doubt the precise pathways to such substances will be revealed in the course of time.

The first chapter describes the labelling experiments establishing the monoterpenoid origin of the non-tryptophan moiety of three major classes of indole alkaloids. The second chapter classifies indole alkaloids according to their structural types, and the structures of representative examples are presented. The following describe the biosynthesis of the *Corynanthe–Strychnos*, *Aspidosperma–Hunteria*, and *Iboga* alkaloids. The last section of Chapter 5, and Chapters 6 and 7 discuss the biosynthesis of other complex indole alkaloid systems such as the ergot alkaloids and binary indole alkaloids. A table of incorporation data and a table of occurrence of indole alkaloids in plants and micro-organisms according to their structural types are presented in the last two chapters. With over 1500 chemical formulae about 900 references to the original literature, it is hoped that the monograph will provide a fairly comprehensive account of the subject and prove useful both to graduate students and research scholars interested in this fascinating field of chemistry. The literature is covered up to June 1981.

We are grateful to Professor A. R. Battersby and Professor D. Arigoni for supplying unpublished results. We are also indebted to Dr Zaheer M. Khan for a supply of reprints. We wish to thank Mr Amin Siddiqui and Miss Atiya Abbasi for their painstaking efforts in respect of the drawing of structures. Finally we wish to acknowledge our indebtedness to a number of useful reviews by Professor A. I. Scott, Professor M. Hesse, and Professor G. Cordell which greatly facilitated our work.

Karachi and   A–U–R
Chicago   A. B.
May 1982

CONTENTS

MONOTERPENOID ORIGIN OF INDOLE ALKALOIDS

1.1 Introduction

The indole alkaloids comprise a large and complex group of naturally occurring organic compounds possessing the indole or dihydroindole (indoline) nucleus. They include such physiologically active compounds as strychnine [1], a convulsant poison, reserpine [2], a hypotensive sedative agent, and lysergic acid, the dimethylamide derivative [3]of which is the powerful hallucinogen LSD capable of inducing symptoms similar to

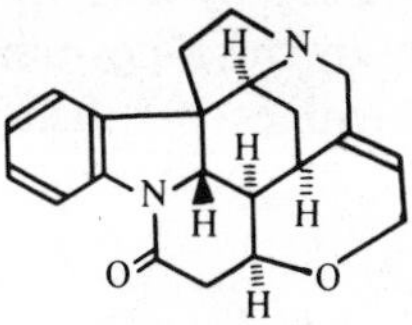

[1]

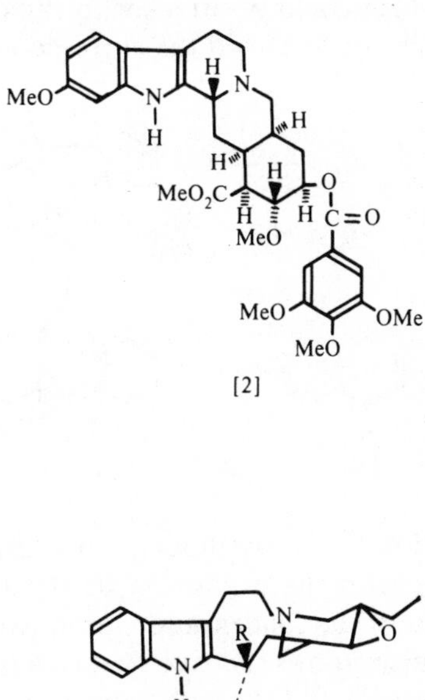

[2]

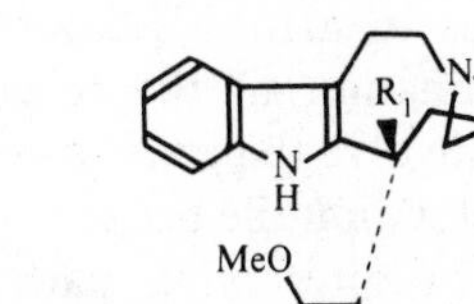

[3]

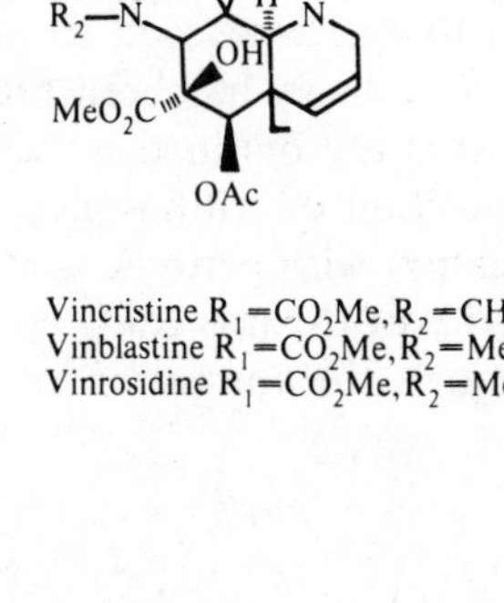

Vinleurosine R=CO$_2$Me [7]

Vincristine R$_1$=CO$_2$Me,R$_2$=CHO,R$_3$=OH,R$_4$=Et [4]
Vinblastine R$_1$=CO$_2$Me,R$_2$=Me,R$_3$=OH,R$_4$=Et [5]
Vinrosidine R$_1$=CO$_2$Me,R$_2$=Me,R$_3$=Et,R$_4$=OH [6]

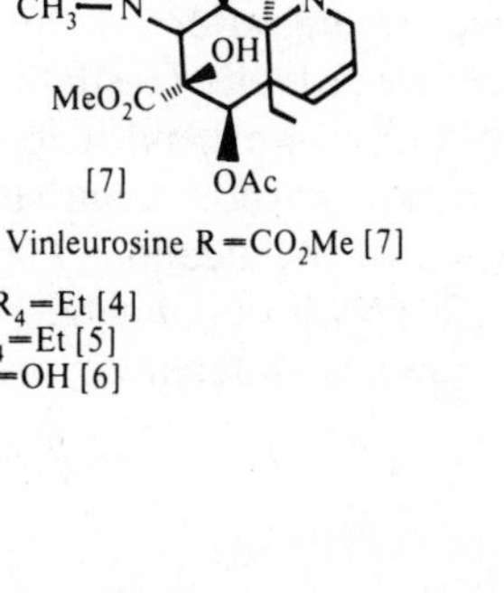

schizophrenia. An illustration of the complexity of the structures encountered in this field are the antileukaemic alkaloids vincristine [4], vinblastine
[5], vinrosidine [6], and vinleurosine [7]. These alkaloids form some of
the most potent drugs available to man for the treatment of a variety of
cancers.

In spite of the superficial complexity of indole-alkaloid structures, when
examined more closely it can be seen that the majority are formally derived
from a Mannich condensation of tryptamine [8] with an aliphatic aldehyde
having nine or ten carbons at the α- or β-positions of the indole nucleus.
Perkin and Robinson[1] were the first to suggest that the aromatic portion
present in the indole alkaloids is derived from tryptophan which has undergone decarboxylation to tryptamine. Experimental evidence for such decarboxylation has been obtained by Battersby and co-workers.[2] Tryptophan or
tryptamine could condense with an appropriate aldehyde to form a Schiff
base which could be attacked intramolecularly by the β-position of the indole nucleus to afford the corresponding indolenines [9] (Scheme 1.1).

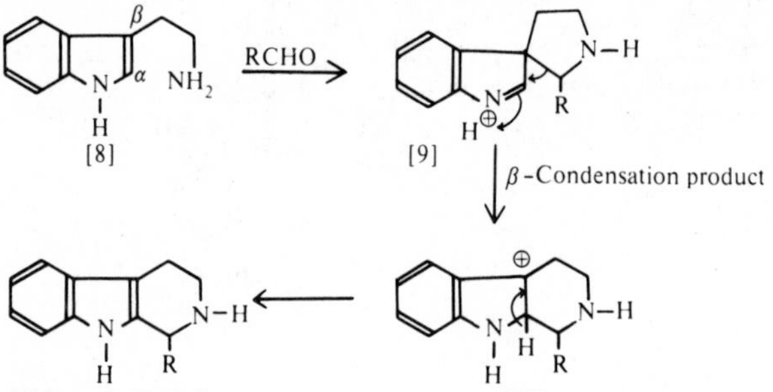

SCHEME 1.1

Previously it was thought that similarly the aldehyde could condense
directly with the α-position of the indole nucleus to afford β-carboline
systems or 'α-condensation products'. An examination of the reaction
mechanism shows that such a direct attack would involve the generation of
an intermediate [15] with a disturbed π-electron cloud on the benzene ring
which would be energetically unfavourable. An alternative pathway
avoiding the generation of such a destabilized structure would be first the
formation of the indolenine [13] from a β-condensation, which could then
rearrange by migration of the R_2 group to the α-position (Scheme 1.2).
Evidence towards the feasibility of such a pathway has been provided by
Jackson and co-workers[3] whose experiments are outlined in Scheme 1.3.
The asterisked carbon atom shows the position of tritium incorporation.
Cyclization of the alcohol [17] and oxidation with periodic acid gives the
ketone [20] with only half the radioactivity of the preceding compounds.
The reaction is thought to proceed through the formation of the sym-

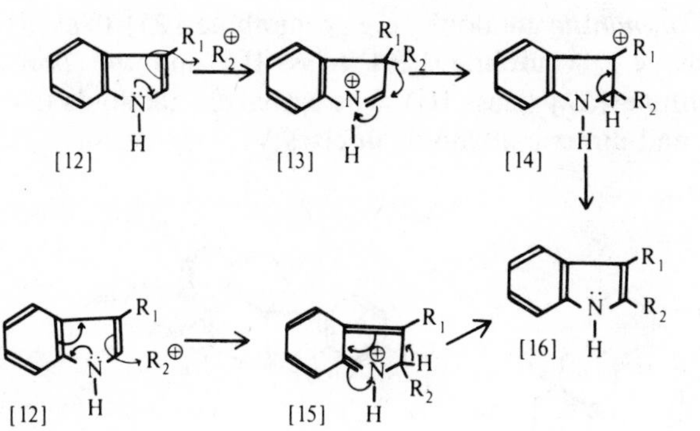

SCHEME 1.2

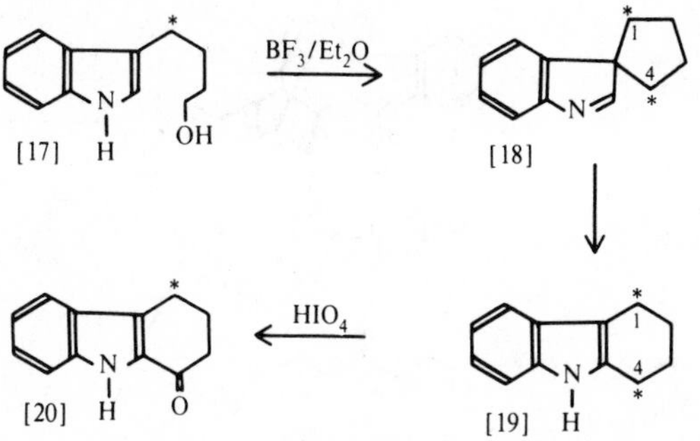

SCHEME 1.3

metrical spirocyclic indolenine [18] which would make positions 1 and 4 equivalent and thus distribute the radioactivity equally in the tetrahydro-carbazole [19]. The formation of the β-carboline alkaloids may be thus as shown in Scheme 1.4.

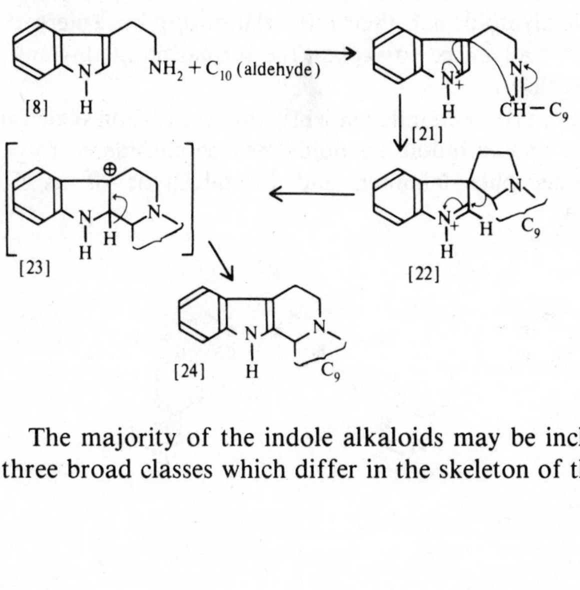

SCHEME 1.4

The majority of the indole alkaloids may be included under essentially three broad classes which differ in the skeleton of the C_{10} unit. These may be exemplified by the *Corynanthe* alkaloids, e.g. yohimbine [25] (class I); *Aspidosperma* alkaloids, e.g. vindoline [26] (class II); and the *Iboga* alkaloids, e.g. catharanthine [27] (class III). The carbazole alkaloids may be included in class IV and dimeric alkaloids in class V.

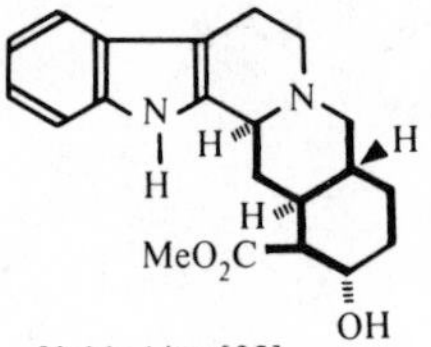

Yohimbine [25]
Class I alkaloid

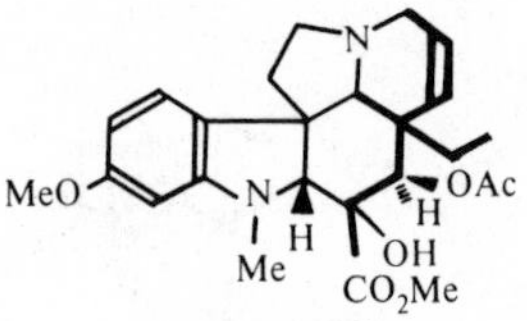

Vindoline [26]
Class II alkaloid

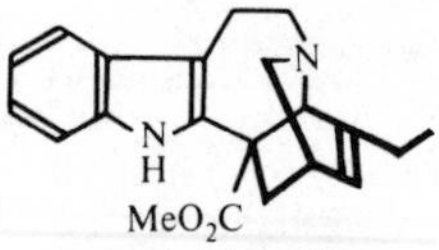

Catharanthine [27]
Class III alkaloid

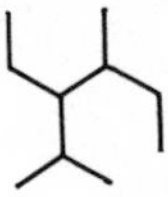

Class I skeleton 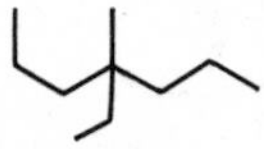   Class II skeleton 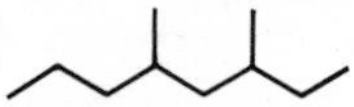   Class III skeleton

The origin of the C_{10} unit which condenses with tryptophan has been the subject of much speculation for many years. It is only during the last few years, however, that a clearer understanding of the mode of biosynthesis of various classes of indole alkaloids and their interrelationship has emerged. Of the various theories[4-14] advanced to explain the formation of this unit, four deserve special mention.

Barger[4] and Hahn[5] have proposed independently that the carbon skeleton of yohimbine and other related indole alkaloids may be considered to be derived from tryptamine, phenylalanine, and formaldehyde or its C_1 equivalent (Scheme 1.5).

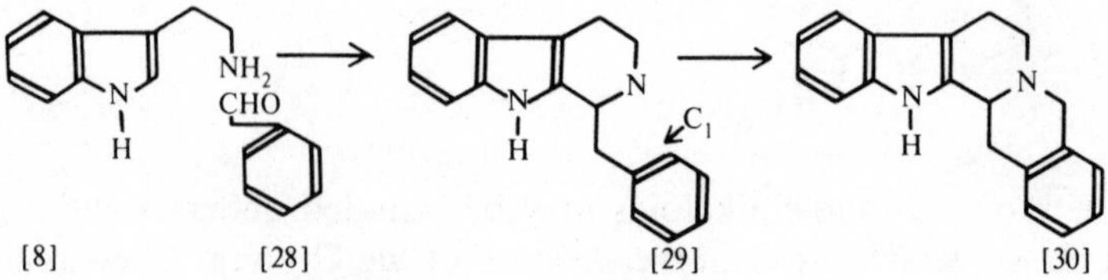

[8]   [28]   [29]   [30]   SCHEME 1.5

Woodward has proposed[6] that the *Strychnos* alkaloids may arise from an intermediate such as [31] by β-condensation of tryptamine with 3,4-dihydroxyphenylalanine and a C_1 unit. This intermediate could be oxidatively cleaved at the carbons bearing the phenolic -OH groups and recyclized to afford the seven-membered ring. N-Acetylation of the indole nitrogen followed by cyclization could then afford strychnine [1] (Scheme 1.6).

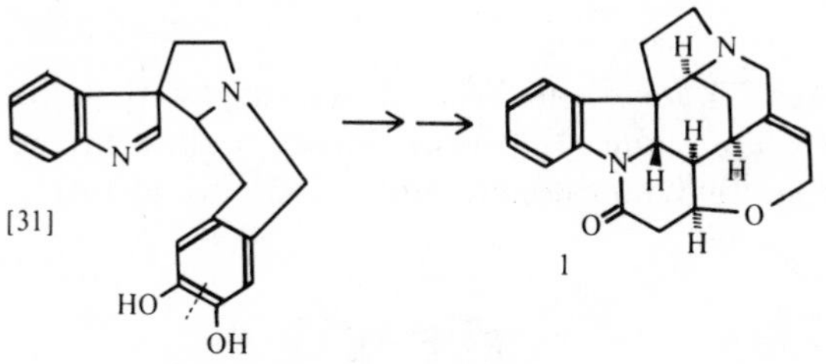

[31]

1

SCHEME 1.6

The 'prephenic acid hypothesis' envisages the involvement of prephenic acid in the biosynthesis of indole alkaloids (Scheme 1.7). Labelling experiments, however, conclusively demonstrated that a one-carbon unit was not involved in the biosynthetic process. All the above theories were therefore rejected on the basis of incorporation data.[15-19]

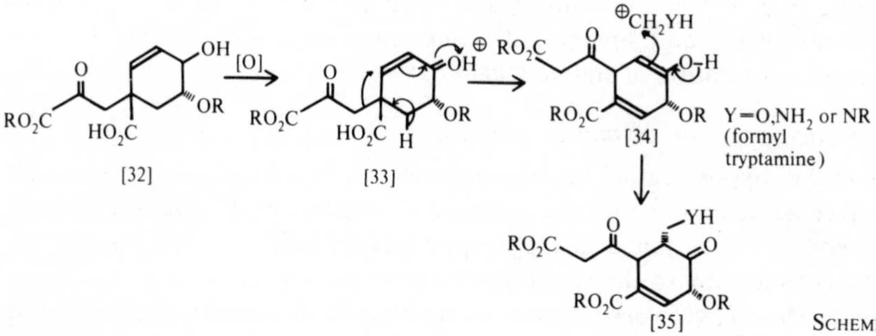

[32] [33] [34] Y = O, NH_2 or NR (formyl tryptamine)

[35] SCHEME 1.7

1.2 Monoterpenoid origin

The first independent proposals of the monoterpenoid origin of the 'non-tryptophan' portion of indole alkaloids were made by Wenkert[8] and Thomas.[17] The striking similarity between several non-alkaloidal (and generally non-nitrogenous) glucosides such as verbenalin [36], genipin [38], and asperuloside [39] on the one hand and the 'non-tryptophan derived portion' of such alkaloids as corynantheine [40] and ajmalicine [41] on the other, led to the suggestion that they may have a common precursor. Moreover, these glycosides have the same stereochemistry at the corresponding site as at C-15 in corynantheine, ajmalicine, etc. The ester function is also at the corresponding position.

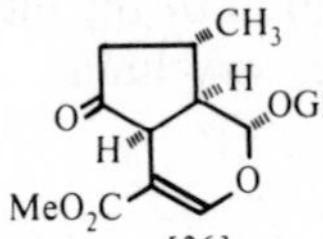

These similarities prompted Thomas and Wenkert to suggest that the non-tryptophan portion of these alkaloids (shown in heavy lines in the structures) was formed from two mevalonate units [42] to afford a

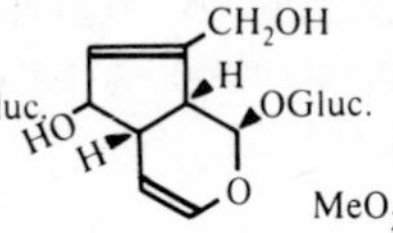

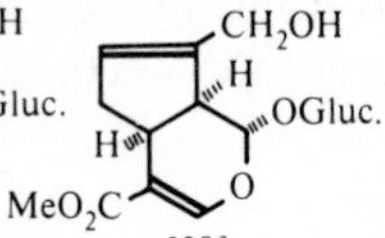

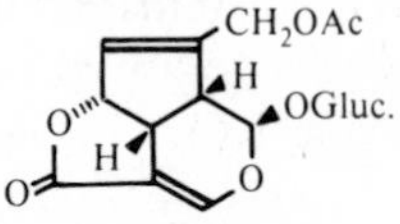

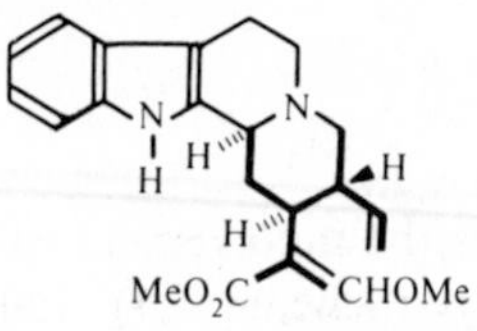

cyclopentane monoterpene which was subsequently cleaved and incorporated into tryptamine. Subsequent work on incorporation of geraniol, mevalonic acid, and loganin, etc. has confirmed this scheme for the biogenetic formation of indole alkaloids.

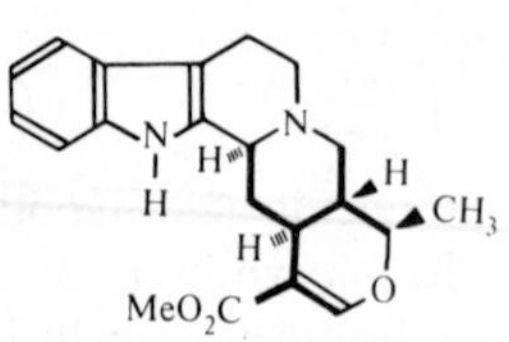

1.2.1 Incorporation of mevalonate unit

The earlier hypotheses of the biogenesis of the non-tryptophan portion all involved a one-carbon unit. Incorporation experiments by several groups of workers[9, 13, 14, 18-20] failed to support the incorporation of such one-carbon units, thus leading to the rejection of earlier theories.

Battersby and co-workers, after obtaining negative results with respect to any of the earlier theories, obtained low incorporation of activity of dl-[2-^{14}C]-mevalonate into cephaeline and ajmaline [48] indicating that indole alkaloids may be of monoterpenoid origin.[20] Evidence of this also came from the experiments of three independent groups led by Scott, Arigoni, and Battersby.[21-23] Scott observed that on administration of dl-[2-^{14}C]-mevalonic acid lactone [49] to the roots of Catharanthus roseus plants, the vindoline isolated was radioactive. Reduction to vindolinol [51] and treatment with periodic acid released formaldehyde which was crystallized as the dimedone derivative (Scheme 1.8). The methoxycarbonyl carbon was shown to contain a quarter of the total radioactivity. This was

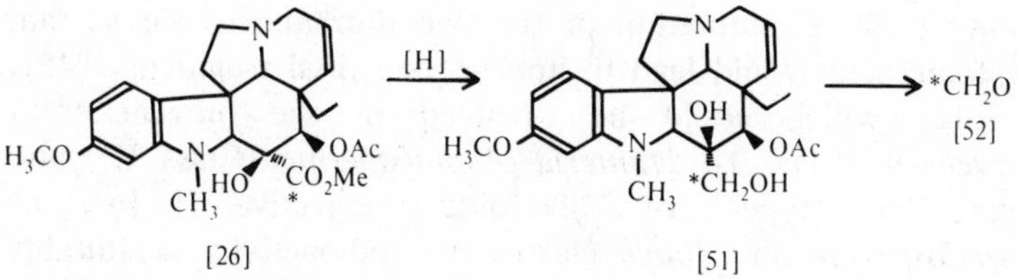
[H]
$*CH_2O$
[52]
H_3CO
OAc
HO
CO_2Me
CH_3
OH
$*CH_2OH$
[26]
[51]

SCHEME 1.8

exciting evidence in favour of the mevalonoid origin of indole alkaloids.[21] Almost identical results were obtained by Arigoni and co-workers who obtained similar incorporation of the label into C-22 of vindoline [26] and reserpine [2] when *dl*-[2-^{14}C]-mevalonate was fed into shoots of *Vinca rosea* plants.[22] Further confirmation came from a brilliant series of experiments by Battersby and co-workers who incorporated sodium (±)-[2-^{14}C]-mevalonate into *V. rosea* plants and obtained radioactive vindoline [26], catharanthine [27], ajmalicine [41], and serpentine [50]. Here again an incorporation of a quarter of the total activity had taken place at C-22. This was consistent with the rearrangement mechanism proposed by Thomas[17] (Scheme 1.9) or the alternative mechanism proposed by

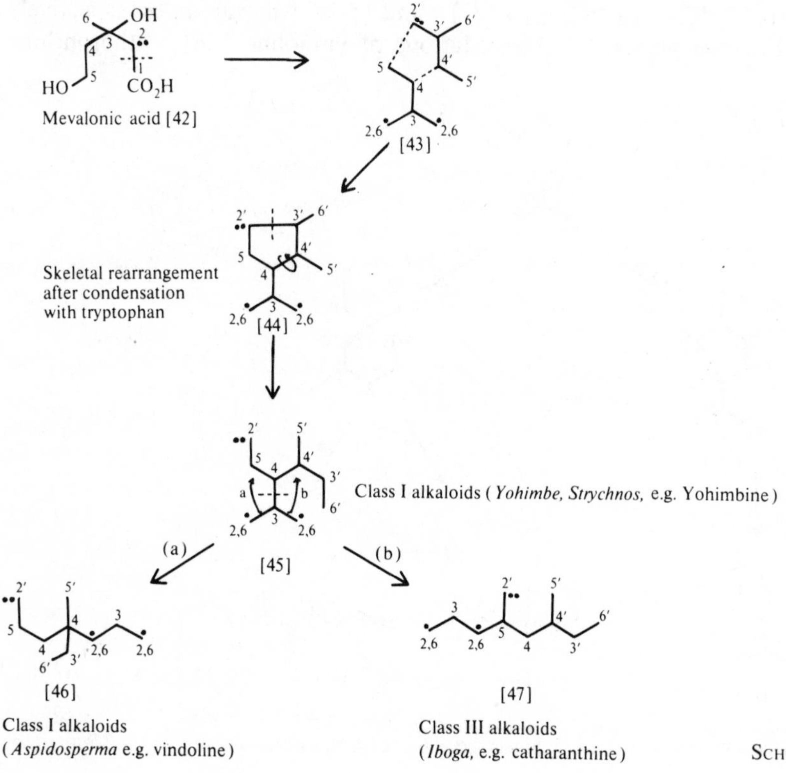
Mevalonic acid [42]
[43]
Skeletal rearrangement
after condensation
with tryptophan
[44]
Class I alkaloids (*Yohimbe, Strychnos,* e.g. Yohimbine)
[45]
(a)
(b)
[46]
[47]
Class I alkaloids
(*Aspidosperma* e.g. vindoline)
Class III alkaloids
(*Iboga,* e.g. catharanthine)

SCHEME 1.9

Scott[21] (Scheme 1.10). Examination of the two mechanisms shows that labelled mevalonic acid would lead to appropriate final fragments [45], [46], and [47] which could be involved in the biogenesis of *Yohimbe–Strychnos* (Class I), *Hunteria–Aspidosperma* (Class II), and *Iboga* (Class III) types of alkaloids respectively. In the *Hunteria–Aspidosperma* and *Iboga* classes the radioactivity in suitably labelled mevalonic acid would be incorporated at different positions depending on which of the two mechanisms was operating, thus affording a way of establishing the correct mechanism.

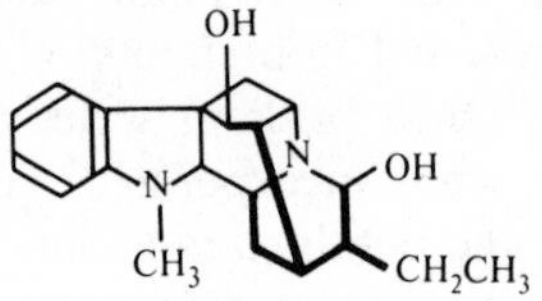

Ajmaline [48] 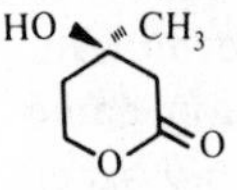   Mevalonic acid lactone [49] 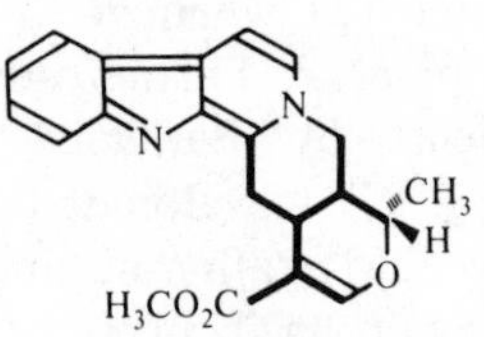   Serpentine [50]

In an exhaustive series of experiments, Battersby and co-workers fed sodium [2-^{14}C]-, [3-^{14}C]-, [4-^{14}C]-, and [5-^{14}C]-mevalonates respectively into *V. rosea* plants.[23, 24] Degradations of vindoline [26], catharanthine

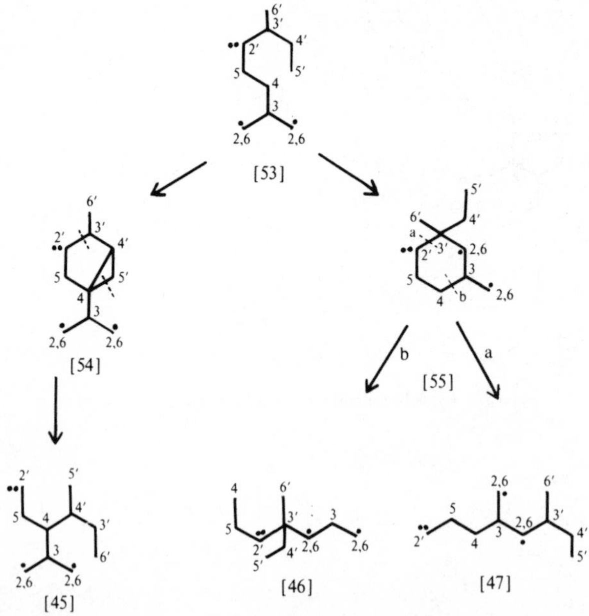

Yohimbe–Strychnos
(Class I alkaloid skeleton)

Hunteria–Aspidosperma
(Class II alkaloid skeleton)

Iboga
(Class III alkaloid skeleton)

SCHEME 1.10

[27], ajmalicine [41], and serpentine [50] showed that incorporation of radioactivity at appropriate positions had taken place as predicted in Thomas and Wenkert's route. The incorporation of 23 per cent of the original activity in C-22 of catharanthine was proof against the involvement of Scott's mechanism of rearrangement as this would predict 50 per cent of the total activity at C-22 starting from sodium ($\pm$)-[2-^{14}C]-mevalonate. Degradation of labelled catharanthine [27] and dehydroaspidospermidine gave results[20, 25] in agreement with the head to tail combination of the two C_5 units.

The host of feeding experiments which followed clearly established the incorporation of mevalonate units in the three main types of indole alkaloidal skeletal systems. The results of these incorporation experiments are summarized in Table 1.1.

Table 1.1

Alkaloids	Mevalonate precursor	Position labelled	% Incorporated	% Expected	Reference
Ajmalicine	[2-^{14}C]	16, 17	22	26(C-1)	24
	[2-^{14}C]	22	24	25	23
	[3-^{14}C]	19	40	50	25
Catharanthine	[2-^{14}C]	3, 14, 15, 21	48	50(C-3)	23
		16, 17	29	25(C-17)	23
		22	23	25	23
		18, 19, 20, OMe	0	0	23
	[3-^{14}C]	18, 19	44	50(C-19)	24
	[4-^{14}C]	20	48	50	24
	[5-^{14}C]	18, 19, 20	0	0	30
1,2-Dehydro-aspidospermidine	[2-^{14}C]	3	65	67	23
	[3-^{14}C]	18, 19, 20	0	0	
		19	47	50	24
		18, 20	0	0	
Perivine	[4-^{14}C]	20	44	50	24
Reserpinine	[2-^{14}C]	22	26	25	24
Serpentine	[2-^{14}C]	3	43	50	30
	[3-^{14}C]	19	42	50	24
	[4-^{14}C]	18	45	50	24
	[5-^{14}C]	14	45	50	24
Vindoline	[2-^{14}C]	22	22, 23	25	21, 22, 31, 32
		18, 19, 20, OMe	0	0	22
		N-Me, OAc	45, 47	50	22, 24
	[3-^{14}C]	18, 19, 20, 22-OMe	0	0	25
	[4-^{14}C]	20	45	50	24
	[5-^{14}C]	18, 19, 20	0	0	30

The usual biosynthetic precursor of mevalonate is acetate which would thus be expected to act as a specific precursor to indole alkaloids. This is strangely not borne out by labelling experiments which showed randomization of acetate radioactivity in *C. roseus*.[20] A possible alternative precursor to mevalonate is leucine, particularly since it has been shown to be convertible to mevalonic acid via β-methylcrotonyl coenzyme A[27]. However, feeding of labelled leucine to *C. roseus* has not resulted in specific incorporations,[28] leaving the origin of the mevalonate unit unanswered. More recently DL-[2-^{14}C]-alanine was found to be preferentially incorporated into the isopentenylpyrophosphate-derived moiety of geraniol.[29]

1.2.2 Incorporation of geraniol derivatives

The above incorporations of mevalonic acid suggested a terpenoid origin of indole alkaloids, since mevalonic acid is known to lead to mono-, di-, tri-, and sesquiterpenoidal systems. As the pathway to these terpenoids from mevalonic acid involves the initial generation of geraniol (i.e. C-10 unit) by the condensation of isopentenyl diphosphate [59] and 3,3-dimethylallyl diphosphate [60] (both decarboxylated isomeric products from mevalonic acid) as shown in Scheme 1.11 it was natural to consider the intermediacy of geraniol in indole-alkaloid biosynthesis.

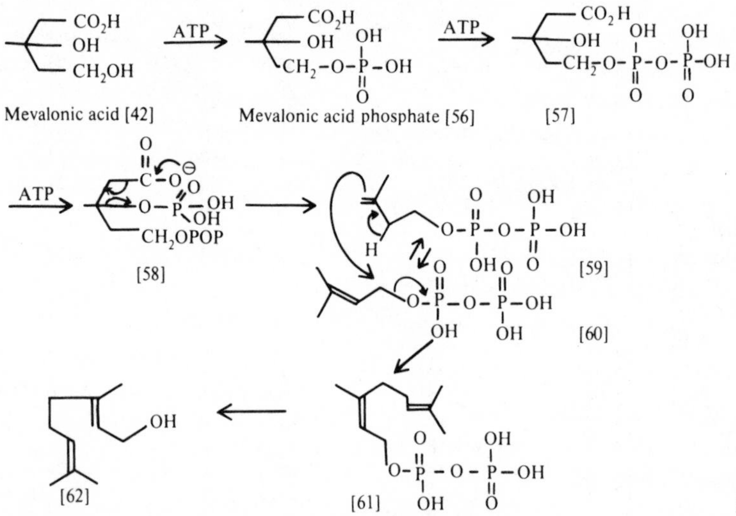

SCHEME 1.11

In order to determine whether geraniol was a precursor, Battersby and co-workers fed [2-^{14}C]-geraniol pyrophosphate [61] to *V. rosea* plants[30] and showed that all the radioactivity was incorporated into C-4 of catharanthine and C-5 of vindoline as expected. Similar experiments carried out independently by Arigoni[25] and Scott[33] confirmed the intermediacy of geraniol in the biosynthetic pathway for the *Corynanthe*, *Aspidosperma*, and *Iboga* classes of bases.

Following these positive feeding results with geraniol, a series of experiments was carried out by Arigoni, Battersby, Leete, Scott, and Schmid using appropriately-labelled geraniol derivatives. Isolation of various alkaloids and determination of the extent of incorporation afforded results entirely consistent with those expected at the predicted positions.[24, 25, 30, 31, 33-35] The results of various feeding experiments with geraniol are summarized in Table 1.2.

Table 1.2

Alkaloids	Mevalonate precursor	Position labelled	% Incorporated	% Expected	Reference
Ajmalicine	[2-^{14}C]-geranyldi-phosphate	15, 16, 17, 18, 19, 20	0	0	30
Catharanthine	[2-^{14}C]-geranyl-phosphate	20	101	100	30
Perivine	[2-^{14}C]-geraniol	20	100	100	24
Vindoline	[2-^{14}C]-geraniol	20	97–99	100	25–30, 32
	[3-^{14}C]-geraniol	19	99	100	35

1.2.3 Incorporation of iridoids

The positive results obtained from the feeding experiments with mevalonate and geraniol provided strong evidence in favour of the theory proposed by Wenkert and Thomas. Since their biogenetic hypothesis had been based on the remarkable similarities between the non-tryptophan portion of the indole alkaloids and certain monoterpene glucosides such as verbenalin [36], acubin [37], genipin [38], asperuloside [39] etc., it followed naturally to investigate whether these glycosides played a more direct role in the biosynthetic pathway to the major groups of indole alkaloids.

Several glycosides were considered as possible intermediates, namely loganin [63], monoterpeinemethyl ester [64], verbenalin [36], and genipin [38]. Many other possible candidates were eliminated on grounds of structural and stereochemical disparities with the non-tryptophan portion of indole alkaloids.

Verbenalin [36] and dihydroverbenalin [65] were labelled by tritium exchange and were found to be totally ineffective as precursors of the indole alkaloids in *V. rosea*.[36] Monoterpeine methylester [64] labelled at the ester-methyl group similarly was found not to be incorporated into the alkaloids. However, when loganin [63] was labelled similarly at the methyl-ester group and the [O-methyl-³H] loganin administered to *V. rosea* plants, good incorporation of activity into all three types of indole alkaloids was obtained. All the radioactivity was found to be introduced in the ester methyl groups in catharanthine [27], vindoline [26], ajmaline [48], and serpentine [50]. This was strong evidence in favour of loganin being a key intermediate in the biosynthetic pathway to indole alkaloids. It had been first proposed in 1964 by Thomas who had suggested that the glucoside incorporated could be loganin, which may undergo cleavage to secologanin prior to incorporation.[37] More labelling experiments carried out by Battersby and Arigoni with loganin labelled at different positions afforded the three major classes of indole alkaloids bearing all the radioactivity at the expected positions (Scheme 1.12). Loganin was thus proved to be a precursor of representative examples from the three major classes of indole alkaloids comprising *Yohimbe*, *Aspidosperma*, and *Iboga* classes.

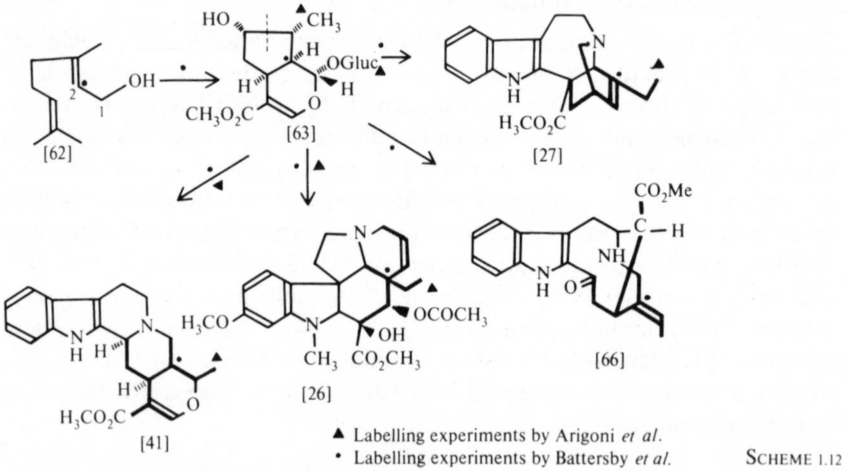

▲ Labelling experiments by Arigoni *et al.*
• Labelling experiments by Battersby *et al.*   SCHEME 1.12

The structure and stereochemistry of loganin [63] has been independently established by Battersby[38] and Arigoni.[39] Biological formation and role of the C_{10} cyclopentane system of loganin was also independently established by Battersby[38] and Arigoni[39] by feeding suitably labelled geraniol to *Menyanthes trifoliata* and isolating the corresponding labelled loganin, followed by separate feeding of this loganin to *V. rosea* shoots.[40, 41] Evidence of the mevalonoid origin of loganin was obtained by

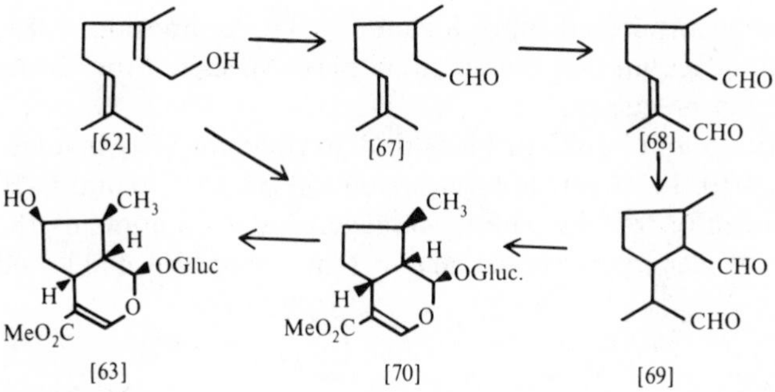

[62] [67] [68] [63] [70] [69]

SCHEME 1.13

Coscia.[42] Loganin itself was observed in *V. rosea* by radiochemical dilution[24, 41] and by isolation.[42] Loganin has also been isolated from *Patrinia villosa*.[44]

The formation of loganin from geraniol has been the subject of intense investigation in the last few years. Battersby had proposed[45, 46] a tentative pathway for the formation of loganin which is outlined in Scheme (1.13).

The above scheme has, however, been shown to be defective in several important respects. [1-³H]-Citronellol [71] failed to be incorporated to any significant extent into loganin or indole alkaloids.[47, 48] Similarly, 10-hydroxy-[1-³H]-citronellol [72], [1-³H]-citronellal [67] and [8-¹⁴C]-10-hydroxylinalool [73] also afforded negative incorporation results, thus indicating that citronellal [67] did not lie in the direct geraniol–loganin pathway and that a 2,3-reduction of geraniol does not occur. Incorporation experiments with labelled iridodial also failed to substantiate its direct intermediacy in indole-alkaloid biosynthesis.[49] Accumulating evidence indicates that the dialdehyde [74] is an important intermediate in which the C-8 and C-10 of geraniol [62] become equivalent prior to incorporation into indole alkaloids or monoterpene glycosides.[50-61] Labelled 10-hydroxygeraniol [76] and 10-hydroxynerol [77], on the other hand, were incorporated into loganin [63] and indole alkaloids,[47, 48] the latter compound being incorporated to a greater extent suggesting that the immediate precursor had the Δ2 double bond in the *cis* configuration. The redox interconversion of geraniol and nerol has been demonstrated in *Tanacetum vulgare*[62] and their biosynthesis has been studied in cell-free extracts of the same plant.[63]

The incorporation of labelled deoxyloganin [70] into loganin as well as into the alkaloids catharanthine [27], vindoline [26], serpentine [50], ajmalicine [41], and perivine [66] showed that hydroxylation at C-7 occurs at a fairly late stage in the biogenesis of loganin.[64] This has received added support from the isolation of the glycosides of deoxyloganin [70] from *C. roseus* and *Strychnos nux vomica* and the demonstration of its existence in *M. trifoliata*. Moreover, the labelled deoxyloganic acid [85] was also shown to be incorporated into loganin [63], geniposide [38], verbenalin [36], and acubin [37] in various plant species, thus firmly establishing that it is a precursor.

The conversion of geraniol [62] or nerol to deoxyloganin [70] involves the oxidation of C-1 to the aldehyde level, oxidation of the C-9 and C-10 methyl groups, saturation of $\Delta 2$ olefinic linkage, and formation of the cyclopentane ring. It has been shown above that citronellol [71] and

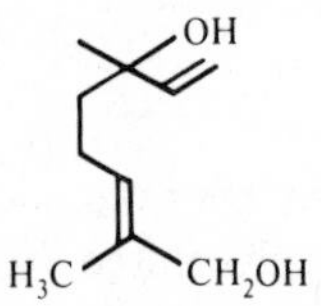

[73] 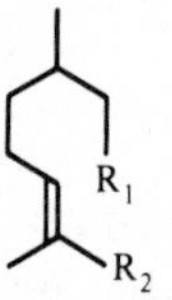 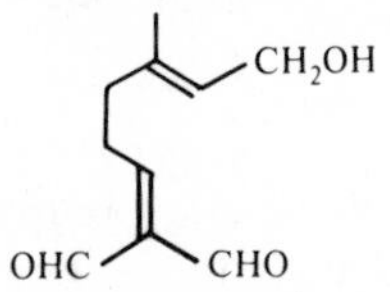   [74]

Citronellol $R_1=CH_2OH$, $R_2=CH_3$ [71]
lo-Hydroxycitronellol $R_1=CH_2OH$, $R_2=CH_2OH$ [72]
Citronellal $R_1=CHO$, $R_2=CH_3$ [67]

iridodial [69] do not lie in the direct pathway as had been tentatively proposed by Battersby.[45] The intermediacy of the dialdehyde [74] or a close analogue and that of deoxyloganin [70] appeared highly probable. Battersby and co-workers, in an effort to clarify the pathway between geraniol and loganin, have prepared several C_{10} terpenes having varying oxidation levels at C-1 and C-10 position, starting from commercially available citral. Feeding experiments from corresponding radiolabelled substances established that 10-hydroxygeranial [78], 10-hydroxyneral [79], 10-oxo-geranial [80], and 10-oxoneral [81] are probable biointermediates between geraniol/nerol and loganin. The monohydroxyaldehydes [94] and [95] were found to be incorporated to a lower degree into loganin and indole alkaloids, which may indicate that they do not lie directly in the biosynthetic pathway.[65] The likely biosynthetic path between geraniol and secologanin is shown in Scheme (1.14) which envisages the oxidation of C-10 of geraniol [62] to 10-hydroxygeraniol [76] as a primary step. Good incorporation with 10-hydroxygeraniol [76] and the isomeric 10-hydroxynerol [77] into loganin [63] and representative alkaloids provided support for this.

Oxidation of 10-hydroxygeraniol [76] or 10-hydroxynerol [77] could afford 10-hydroxygeranial [78] or 10-hydroxyneral [79]. The oxidation of C-10 alcoholic groups could then lead to the isomeric dialdehydes [80] and [81]. These dialdehydes could be progressively oxidized to the dialdehyde alcohol [82] or the trialdehyde [83] in which the terminal carbon atoms have become equivalent.

The structural relationships between these molecules strongly suggested that deoxyloganin [70] is the immediate precursor of loganin. This was

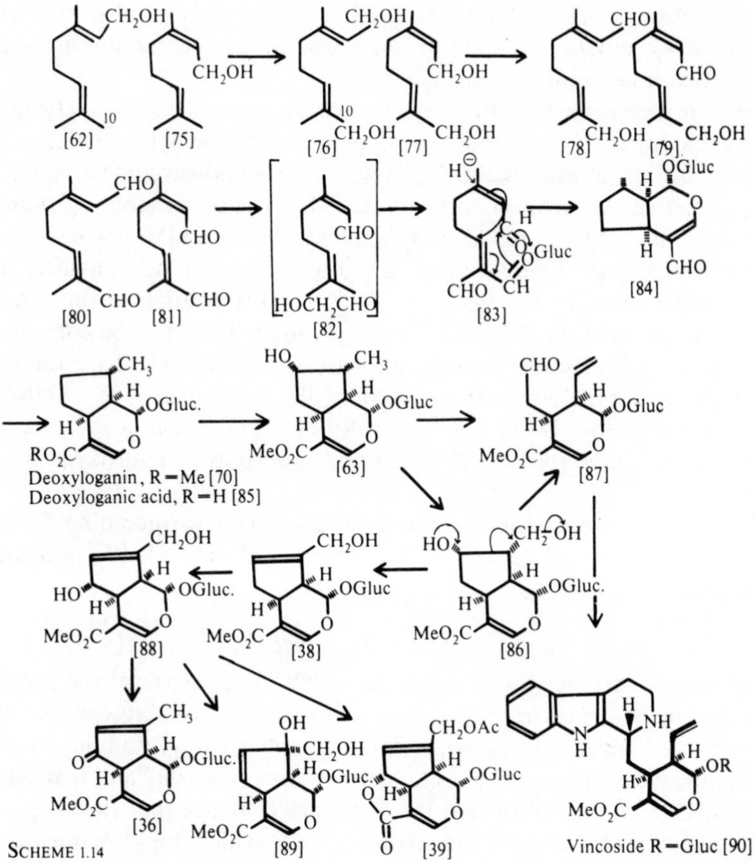
CH2OH
CH2OH
10
[62]
[75]
CH2OH
CH2OH
10
CH2OH
CH2OH
[76]
[77]
CHO
CHO
CH2OH
CH2OH
[78]
[79]
CHO
CHO
CHO
CHO
[80]
[81]
CHO
HOCH2CHO
[82]
H
H
OGluc
CHO
CH
[83]
OGluc
H
O
H
CHO
[84]
CH3
H
OGluc.
H
O
RO2C
Deoxyloganin, R=Me [70]
Deoxyloganic acid, R=H [85]
HO
CH3
H
OGluc
H
O
MeO2C
[63]
CHO
H
OGluc
H
O
MeO2C
[87]
CH2OH
H
OGluc.
HO
H
O
MeO2C
[88]
CH2OH
H
OGluc
H
O
MeO2C
[38]
HO
CH2OH
H
OGluc.
H
O
MeO2C
[86]
CH3
H
OGluc.
O
H
O
MeO2C
[36]
OH
CH2OH
H
OGluc.
H
O
MeO2C
[89]
CH2OAc
H
OGluc
O
H
O
C
O
[39]
N
H
H
NH
H
OR
H
O
MeO2C
Vincoside R=Gluc [90]

SCHEME 1.14

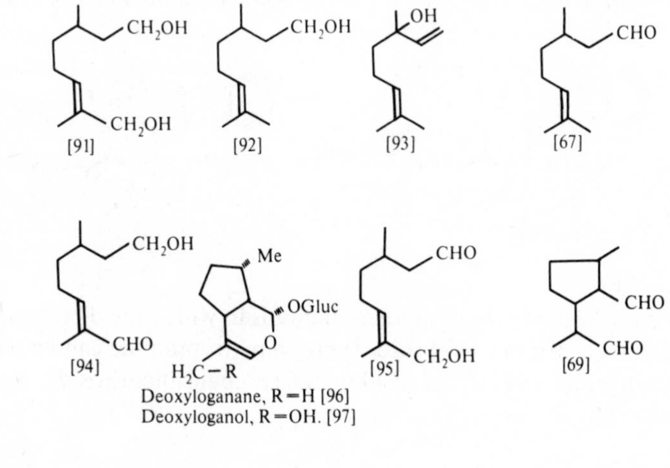
CH2OH
CH2OH
[91]
CH2OH
[92]
OH
[93]
CHO
[67]
CH2OH
CHO
[94]
Me
OGluc
O
H2C−R
Deoxyloganane, R=H [96]
Deoxyloganol, R=OH. [97]
CHO
CH2OH
[95]
CHO
CHO
[69]

proved by the work of Inouye and others.[65] It appeared also that deoxy-loganol [84] and its corresponding aglycone were intermediates in the path-way between geraniol and loganin.

Feeding experiments with their tritiated derivatives conclusively showed that deoxyloganol [84] and deoxyloganol aglucone act as precursors to loganin and the indole alkaloids. These results demonstrated the validity of the reductive cyclization process to the cyclopentanoid systems as postulated by Arigoni, and also explain how the C-9 and C-10 carbon atoms of the isopropylidene side chain in geraniol—nerol become equivalent in the biosynthetic pathway to loganin.[65] Further hydroxylation at the exocyclic methyl group affords 10-hydroxyloganin [86] which must be converted to secologanin [87] before condensing with tryptophan to afford the indole alkaloids. Other iridoids such as genipin [38], scandoside [88], verbenalin [36], gardenoside [89], and asperuloside [39] could also arise from 10-hydroxyloganin [86] by dehydration,[66] appropriate hydroxylation, and oxidation.

Additional support for the above proposal was forthcoming from the non-incorporation of various monoterpenes [67, 69, 91—95] into indole alkaloids.[48, 49]

1.2.4 Cleavage of the cyclopentane framework

What happens to loganin once it is formed? The cleavage of the cyclopen-tane has to occur at some stage to give rise to a suitably functionalized aliphatic unit which can condense with tryptophan (or tryptamine) to afford the simple indole bases. The occurrence of vincoside [90] and isovincoside [98] in *C. roseus*[67, 2] provided the necessary evidence that the cleaved (or 'seco') iridoid would have the structure [87] which could condense with tryptamine to afford these alkaloids. *In vitro* condensation of secologanin [87] with tryptamine afforded a mixture of vincoside [90] and the 3-epimer isovincoside [98].[2, 67, 68] (Isovincoside is identical to 'strictosidine'

[90]   [98]

isolated from *Rhazya stricta*.) Other alkaloids in which the direct condensa-tion of the 'seco' iridoid [87] into tryptophan/tryptamine can be recogniz-ed are cordifoline [99],[69] ipecoside [100],[70] pauridianthoside, and doli-chantoside.[72]

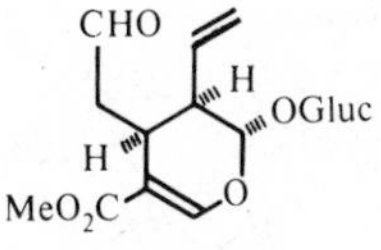

[87] 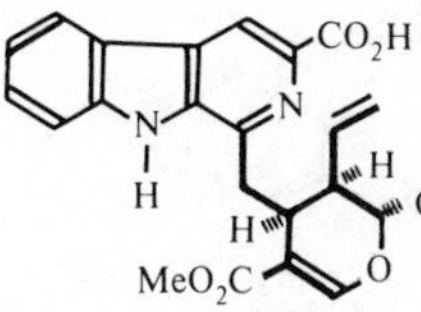 [99] 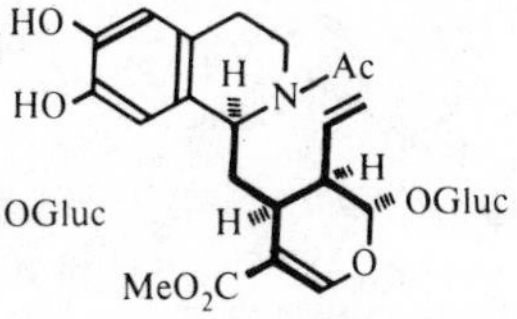 [100]

Secologanin [87] has been isolated from *Lonicera morrowii*,[73] fresh *Rhazya orientalis* leaves,[74] *Cornus officinalis* and *Cornus mass*.[75] The existence of secologanin has also been demonstrated in *C. roseus*[64, 61, 84, 85] and *M. trifoliata*[74] by dilution analysis. The secologanin unit can be recognized in a masked form in the secoiridoids sweroside [103],[76-81] foliamenthin [102],[79] menthiafolin [104],[79] and dihydromenthiafolin [105].[79, 81] Hydrolysis of menthiafolin [104] afforded [103] which proved invaluable since it represents secologanin masked by lactol formation. Ring opening and methylation afforded secologanin which could be used in in-

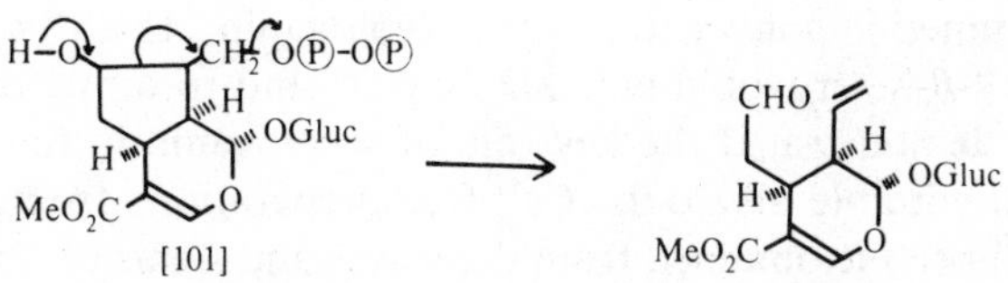

[101] [87]

corporation experiments after suitable labelling. Secologanin was shown to be a precursor of the glucosides foliamenthin [102], menthiafolin [104], and sweroside [103] in *M. trifoliata*.[82] Moreover, sweroside [103] was also shown to be incorporated into vindoline[93] which is best understood by the interconversion of secologanin and sweroside (Scheme 1.15).

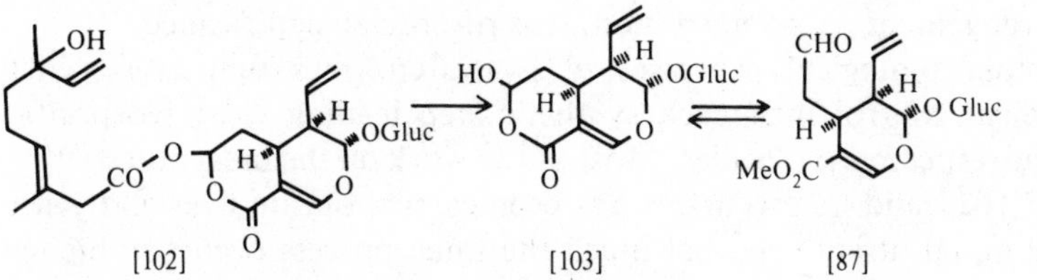

[102] [103] [87] SCHEME 1.15

The terpenoid origin of secologanin and other related secoiridoids has been established by labelling experiments. Labelled mevalonates were incorporated into secologanin [87] and secologanic acid [106] in *C. roseus*.[84] More interesting was the incorporation of loganin [63] into secologanin [87] in *C. roseus*.[67, 85]

[106]   [107]   [104]

It has been shown that enzyme extracts obtained from *C. roseus* can efficiently esterify loganic acid [107] and secologanic acid [106] to the corresponding methyl esters, loganin [63] and secologanin [87] respectively.[83, 86-88] Similarly labelled loganic acid [107] was found to be effectively incorporated into loganin [63], secologanin [87], and secologanic acid [106], the esterification being effected in the plant by the methyl transferase enzyme present.

[OMe-³H]-Secologanin [87] has been fed to *C. roseus* plants[85, 89] and the various indole alkaloids isolated were labelled at the expected positions, thus establishing that secologanin [87] is a general precursor to the major groups of indole alkaloids.[130]

The biomimetic conversions of secologanin to 18-oxoyohimban alkaloids,[90] 19-β-heteroyohimbine alkaloids,[91] and to dihydromancunine[92] have further demonstrated the key role of secologanin in the biosynthetic pathway to the indole alkaloids. Cell-free preparations from young seedlings and callus tissue obtained from the leaves and stems of *C. roseus* were shown to contain soluble enzymes which catalysed the condensation of tryptamine with secologanin to afford the various indole alkaloids.[93, 94]

1.2.5 Stereochemistry of loganin and secologanin

No mention has been made so far of the stereochemistry of the biosynthetic process leading to secologanin [87] and its precursors. Since it is this relative stereochemistry that will be carried through into the indole alkaloids at the asymmetric carbon atoms unless it is modified during subsequent rearrangement or isomerization, it is of crucial significance.

As in other biological processes, it is the enzymes that exert strict stereochemical control over each synthetic step leading to stereospecific syntheses of respective molecules. Most of the work on the biosynthesis[95-101] of geraniol [62] and its precursors has been carried out on liver and yeast systems, although it is likely that much the same process occurs in higher plants.

Three molecules of acetyl coenzyme A [108] condense to form, initially, acetoacetyl coenzyme A [109] which undergoes further condensation with acetyl coenzyme A to afford 3-hydroxy-3-methylglutaryl coenzyme A (HMGCoA) [110].[102, 103] This compound is then reduced via the hemithioacetal intermediate [111][104, 105] to R(+)-mevalonic acid [42][106] by

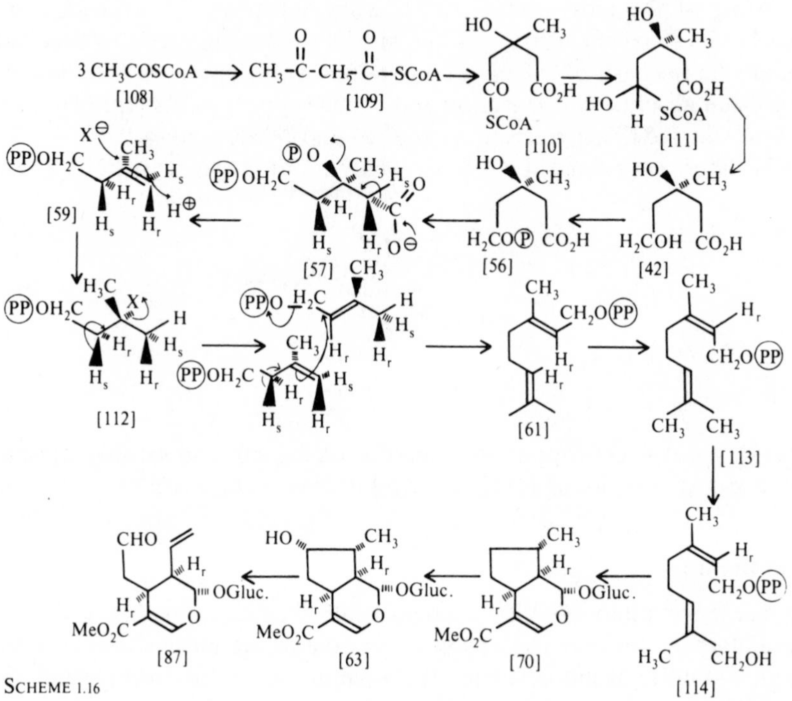

$3 CH_3COSCoA \longrightarrow CH_3\text{-}\overset{O}{\overset{\|}{C}}\text{-}CH_2\text{-}\overset{O}{\overset{\|}{C}}\text{-}SCoA \longrightarrow$ [108] [109] [110] [111] [56] [42] [57] [59] [112] [61] [113] [114] [87] [63] [70]

SCHEME 1.16

TPNH[107] (Scheme 1.16). Mevalonic acid is then converted to the 5-phosphate [56] which is further phosphorylated to the 5-pyrophosphate [57].[108-111] This undergoes decarboxylative *trans* elimination[97, 112] to give isopentenyldiphosphate [59].[43, 114] The next step is its conversion to the double-bond isomer, dimethylallyldiphosphate [60]. This involves the stereoselective loss[115] of the pro-4 S hydrogen[116] and the stereoselective addition of a proton to the 're' side of the olefin[117] to afford dimethylallyl-diphosphate [60].[118] This may take place by an enzyme-controlled addition–elimination sequence or by a concerted enzyme-directed protona-tion–deproptonation mechanism. Isopentenyldiphosphate [59] then couples with dimethylallyldiphosphate [60] by an enzyme-controlled mechanism which results in the stereospecific loss of the pro-4 S proton from isopentenyldiphosphate [59] to afford geranyldiphosphate [61][96, 97, 119, 123, 124]. It is notable that the two methyl groups of dimethylallyldiphosphate [60] are not biosynthetically equivalent as they are derived from C-2 and C-6 of mevalonic acid [42]. Thus in geranyl-diphosphate [61], C-10 methyl is derived from C-2 of mevalonic acid and C-8 and C-9 methyl groups from C-6 of mevalonic acid. This has been shown to be so in a mould metabolite[120, 121] and a triterpenoid.[122]

Most of the biosynthetic studies using loganin [63] were carried out before its stereochemistry was known. Its stereochemistry was deduced initially on the basis of chemical correlations with substances of known structure and absolute configuration and added support for its representation as shown in [63] was received by optical and NMR studies.[38, 39, 125, 126] This was later confirmed by X-ray analysis of loganin methoxybromide

[115]

[115].[127, 128] A conformational analysis of loganin and secologanin based on a study of their ^{13}C-NMR spectra has been published.[129]

1.3 Summary

It has been established by extensive labelling experimentation that the monoterpene loganin [63] plays a key role in the biosynthesis of several major families of indole alkaloids. Loganin itself arises in the plants from mevalonate [42] which is transformed by a series of steps to isopentenyl-diphosphate [59] and dimethylallyldiphosphate [60]. Combination of [59] and [60] leads initially to geraniol [62], then to loganin [63] and finally to secologanin [87] before combination with tryptophan. The sequence of steps leading to the *Corynanthe*-type alkaloids is shown in Schemes 1.11–1.16.

References

1. PERKIN, W. H. and ROBINSON, R. *J. Chem. Soc.* **115**, 933 (1919).
2. BATTERSBY, A. R., BURNETT, A. R., and PARSONS, P. G. *J. Chem. Soc.* 1193 (1969).
3. JACKSON, J. H. and SMITH, P. *Chem. Commun.* 265 (1967).
4. BARGER, G. and SCHOLZ, G. *Helv. Chim. Acta* **16**, 1343 (1933).
5. HAHN, G. and WERNER, H. *Justus Liebigs Ann. Chem.* **520**, 123 (1933).
6. WOODWARD, R. B. *Nature (Lond.)* **162**, 155 (1949); *Angew. Chem.* **68**, 13 (1956).
7. WENKERT, E. and BRINGI, N. Y. *J. Am. Chem. Soc.* **81**, 1447 (1959).
8. WENKERT, E. *J. Am. Chem. Soc.* **84**, 98 (1962).
9. STOLLE, K., GROGER, D., and MOTHES, K. *Chem. Ind.* (Lond.) 2065 (1965).
10. LEETE, E., GOSHAL, S., and EDWARDS, P. N. *J. Am. Chem. Soc.* **84**, 1068 (1962).
11. LEETE, E. and EDWARDS, P. N. *Chem. Ind. (Lond.)* 1966 (1961).

12. LEETE, E. and GOSHAL, S. *Tetrahedron Lett.* 1179 (1962).
13. BATTERSBY, A. R., BINKS, R., LAWRIE, W., PARRY, G. V., and WEBSTER, B. R. *Proc. Chem. Soc.* 369 (1963).
14. LEETE, E., AHMAD, A., and KOMPIS, I. *J. Am. Chem. Soc.* **87**, 4168 (1963).
15. BATTERSBY, A. R., BINKS, R., LAWRIE, N., PARRY, G. V., and WEBSTER, B. R. *Proc. Chem. Soc.* 369 (1963).
16. STOLLE, K., GROGER, D., and MOTHES, K. *Chem. Ind.* 2065 (1965).
17. THOMAS, R. *Tetrahedron Lett.* 544 (1961).
18. GOEGGEL, H. and ARIGONI, D. *Experientia* **21**, 369 (1965).
19. BARTON, D. H. R., KIRBY, G. W., PRAGER, R. H., and WILSON, E. M. *J. Chem. Soc.* 3990 (1965).
20. BATTERSBY, A. R., BINKS, R., LAWRIE, W., PARRY, G. V., and WEBSTER, B. R. *J. Chem. Soc.* 7459 (1965).
21. MCCAPRA, F., MONEY, T., SCOTT, A. I., and WRIGHT, I. G. *Chem. Commun.* 537 (1963).
22. GOEGGEL, H. and ARIGONI, D. *Chem. Commun.* 538 (1965).
23. BATTERSBY, A. R., BROWN, R. T., KAPIL, R. S., PLUNKETT, A. O., and TAYLOR, J. B. *Chem. Commun.* 46 (1966).
24. BATTERSBY, A. R., BROWN, R. T., KAPIL, R. S., KNIGHT, J. A., MARTIN, J. A., and PLUNKETT, A. O. *Chem. Commun.* 23, 810; 888 (1966).
25. LOEW, P., GOEGGEL, H., and ARIGONI, D. *Chem. Commun.* 347 (1966).
26. LEETE, E., AHMAD, A., and KOMPIS, I. *J. Am. Chem. Soc.* **87**, 4168 (1965).
27. RODWELL, V. W. *Metabolic pathways* (eds. D. M. Greenberg) 3rd edn., Vol. III, p. 191. Academic Press, New York (1969).
28. WIGFIELD, D. C. and WEN, B. P. *Bioorg. Chem.* **6**, 511 (1977).
29. TANGE, K., HIRATA, T., and SUGA, T. *Chem. Lett.*, No. 3, 269 (1979).
30. BATTERSBY, A. R., BROWN, R. T., KNIGHT, J. A., MARTIN, J. A., and PLUNKETT, A. O. *Chem. Commun.* 346 (1966).
31. GROGER, D., STOLLE, K., and MOTHES, K. *Arch. Pharmac.* **300**, 393 (1967); *Z. Naturforsch.* **21b**, 206 (1966).
32. MONEY, T., WRIGHT, I. G., MCCAPRA, F., HALL, E. S., and SCOTT, A. I. *J. Am. Chem. Soc.* **90**, 4144 (1968).
33. HALL, E. S., MCCAPRA, F., MONEY, T., FUKUMOTO, F., HANSON, J. R., MOOTO, B. S., PHILLIPS, G. T., and SCOTT, A. I. *Chem. Commun.* 348 (1966).
34. SCHLATTER, G., WALDNER, E., GROGER, D., MAIER, W., and SCHMID, H. *Helv. Chim. Acta.* **52**, 776 (1969).
35. LEETE, E. and VEDA, S. *Tetrahedron Lett.* 4915 (1966).
36. BATTERSBY, A. R., BROWN, R. T., KAPIL, R. S., MARTIN, J. A., and PLUNKETT, A. O. *Chem. Commun.* 23, 890 (1966).
37. THOMAS, R. *Biogenesis of antibiotic substances* (eds. Z. Vanek and Z. Hostalek) (Symposium, June 1964) p. 163. Academic Press, New York (1965).
38. BATTERSBY, A. R., KAPIL, R. S., and SOUTHGATE, R. *Chem. Commun.* 131 (1968).
39. BRECHBUHLER-BADER, S., COSCIA, C. J., LOEW, P., VON SZIZEPANSKI, CH., and ARIGONI, D. *Chem. Commun.* 3, 136 (1968).
40. BATTERSBY, A. R., KAPIL, R. S., MARTIN, J. A., and MO, L. *Chem. Commun.* 133 (1968).
41. ARIGONI, D. and LOEW, P. *Chem. Commun.* 137 (1968).
42. COSCIA, C. J. and GUARNACCIA, R. *Chem. Commun.* 138 (1968).
43. BATTERSBY, A. R., BURNETT, A. R., HALL, E. S., and PARSONS, P. G. *Chem. Commun.* 1582 (1968).

44. TAGUCHI, H., YOKOKAWA, T., and ENDO, T. *Yakugaku Zasshi.* **93**, 607 (1973).
45. BATTERSBY, A. R. 4th Int. *Symposium on the Chemistry of Natural Products*, Stockholm, June 1966.
46. BATTERSBY, A. R. *Pure Appl. Chem.* **14**, 117 (1967).
47. ESCHER, S., LOEW, P., and ARIGONI, D. *Chem. Commun.* 823 (1970).
48. BATTERSBY, A. R., BROWN, S. H., and PAYNE, T. G. *Chem. Commun.* 827 (1970).
49. BOWMAN, R. M. and LEETE, E. *Phytochemistry* **8**, 1003 (1969).
50. YEOWELL, D. A. and SCHMID, H. *Experientia* **20**, 250 (1964).
51. SCHMID, H. *Chimia* **22**, 312 (1968).
52. GUARNACCIA, R., BOTTA, L., and COSCIA, C. J. *J. Am Chem. Soc.* **92**, 6098 (1970).
53. COSCIA, C. J., GUARNACCIA, R., and BOTTA, L. *Biochemistry* **8**, 5036 (1969).
54. HUNI, G. E. S., HILTEBRAND, H., SCHMID, H., GROGER, D., JOHNE, S., and MOTHES, K. *Experientia* **22**, 656 (1966).
55. INOUYE, H., UEDA, S., and NAKAMURA, Y. *Tetrahedron Lett.* 3221 (1967).
56. COSCIA, C. J. and GUARNACCIA, R. *J. Am. Chem. Soc.* **89**, 1280 (1967).
57. INOUYE, H., UEDA, S., and NAKAMURA, Y. *Chem. Pharmac. Bull. (Tokyo)* **18**, 2043 (1970).
58. BIOLLAZ, M. and ARIGONI, D. *Chem. Commun.* 633 (1969).
59. CORBELLA, A., GARIBOLDI, P., JOMMI, G., and SCOLASTICO, C. *Chem. Commun.* 634 (1969).
60. AUDA, H., JUNEJA, H. R., EISENBRAUN, E. J., WELLER, G. R., KAYS, W. R., and APPEL, H. H. *J. Am. Chem. Soc.* **89**, 2476 (1967).
61. HORODYSKY, A. G., WALLER, G. R., and EISENBRAUN, E. J. *J. Biol. Chem.* **244**, 3110 (1969).
62. BANTHORPE, D. V., MODAWI, B. M., POOTS, I., and ROWAN, M. G. *Phytochemistry* **17**, 115 (1978).
63. BANTHORPE, D. V., BUCKNALL, G. A., HILARY, D. J., DOONAN, S., and ROWAN, M. G. *Phytochemistry* **15**, 91 (1976).
64. BATTERSBY, A. R., BURNETT, A. R., and PARSONS, P. G. *Chem. Commun.* 826 (1970).
65. BATTERSBY, A. R., Private communication.
66. TIETZE, L. F., *Angew. Chem.* **85**, 763 (1973).
67. BATTERSBY, A. R., BURNETT, A. R., and PARSONS, P. G. *Chem. Commun.* 1282 (1968).
68. BATTERSBY, A. R. and PARRY, R. J. *Chem. Commun.* 902 (1971).
69. BROWN, R. T. and ROW, L. R. *Chem. Commun.* 453 (1967).
70. BATTERSBY, A. R. and GREGORY, B. *Chem. Commun.* 134 (1968).
71. LEVESQUE, J., POUSSET, J. L., and CAVA, A. *Fitoterapia* **48**, 5 (1977).
72. COUNE, C. and ANGENOT, L. *Planta Med.* **34**, 53 (1978).
73. SOUZU, I. and MITSUHASHI, H. *Tetrahedron Lett.* 191 (1970).
74. DESILVA, K. T. D., KING, D., and SMITH, G. N. *Chem. Commun.* 908 (1971).
75. JENSEN, S. R., KJAER, A., and NIELSEN, B. J. *Phytochemistry* **12**, 2064 (1973).
76. INOUYE, H., YOSHIDA, H., NAKAMURA, T., and TOBITA, S. *Tetrahedron Lett.* 4429 (1968).
77. INOUYE, H., UEDA, S., and NAKAMURA, Y. *Tetrahdron Lett.* 5229 (1966).
78. LINDE, H. A. and RAGAB, M. S. *Helv. Chim. Acta.* **50**, 991 (1967).
79. BATTERSBY, A. R., BURNETT, A. R., KNOWLES, G. D., and PARSONS, P. G. *Chem. Commun.* 1277 (1968).
80. SOUZU, I. and MITSUHASHI, H. *Tetrahedron Lett.* 2725 (1969).

81. LOEW, P., SZCZEPANSKI, C., VON COSCIA, C. J., and ARIGONI, D. *Chem. Commun.* 1276 (1968).
82. BATTERSBY, A. R. *The alkaloids*, Vol. I, pp. 39–40, Specialist Periodical Reports, Chemical Society, London (1971).
83. INOUYE, H., UEDA, S., and TAKEDA, Y. *Tetrahedron Lett.* 3453 (1968).
84. GUARNACCIA, R. and COSCIA, C. J. *J. Am. Chem. Soc.* **93**, 6320 (1971).
85. BATTERSBY, A. R., BURNETT, A. R., and PARSON, P. G. *J. Chem. Soc.,* (C), 1187 (1969).
86. MADYASTHA, K. M., GUARNACCIA, R., and COSCIA, C. J. *FEBS Lett.* **14**, 175 (1971).
87. COSCIA, C. J., MADYASTHA, K. M., and GUARNACCIA, R. *Fed. Proc. Fed. Am. Soc. exp. Biol.* **30**, Abstr. 1472 (1971).
88. MADYASTHA, K. M., GUARANCCIA, R., and COSCIA, C. J. *Biochem. J.* **128**, 341 (1972).
89. BATTERSBY, A. R., BURNETT, A. R., and PARSONS, P. G. *Chem. Commun.* 1280 (1968).
90. BROWN, R. T., CHAPPLE, C. L., DUCKWORTH, D. M., and PLATT, R., *J. Chem. Soc., Perkin Trans.* I(2), 160 (1976).
91. BROWN, R. T., LEONARD, J., and SLEIGH, S. K. *J. Chem. Soc. Chem. Commun.* 636 (1977).
92. BROWN, R. T., CHAPPLE, C. L., PLATT, R., and SLEIGH, S. K. *Tetrahedron Lett.* 1829 (1976).
93. SCOTT, A. I. and LEE, S-L. *J. Am. Chem. Soc.* **97**, 6906 (1975).
94. SCOTT, A. I., and LEE, S-L. *Rev. Latinoam Quim.* **9**, 131 (1978).
95. BANTHORPE, D. V., CHARLWOOD, B. V., and FRANCIS, M. J. O. *Chem. Rev.* **72**, 115 (1972).
96. COSCIA, C. J., BOTTA, L., and GUARNACCIA, R. *Arch. Biochem. Biophys.* **136**, 498 (1970).
97. POPJAK, C. and CORNFORTH, J. W. *Biochem. J.* **101**, 553 (1966).
98. CLAYTON, R. B. *Q. Rev. (Lond.)* **19**, 168 (1965).
99. CLAYTON, R. B. *Q. Rev. (Lond.)* **19**, 201 (1965).
100. RICHARDS, J. H. and HENDRICKSON, J. B. *Biosynthesis of steroids, terpenoids and acetogenins*, W. A. Benjamin, New York, p. 416 (1964).
101. POPJAK, G. *Natural substances formed biologically from mevalonic acid*, Academic Press, New York (1970).
102. LYNEN, F., HENNING, U., BUBLITZ, C., SORBO, B., and RUEFF, L. K. *Biochem. Z.* **330**, 269 (1958).
103. RUDNEY, H. and FERGUSON, J. J. *J. biol. Chem.* **234**, 1076 (1959).
104. RETEY, J., STETTEN, E., VON COY, O., and LYNEN, F. *Euro. J. Biochem.* **15**, 72 (1970).
105. BLATTMAN, P. and RETEY, J. *Chem. Commun.* 1394 (1970).
106. HANSON, K. R. *J. Am. Chem. Soc.* **88**, 2731 (1966).
107. DURR, I. F. and RUDNEY, H. *J. biol. Chem.* **235**, 2572 (1960).
108. GARCIA-PEREGRIN, E., SUAREZ, M. D., ARAGON, M. C., and MAYOR, F. *Phytochemistry* **11**, 2495 (1972).
109. POTTY, V. H. and BRUEMMER, J. H. *Phytochemistry* **9**, 1229 (1970).
110. BANTHORPE, D. V. and WIRZ-JUSTICE, A. *J. Chem. Soc. Perkin Trans.* **1**, 1769 (1972).
111. VALENZUELA, P., BEYTIA, E., CORI, O., and YUDELEVICH, A. *Arch. Biochem. Biophys.* **113**, 536 (1966).

112. CORNFORTH, J. W., CORNFORTH, R. H., POPJAK, G., and VENGOYAN, L. *J. biol. Chem.* **241**, 3870 (1966).
113. CHAKYIN, S., LAW, J., PHILLIPS, A. H., TCHEN, T. T., and BLOCH, K. *Proc. Natl. Acad. Sci., U.S.A.* **44**, 998 (1958).
114. LYNEN, F., EGGERE, H., HENNING, U., and CASTLE, I. *Angew. Chem.* **70**, 738 (1958).
115. ELIEL, E. L. *Stereochemistry of carbon compounds*, p. 436. McGraw-Hill, New York (1962).
116. CORNFORTH, J. W., CORNFORTH, R. H., DONINGER, C., and POPJAK, G. *Proc. R. Soc. Lond. B.* **163**, 492 (1966).
117. CLIFFORD, K., CORNFORTH, J. W., MALLABY, R., and PHILLIPS, G. T. *Chem. Commun.* 1599 (1971).
118. AGRANOFF, B. W., EGGERER, H., HANNING, U., and LYNEN, F. *J. biol. Chem.* **235**, 326 (1960).
119. FRANCIS, M. J. O., BANTHROPE, D. V., and LE-PATOUREL, G. N. J. *Nature (Lond.)* **228**, 1005 (1970).
120. BIRCH, A. J., KOCOR, M., SHEPPARD, N., and WINTER, J. *J. Chem. Soc.* 1502 (1962).
121. BIRCH, A. J., ENGLISH, R. J., MASSEY-WESTROP, R. A., and SMITH, H. *J. Chem. Soc.* 369 (1958).
122. ARIGONI, D. *Experientia* **14**, 153 (1958).
123. BATTERSBY, A. R., BYRNE, T. C., KAPIL, R. S., MARTIN, J. A., PAYNE, T. G., ARIGONI, D., and LOEW, P. *Chem. Commun.* 951 (1968).
124. GUARNACCIA, R., BOTTA, L., and COSCIA, C. J. *J. Am. Chem. Soc.* **91**, 204 (1969).
125. BATTERSBY, A. R., HALL, E. S., and SOUTHATE, R. *Chem. Commun.* 131 (1968).
126. INOUYE, H., YOSHIDA, T., and TOMITA, S. *Tetrahedron Lett.* 2945 (1968).
127. LENZ, P. J. and ROSSMANN, M. G. *Chem. Commun.* 1269 (1969).
128. BATTERSBY, A. R. and HALL, E. S. *Chem. Commun.* 793 (1969).
129. HECKENDORF, A. H., MATTES, C. K., HUTCHINSON, C. R., HAGAMAN, E. W., and WENKERT, E. *J. org. Chem.* **41**, 2045 (1976).
130. STUART, K. L., KUTNEY, J. P., HONDA, T., LEWIS, N. G., and WORTH, B. R. *Heterocycles* **9**, 647 (1978).

2

BIOGENETIC CLASSIFICATION OF INDOLE ALKALOIDS

In the vast domain of indole alkaloid chemistry, a very large number of highly complex heterocyclic structures exist. When the common structural features in these alkaloids are considered, they can be readily assigned to five broad classes based on the skeleton of the non-tryptophan unit.

In Chapter 1, the intermediacy of loganin and secologanin in the biosynthetic schemes was discussed. After condensation with tryptophan these skeletal systems can be recognized in an unrearranged form as the α- or β-condensation products, in several common types of indole alkaloids, e.g. *Corynanthe*, *Yohimbe*, *Strychnos* groups, and these form the first class of indole alkaloids. In the same class are also included the secodine group of indole alkaloids in which the secologanin group can be found as an opened (but unrearranged) form.

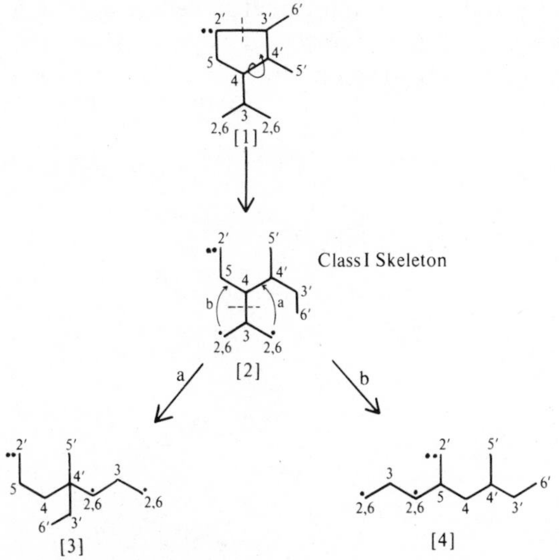

Class II Skeleton

Class III Skeleton

SCHEME 2.1

In the second class of indole alkaloids, the secologanin skeletal system is no longer found in its original form. A cleavage has occurred between C-3 and C-4 and a new bond has been formed between C-2,6 and C-4 (Scheme 2.1). The majority of alkaloids in this class are those of the *Aspidosperma* or *Hunteria* families. There is a growing volume of evidence that the rearrangement of these terpenoidal fragments occurs *after* condensation with tryptophan. Incorporation experiments with vincoside and stricto-sidine showed that these were incorporated into catharanthine, vindoline, serpentine, ajmalicine, and perivine, providing strong evidence that the rearrangement process occurs after condensation of the C-10 aldehyde unit with tryptophan. The mechanisms of these rearrangements will be discussed in the following chapters.

The third class of indole alkaloids is divided into two sub-classes. The first is that of the *Iboga* type. In these alkaloids, the same C—C cleavage has occurred as in the *Hunteria–Aspidosperma* series but the new bond for-mation is between C-2(6) and C-5 (Scheme 2.1). The second sub-class in-cludes those alkaloids which are derived from tryptophan and secologanin but which present a novel C-10 skeleton due to expansive rearrangement.

The fourth class of indole alkaloids comprises (a) non-tryptophan indole alkaloids, i.e., those containing the carbazole nucleus, (b) non-isoprenoid tryptophan alkaloids, and (c) fungal indole alkaloids.

The fifth class of indole alkaloids consists of the bis-indole alkaloids. In these 'dimeric' substances, one or both moieties bear the indole nucleus either as such or in a modified form. The various binary alkaloids have been grouped on the basis of the classes of the individual moieties. The anti-leukaemic alkaloid vincristine, for instance, would fall in the II–III group-ing as one of the moieties bears a class II skeleton while the other moiety has a class III skeleton.

2.1 Class I alkaloids

Vincoside group

Rubenine group

5α - Carboxystrictosidine.

Rubenine

Macrolidine group

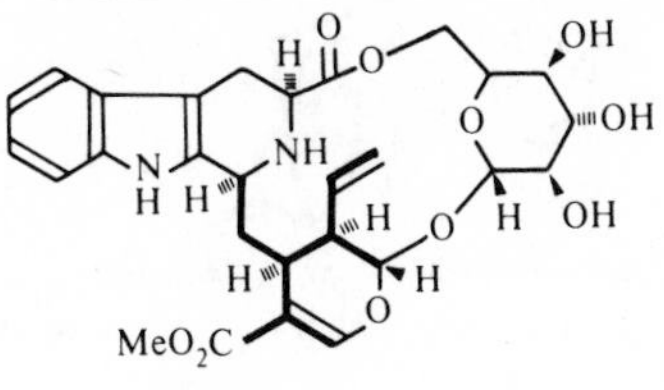

Macrolidine

Cadambine group

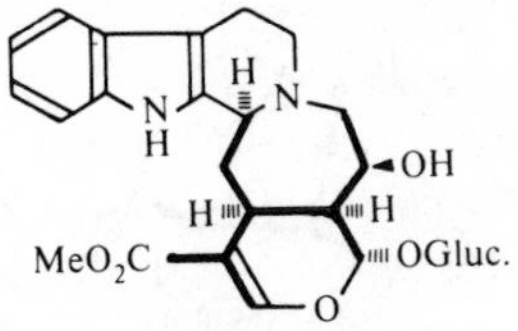

3α - Dihydrocadambine

Strictosamide group

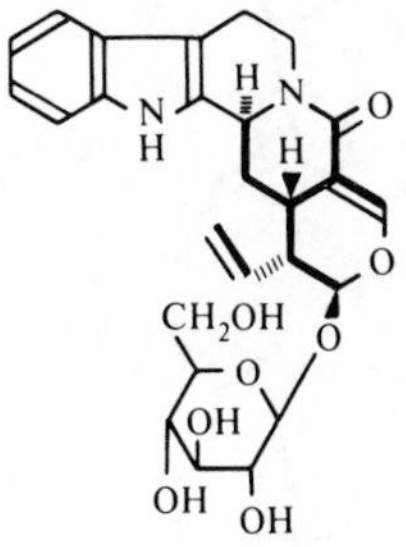

Strictosamide

Cordifoline group

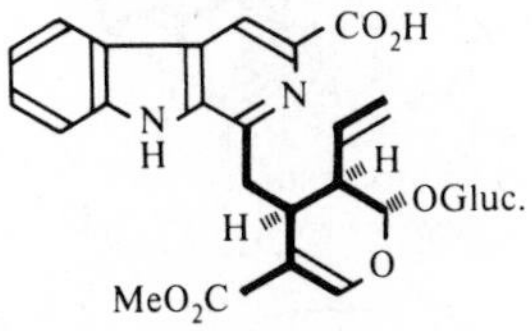

Cordifoline

Lyalidine group

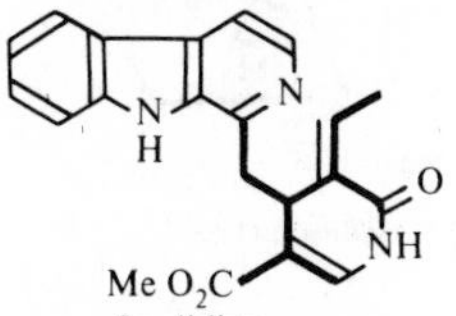

Lyalidine

Talbotine group

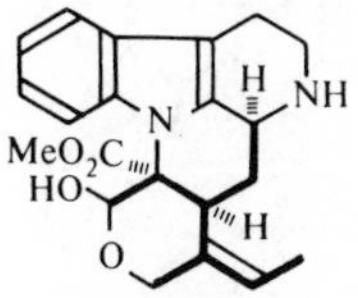

Talbotine

Cadamine group

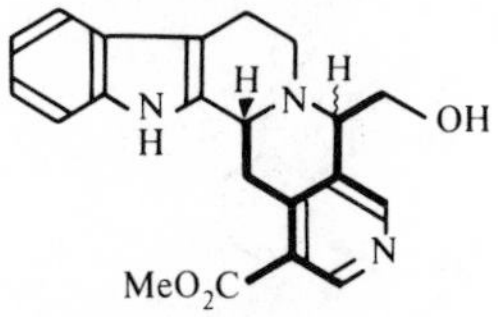

Cadamine

Vallesiachotamine group

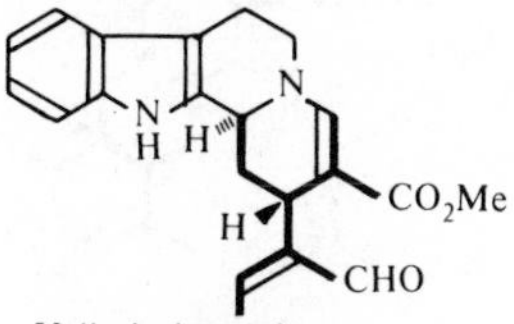

Vallesiachotamine

Adifoline group

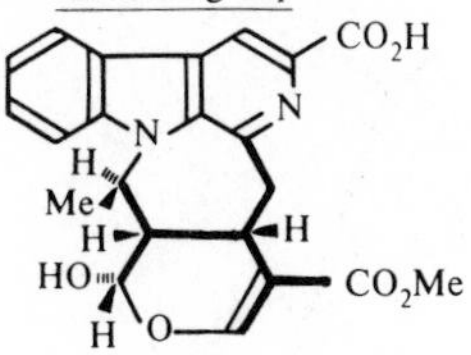

Adifoline

Corynantheine group

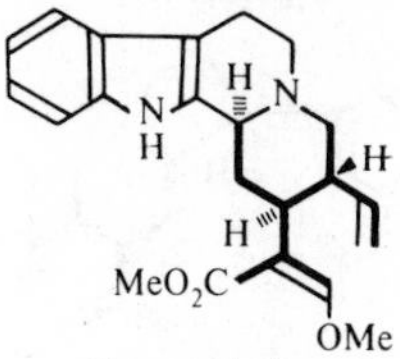

Corynantheine

C-Fluorocurine group

HOH₂C

C-Fluorocurine

Reflexine group

Reflexine

Aspidospermatine group

Condylocarpine CO₂Me

Gardnutine group

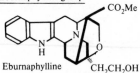

Gardnutine

Vinoxine group

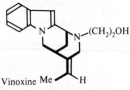

Vinoxine Me

Eburnaphylline group

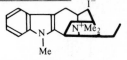

Eburnaphylline CH₂CH₂OH

Sarpagine group CH₂OH

CO₂Me

Polyneuridine

Isoaffisinine group

Isoaffisinine iodomethylate

Ajmalicine group

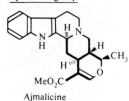

Ajmalicine

Oxogambirtannine group

MeO₂C

Oxogambirtannine

Elegantine group

H₃CO

OCH₃

MeO₂C
Elegantine

Picraline group

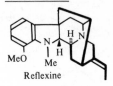

Akuammiline

Picraphylline group

Picraphylline

Aspidodasycarpine group

Aspidodasycarpine

Yohimbine group

Yohimbine

Mavacurine group

Mavacurine

Ajmaline group

Ajmaline

Vobasine group

Vincadiffine

Cinchonamine group

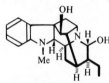

Cinchonamine

Macroline group

Macroline

Corynine group

Corynine

Perakine group

Perakine

Erinine group

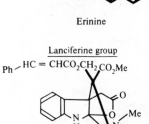

Erinine

Peraksine group

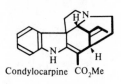

Peraksine

Lanciferine group

$Ph-HC=CHCO_2CH_2CO_2Me$

Lanciferine

Condylocarpine group

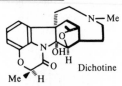

Condylocarpine CO_2Me

Chitosenine group

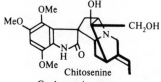

Chitosenine

Dichotine group

Dichotine

Gardneramine group

Gardneramine

Tubifoline group

Tubifoline

Rhyncophylline group

Rhyncophylline OMe

Pandine group

Pandine

Stemmadenine group

Stemmadenine CO_2Me

Vomicine group

Vomicine

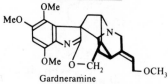
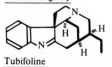

Uleine group

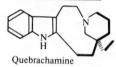

De-N-methyluleine

Secodine group

16,17-Dihydrosecodin-17-ol

Strychnine group

Strychnine

Calebassinine group

C-calebassinine

2.2 Class II alkaloids

Quebrachamine group

Quebrachamine

Aspidospermine group

CO$_2$Me
Vincadifformine

Cathovaline group

CH$_3$ CO$_2$Me
Cathovaline

Apodine group

Apodine CO$_2$Me

Buxomeline group

OH
N
H
CO$_2$Me
Buxomeline

Aspidoalbidine group

Fendleridine

Alalakine group

MeO
MeO
Et Alalakine

Vincatine group

CH$_3$ CO$_2$Me
Vincatine

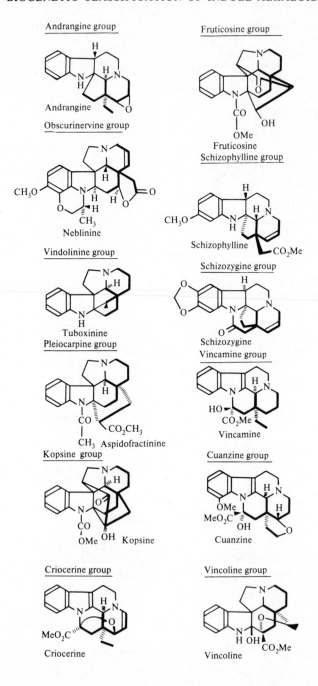

Andrangine group

Andrangine

Obscurinervine group

CH₃O

Neblinine

Vindolinine group

Tuboxinine

Pleiocarpine group

CO₂CH₃

CH₃ Aspidofractinine

Kopsine group

OMe OH Kopsine

Criocerine group

MeO₂C

Criocerine

Fruticosine group

CO
OMe
Fruticosine

Schizophylline group

CH₃O

Schizophylline CO₂Me

Schizozygine group

Schizozygine

Vincamine group

HO
CO₂Me
Vincamine

Cuanzine group

OMe
MeO₂C
OH
Cuanzine

Vincoline group

OH CO₂Me
Vincoline

2.3 Class III alkaloids

Catharanthine group

Catharanthine

MeO₂C
Catharanthine

Rupicoline group

MeO

Rupicoline   CO₂Me

Eglandine group

MeO₂C
Eglandine

Capuronine group

Capuronine

Ervetamine group

CO₂ Me
N—Me

Ervetamine

Ervitsine group

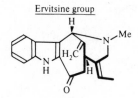

N—Me

H₂C

Ervitsine

Pandoline group

OH

Pandoline   CO₂Me

Ibophyllidine group

Ibophyllidine   CO₂Me

Iboxyphylline group

Me

OH

CO₂Me
Iboxyphylline

Andranginine group

MeO₂C

Andranginine

Nitramidine group

Nitrarine

2.4 Class IV alkaloids (not derived from secologanin)

2.4.1 Non-tryptophan indole alkaloids

Murrayanine group

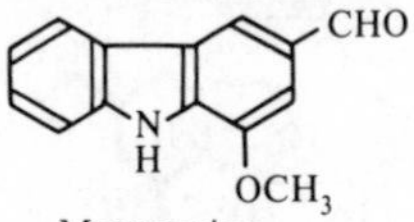

Murrayanine

Currayangine group

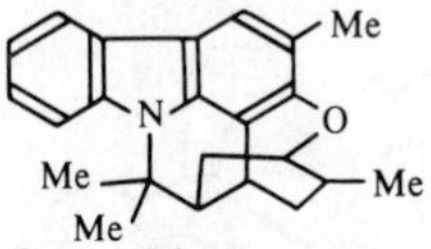

Currayangine

Girinimbine group

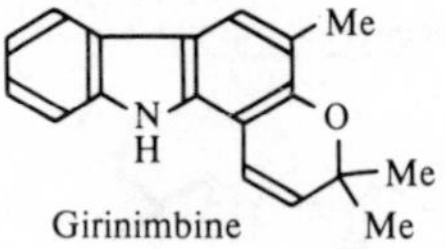

Girinimbine

Murrayazolidine group

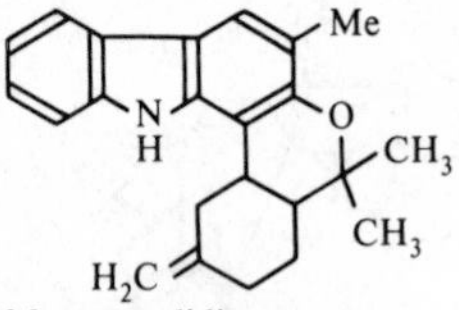

Murrayazolidine

Mahanimbidine group

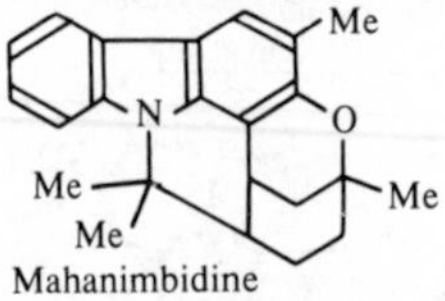

Mahanimbidine

Bicyclomahanimbine group

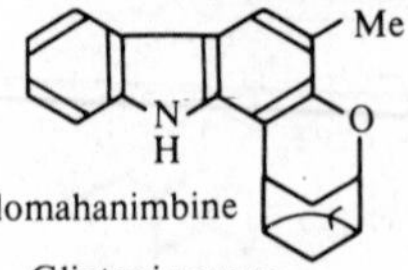

Bicyclomahanimbine

Cyclomahanimbine group

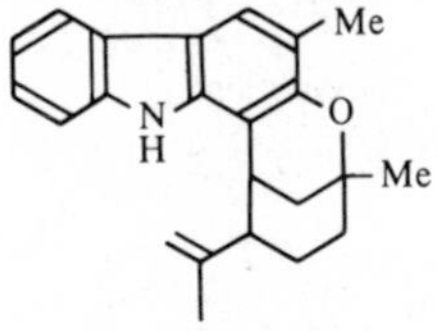

Cyclomahanimbine

Gliotoxin group

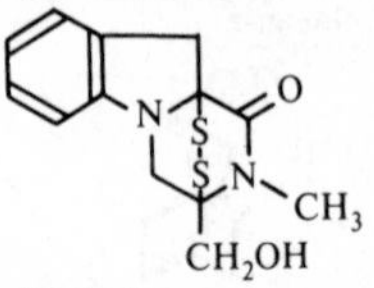

Gliotoxin

Subincanine group

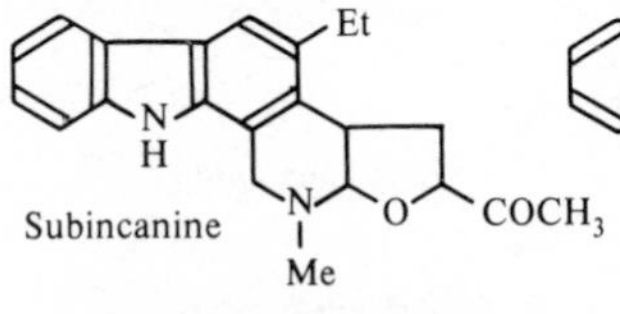

Subincanine

Paxilline group

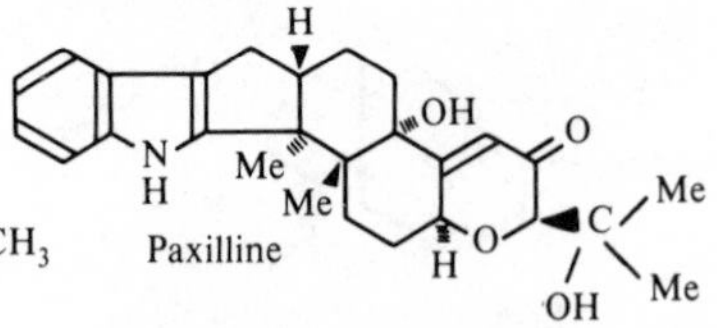

Paxilline

Couroupitine A group

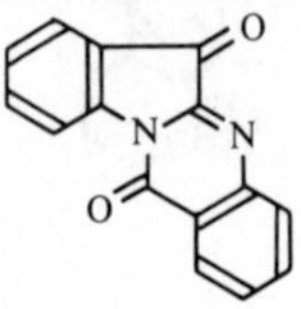

Couroupitine A

Physovenine group

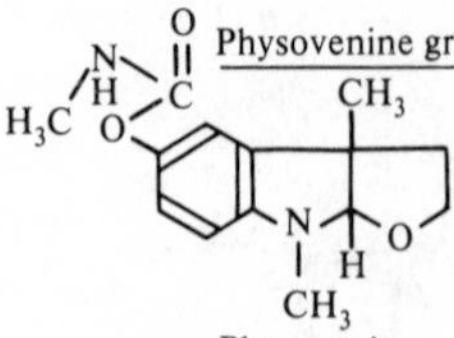

Physovenine

2.4.2 Non-isoprenoid tryptophan alkaloids

Tryptophan group

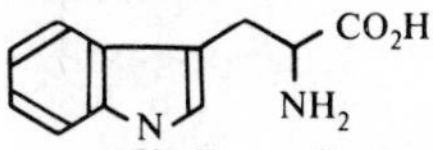

Tryptophan

Harmaline group

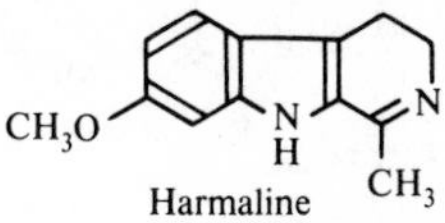

Harmaline

Amphibine group

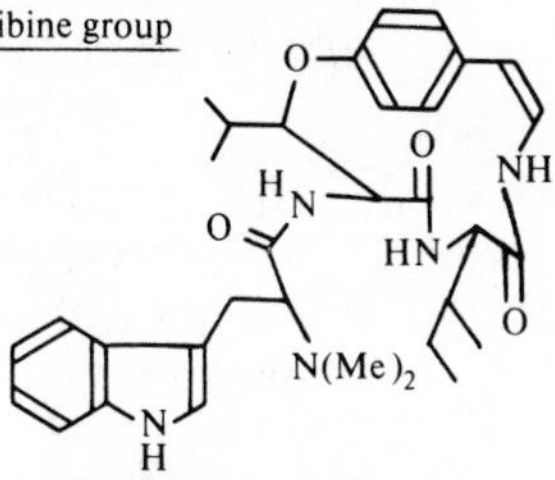

Amphibine A

Indolopyridine

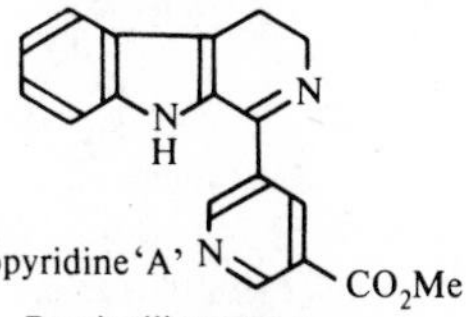

Indolopyridine 'A'

Peduncularine group

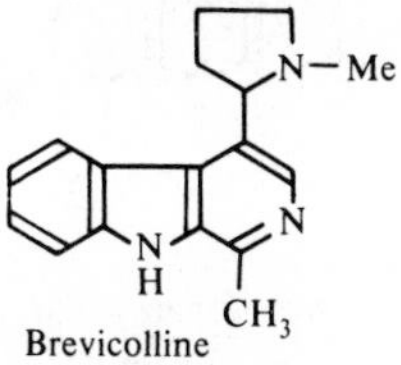

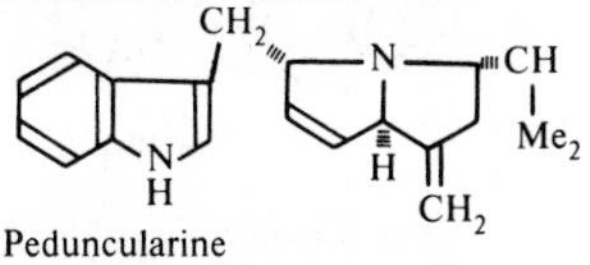

Peduncularine

Brevicolline group

Brevicolline

Indolmycin group

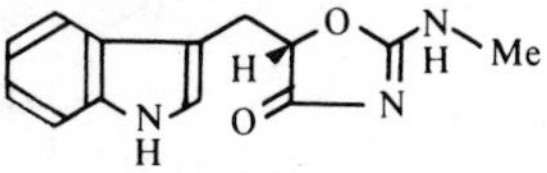

Indolmycin

Perlolyrine group

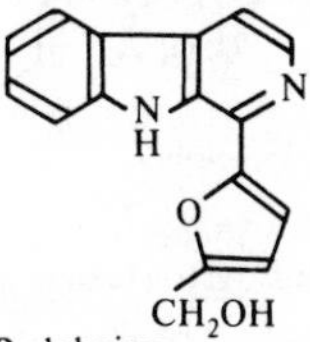

Perlolyrine

Indoloquinolizidine group

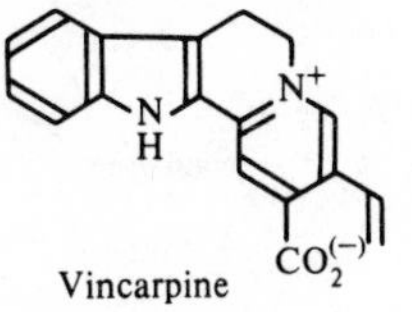

Vincarpine

Naufoline group

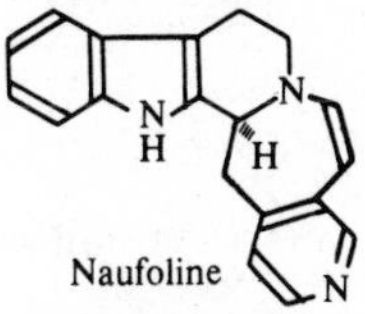

Naufoline

Nitrarine group

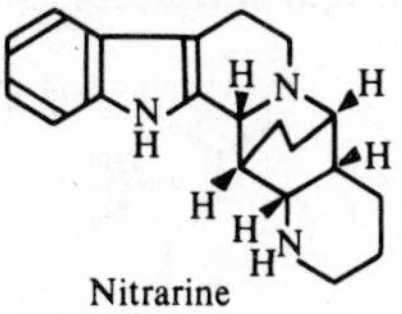

Nitrarine

Canngunine group

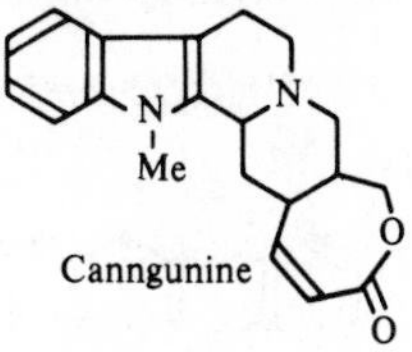

Canngunine

Parvine group

Parvine

Physostigmine group

Physostigmine

Rutecarpine group

Rutecarpine

Geneserine group

Geneserine

Elaeocarpidine group

Elaeocarpidine

Dendrodoine group

Dendrodoine

Naucleonine group

Naucleonine

Nauclefine group

Nauclefine

Coumaryl tryptamine group

N_b-(λ-coumaryl)-tryptamine

Naulafine group

Naulafine

2.4.3 Isoprenoid tryptophan alkaloids (fungal indole alkaloids)

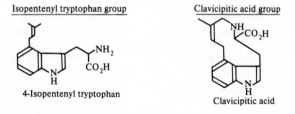

Isopentenyl tryptophan group

4-Isopentenyl tryptophan

Clavicipitic acid group

Clavicipitic acid

Borreline group

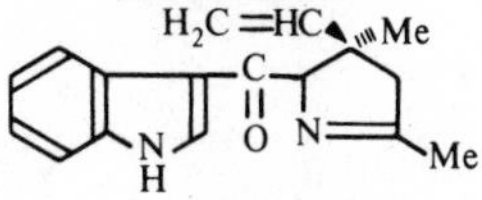

Borreline

Borrecarpine group

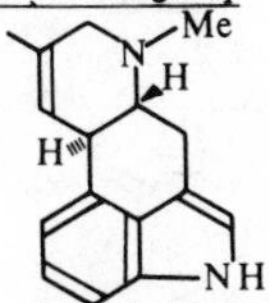

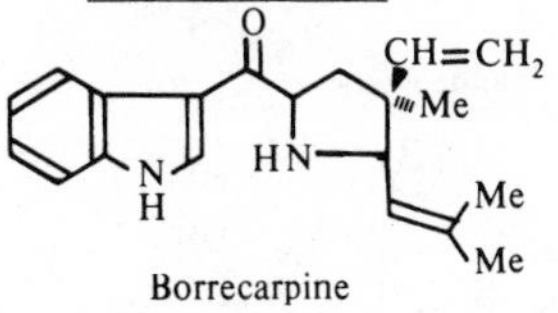

Borrecarpine

Chanoclavine group

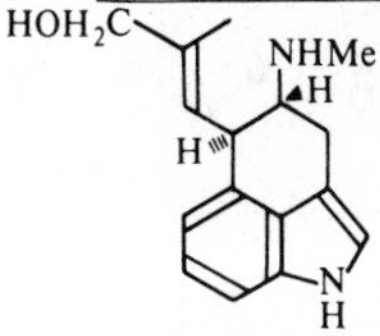

Chanoclavine-I

Ergotoxin group

Ergocristine

Rugulovasine group

Rugulovasine

β - Cyclopiazonic acid group

β - Cyclopiazonic acid

Agroclavine group

Agroclavine

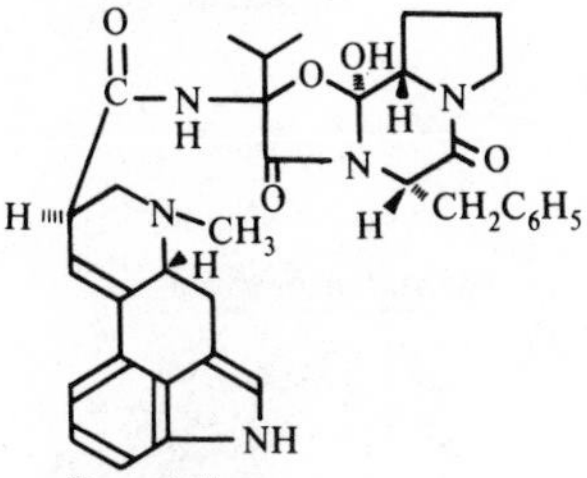

Cycloclavine group

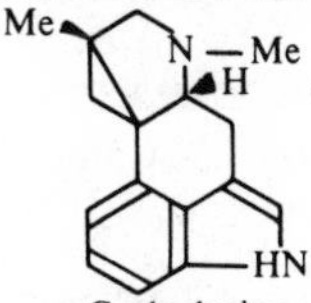

Cycloclavine

Lysergic acid group

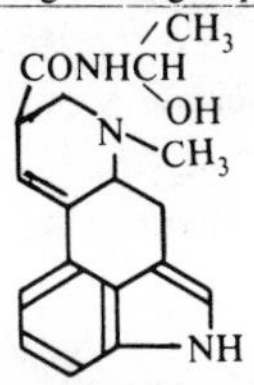

Lysergic acid-α-
hydroxyethyl amide

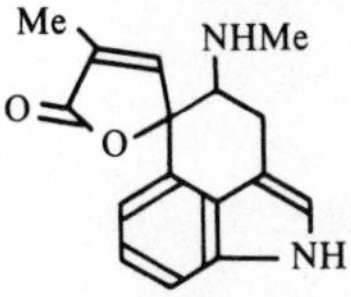

Cyclopiazonic acid group

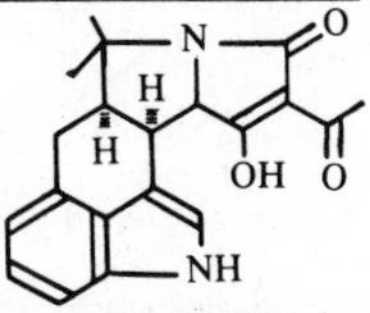

Cyclopiazonic acid

Neoechinulin group

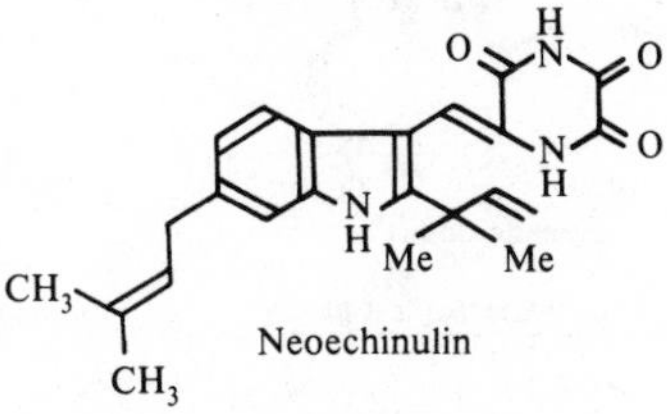

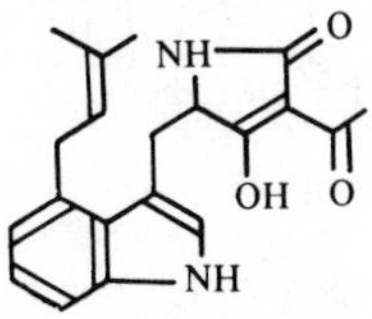

Neoechinulin

Alysinopsin group

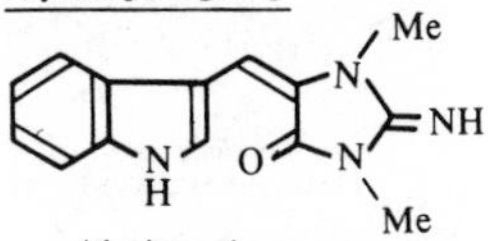

Alysinopsin

Olivacine group

Olivacine

Akaferine group

Akaferine

Chaetoglobosin group

Chaetoglobosin A

Oxaline group

Oxaline

Cryptoechinulin group

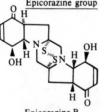

Cryptoechinulin B

Aristotelone group

Aristotelone

Epicorazine group

Epicorazine B

Austamide group

12,13-Dihydro-12-hydroxyaustamide

Dioxopiperazine-I group

Dioxopiperazine-I

Fumitremorgin group

Fumitremorgin

Dioxopiperazine- II group

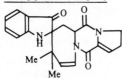

Dioxopiperazine-II

Roquefortine group

$H_2C \rightarrow HCCMe_2$

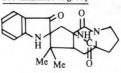

Roquefortine

Dioxopiperazine - III group

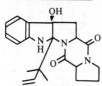

Dioxopiperazine-III

Brevianamide A group

Brevianamide A

Brevianamide E group

OH

Brevianamide E

Aristoteline group

Aristoteline

Aristone group

Aristone

Paspalicine group

Paspalicine

Paspaline group

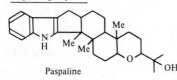

Paspaline

Sporidesmin group

Sporidesmin A

Tryptoquivaline group

AcO

Tryptoquivaline

Surugatoxin group

Surugatoxin

2.5 Class V alkaloids (binary indole alkaloids)

0-II group (0 = non-indolic moiety)

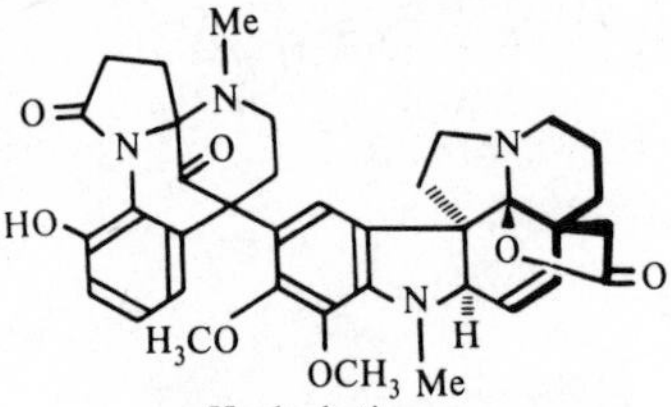

Haplophytine

0-IV group

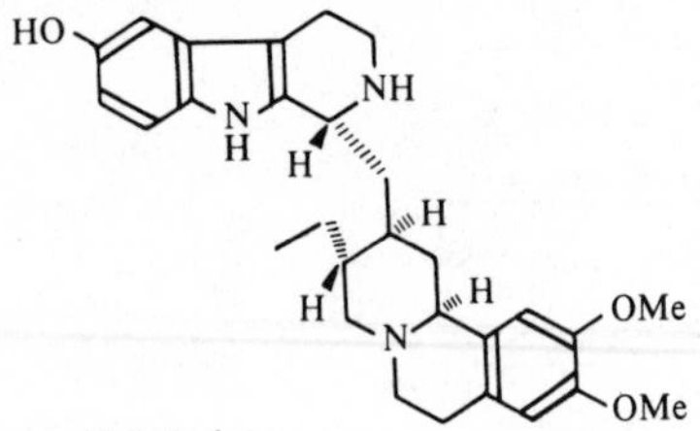

Tubulosine

I - I group

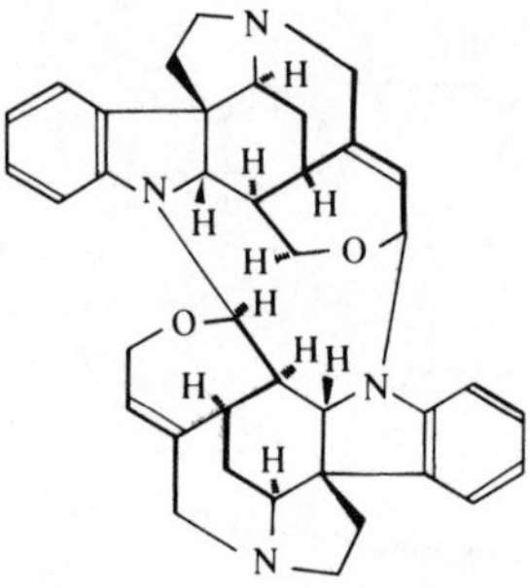

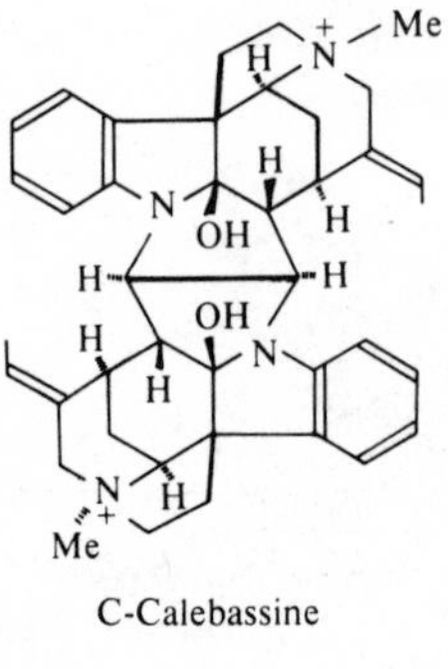

C-Calebassine

I - I group (cont.)

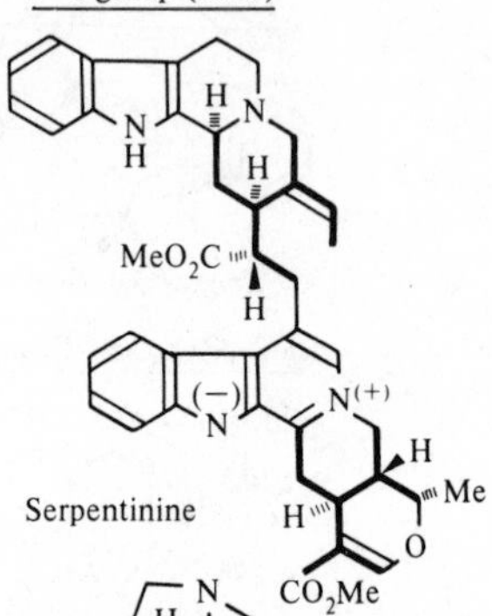

Serpentinine

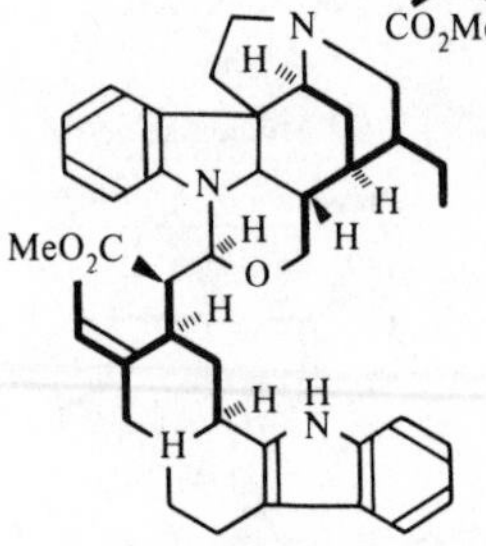

Geissospermine

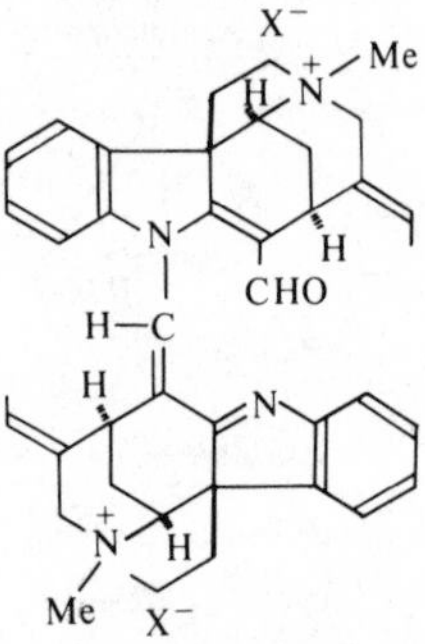

Ultracurine A

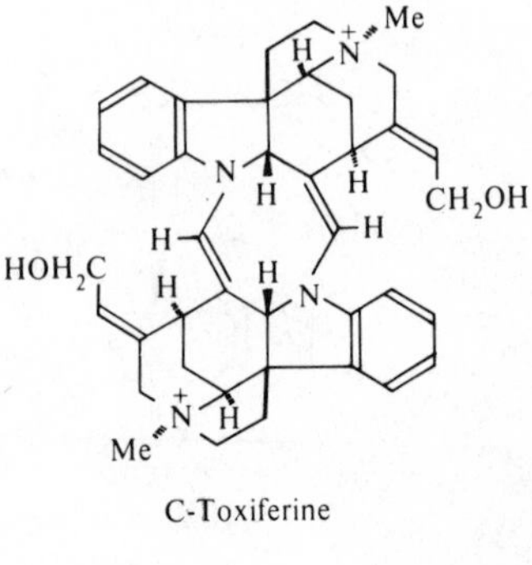

C-Toxiferine

I – II group

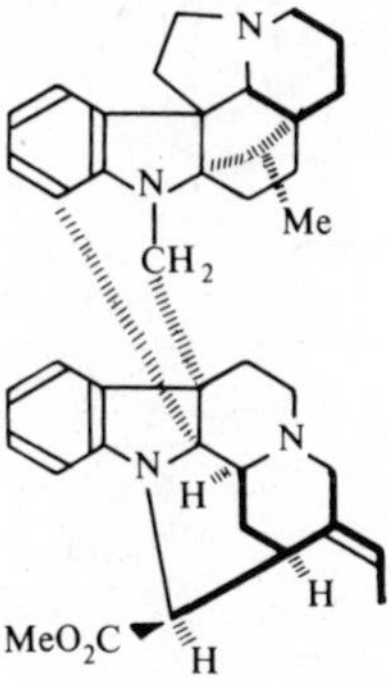

14′,15′- dihydropycnanthine

I – III group

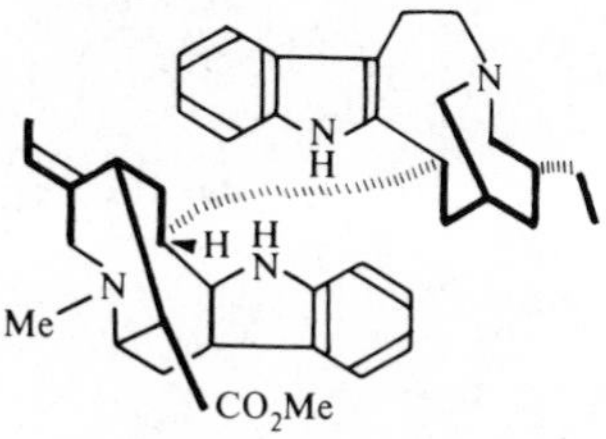

Dehydroxycopuvosine

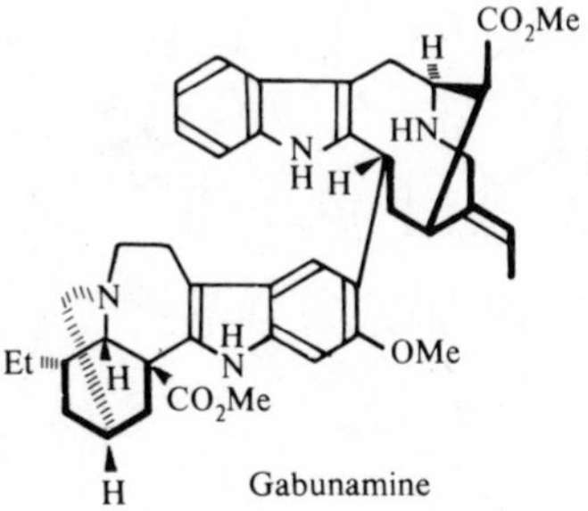

Gabunamine

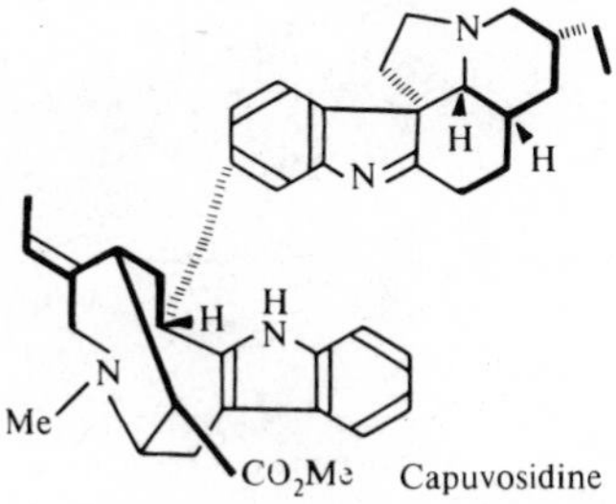

Capuvosidine

I – III group (cont.)

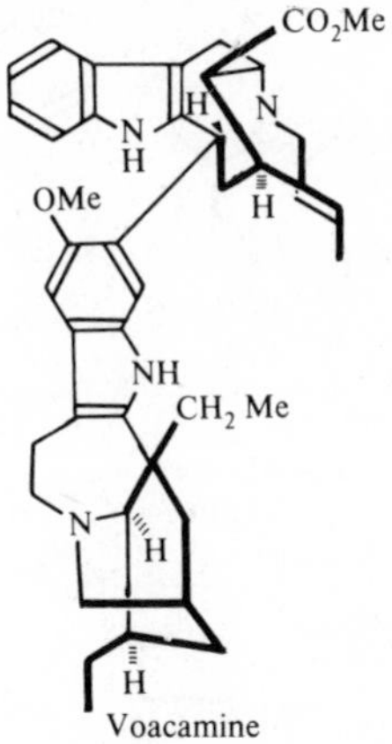

Voacamine

I – IV group

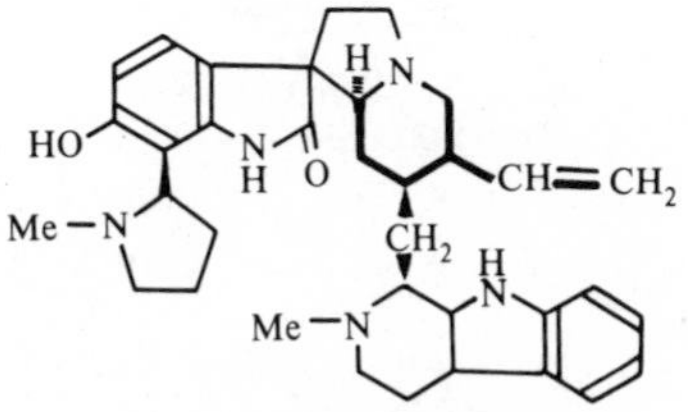

Strychnophylline

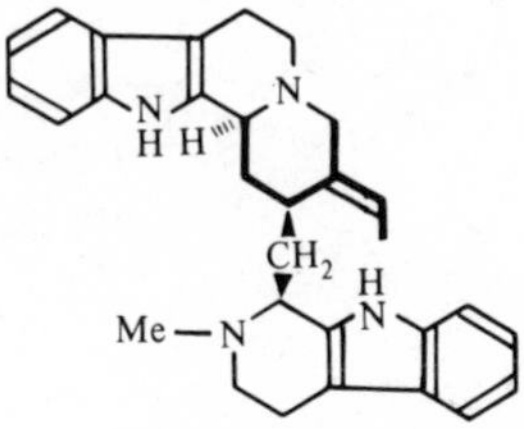

Usambarine

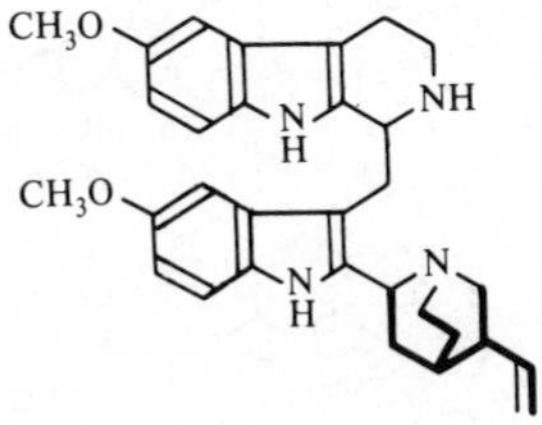

Cinochophyllamine

II – II group

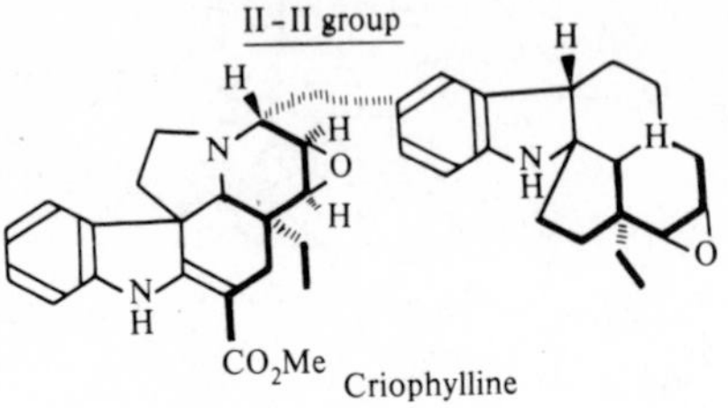

Criophylline

II – III group (cont.)

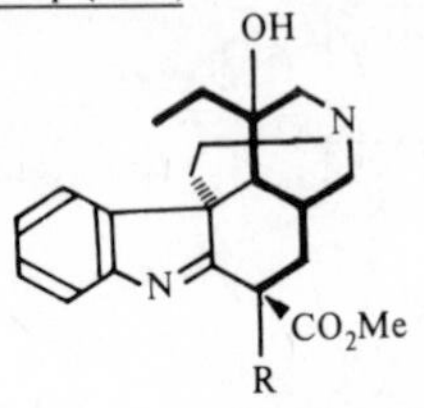

Vincathicine (R = Vindoline)

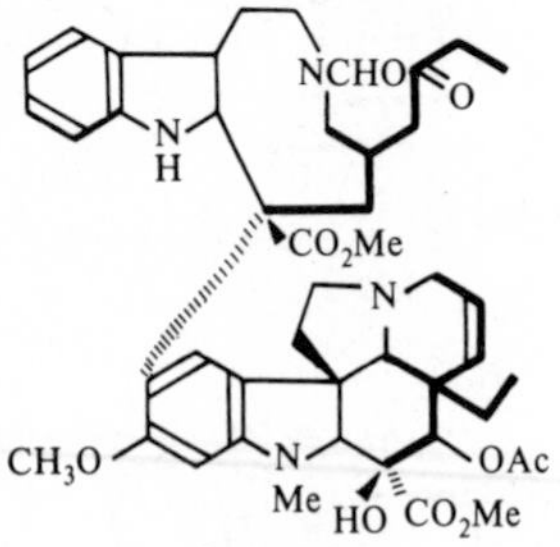

Catharinine

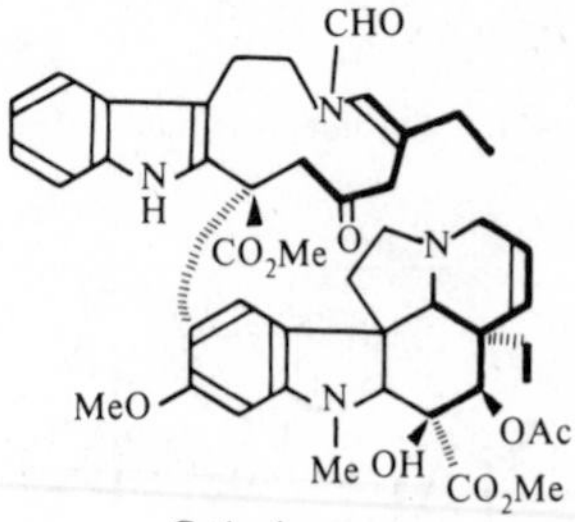

Catharine

III – III group

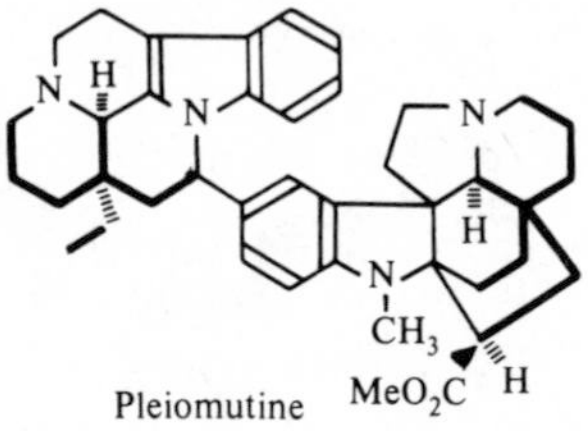

Pleiomutine

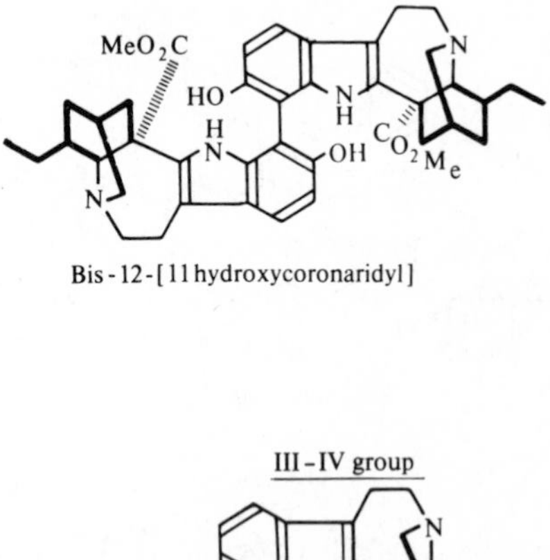

Bis - 12 - [11 hydroxycoronaridyl]

II – III group

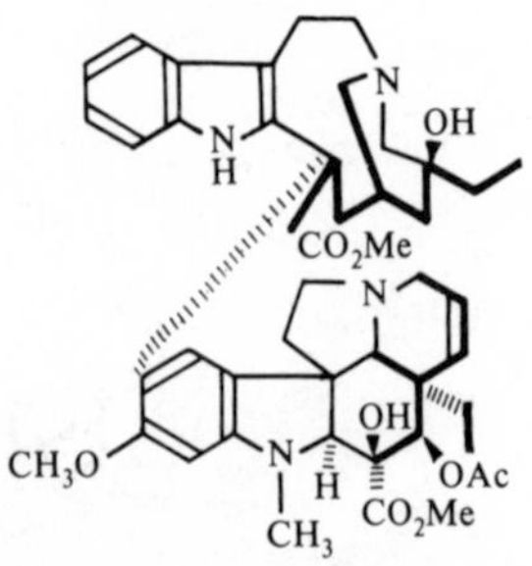

Vinblastine

III – IV group

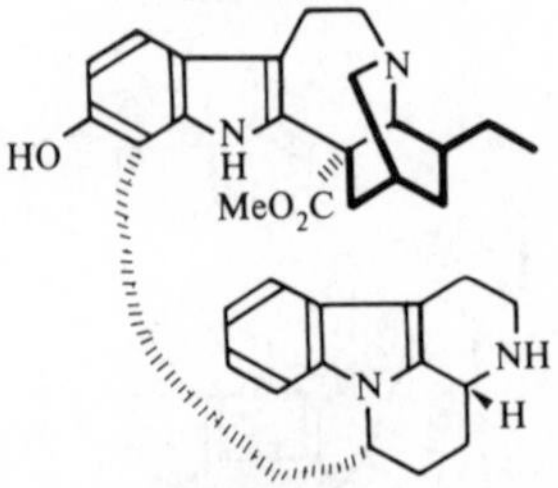

Bonafousine

III – IV group (cont.)

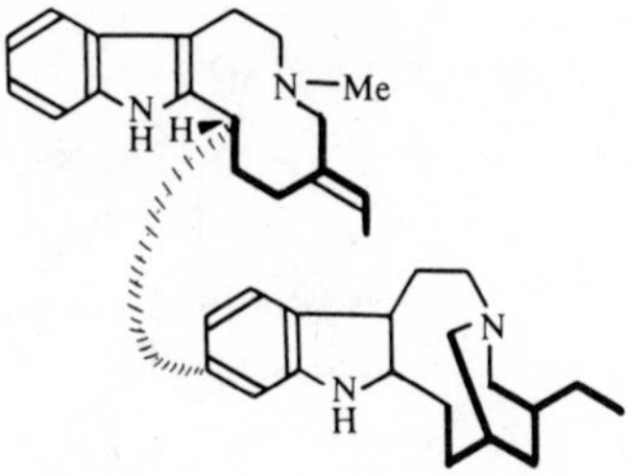

IV – IV group (cont.)

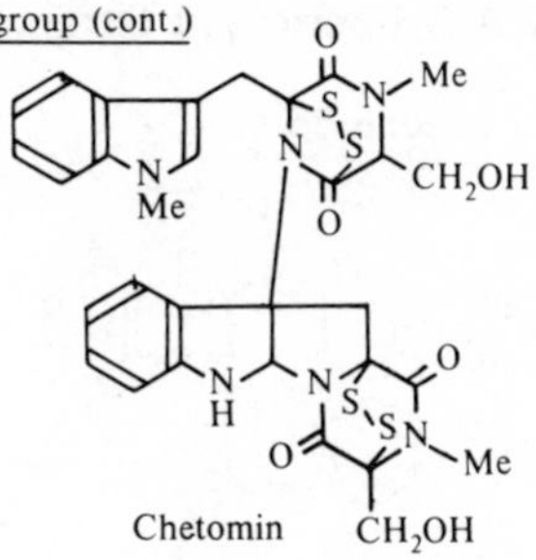

Chetomin

IV – IV group

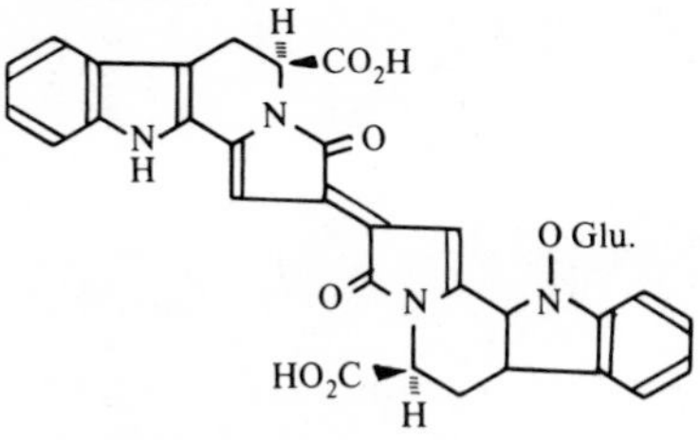

Trichotomine G₁

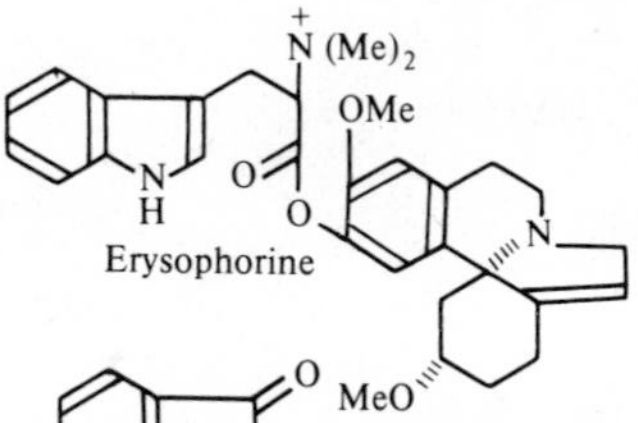

Erysophorine

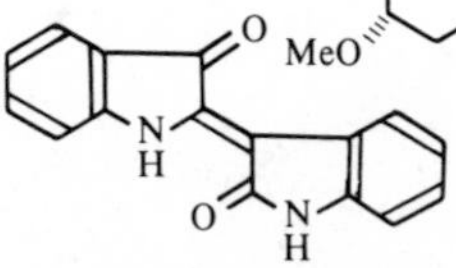

Couropitine B (Indirubin)

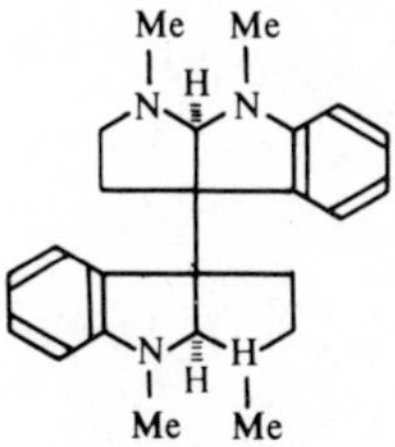

Foliocanthine

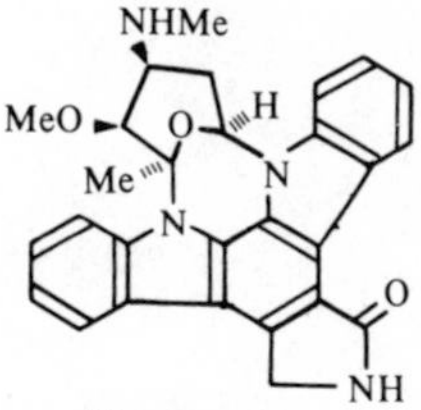

Staurosporine

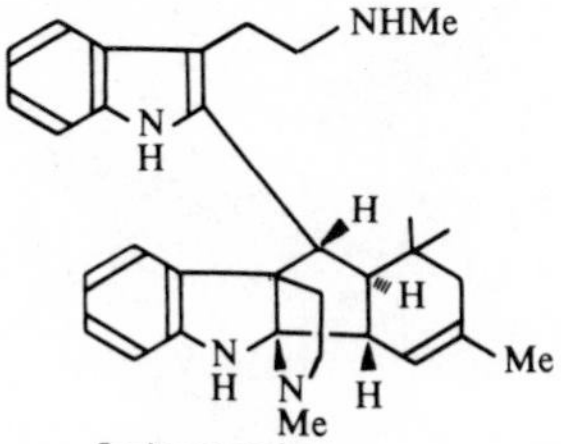

Isoborreverine

2.6 Trimeric and tetrameric indole alkaloids

IV–IV–IV group

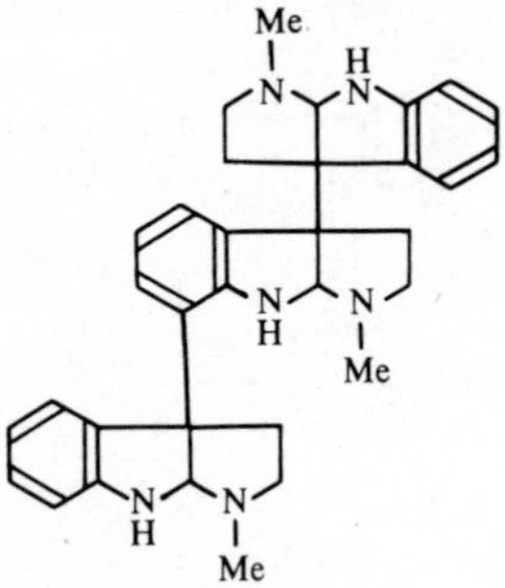

Hodgkinsine

IV–IV–IV–IV group

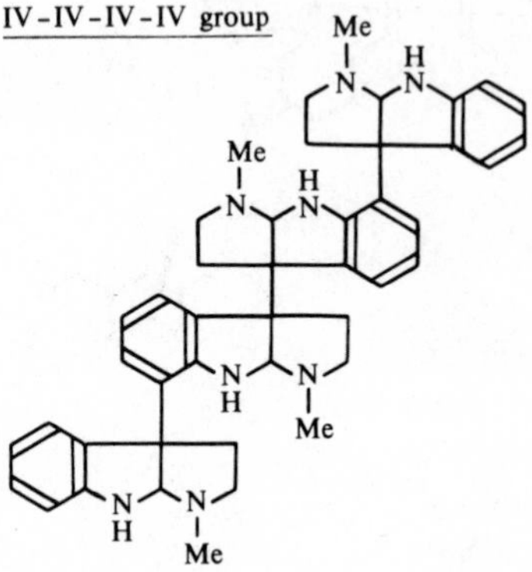

Quadrigemine - A

3

BIOSYNTHESIS OF CLASS I ALKALOIDS
(*Corynanthe–Strychnos Type*)

A very large number of indole alkaloids belong to class I in which an unrearranged secologanin unit [1] can be recognized in combination with tryptamine/tryptophan. The majority of the alkaloids in this class possess S chirality at C-15. In vallesiachotamine [2] and certain other closely related alkaloids, the opposite chirality is present at this centre.

The subsequent elaboration of these alkaloids to the vast diversity of structures found within each indole alkaloid family makes fascinating reading for organic chemists. The complex functionalities built up in a step-wise manner serve to illustrate the intricate, yet beautifully simple synthetic, processes operative in nature. The majority of the reactions through which these elaborations occur can be envisaged to proceed through enamine or dienamine chemistry. Within each class there are certain alkaloids of the correct activity and oxidation level which serve as 'link' alkaloids, linking one class with the other. During the last few years an enormous amount of work has been conducted in the detection of such 'link' alkaloids. Feeding experiments with such alkaloids are then being used to unravel the complex intra- and inter-class pathways.

The stepwise formation of secologanin [1] from mevalonate was described in Chapter 1. Condensation of secologanin [1] with tryptophan should afford the alkaloid [3], a typical example of the strictosidine group. 5α-Carboxystrictosidine, has been isolated and its absolute stereochemistry has been determined.[1] The first non-glycosidic alkaloid to be derived from tryptophan and secologanin is adirubine [4].

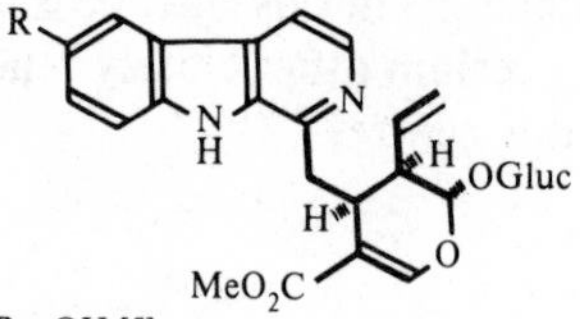

R=OH [5]
R=H [6]

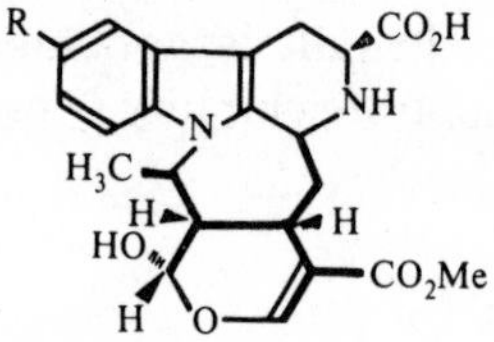

R=OH [7]
R=H [8]

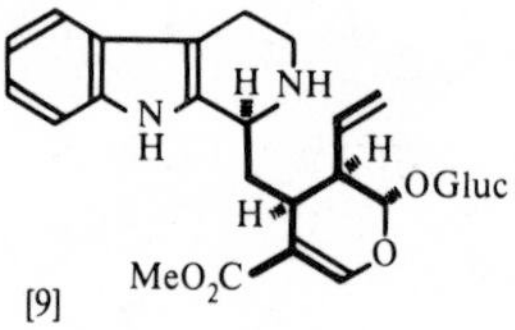

[9]

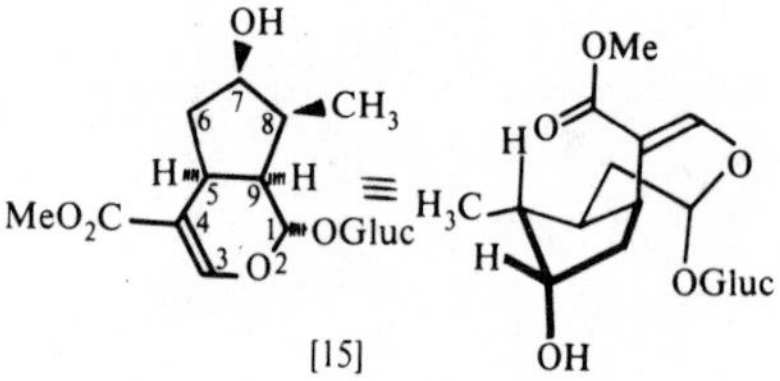

[15]

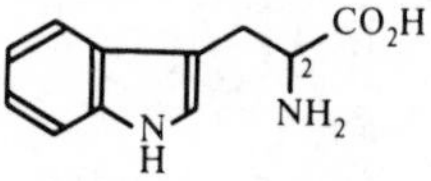

[16]

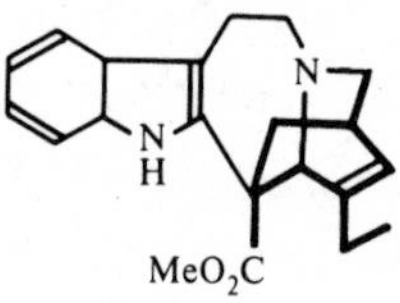

[17]

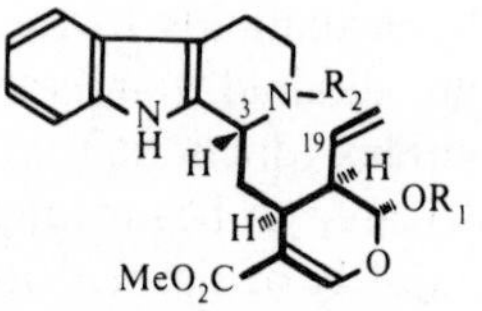

R₁=gluc., R₂=H [10]
R₁= gluc (OAc)₅, R₂=H [11]
R₁=gluc., R₂=Ac [12]

The first examples of secoiridoid units in combination with tryptophan, however, were the alkaloids cordifoline [5][2] and adifoline [7][3] and their deoxyderivatives [6] and [8] respectively, isolated from *Adina cordifolia*. Subsequently Smith and co-workers isolated the tryptamine–secologanin derivative strictosidine [9] from *Rhazya stricta*,[4] *Rhazya orientalis*, and *Catharanthus roseus*.[5, 6] Battersby later showed by dilution analysis that both the C-3 epimer vincoside [10] and strictosidine [9] were present in *C. roseus*.[6] These efforts finally led to the isolation of vincoside [10], strictosidine [9], vincoside pentaacetate [11],[7] and N-acetylvincoside [12][6, 8] from *C. roseus*. More recently, pauridianthoside [13] from *Pauridiantha lyalli*[9] and dolichantoside [14] from *Strychnos gossweileri Exell*[10] have been isolated, in which the secologanin unit can be recognized intact. Strictosamide has also been isolated from the root bark of *Nauclea latifolia*[11] and 5α-carboxystrictosidine from *Vinca elegantissima*.[12] Labelled loganin [15] and tryptophan [16] were found to be incorporated into strictosidine [9] and vincoside [10] in *C. roseus*.[5, 8, 13]

Pauridianthoside [13]   Dolichantoside [14]

The earlier incorporation work suggested that when a mixture of labelled vincoside [10] and strictosidine [9] was fed into *C. roseus*, both glycosides were found to be incorporated into catharanthine [17], vindoline [18], serpentine [19], ajmalicine [20], and perivine [21].[5, 6, 14] However, when the individual isomers were fed, only vincoside seemed to be incorporated, strictosidine [9] and dihydrostrictosidine [22] being ineffective.[6, 8] As the majority of *Corynanthe* alkaloids have the opposite stereochemistry to vincoside at C-3, this raised a mechanistic problem regarding the mode of epimerization at this centre. More recent work of Stockigt has established that it is strictosidine and *not* vincoside that is incorporated into the three major classes of indole alkaloids.[15] This has subsequently been independently confirmed by Battersby,[16] Scott,[17] and Brown.[18] Moreover strictosidine has been shown to be incorporated into indole alkaloids with 3β-H stereochemistry as well as those with 3α-H configuration.[19]

It was proposed by Van Tamelen[20, 21] that 5α-carboxystrictosidine and 5α-carboxyvincoside may play a significant part in the biosynthesis of sarpagine- and ajmaline-type alkaloids. Feeding of the labelled 5α-carboxy compounds has, however, shown negligible incorporation into the indole

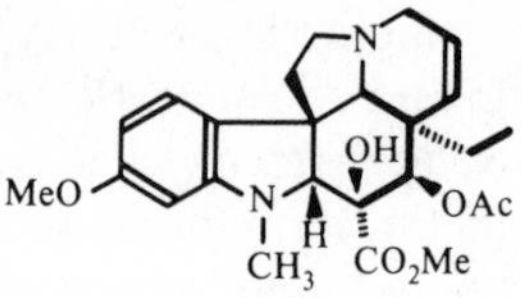

[18]

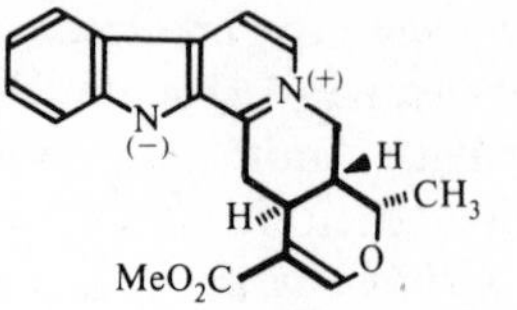

[19]

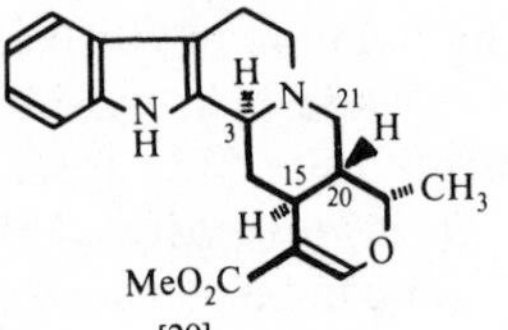

[20]

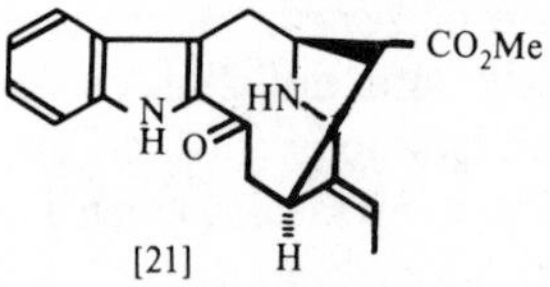

[21]

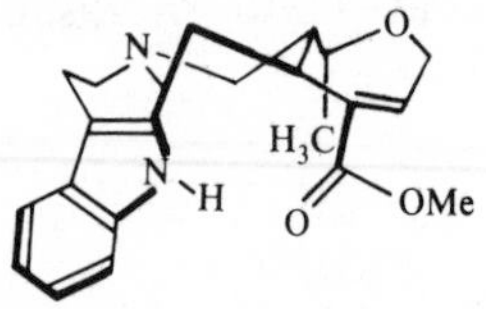

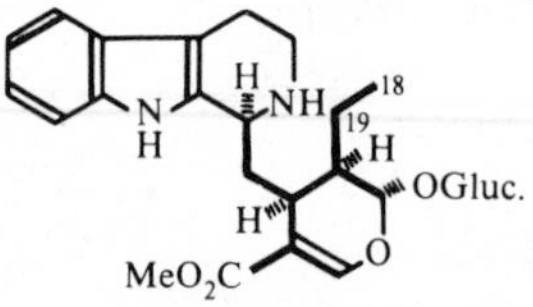

[22]

alkaloids.[22] An enzymically catalysed condensation of tryptamine and secologanin resulted in the formation of strictosidine [3α(S) configuration] which was further converted to ajmalicine, 19-epiajmalicine, and tetrahydroalstonine, confirming that the key intermediate is strictosidine and not vincoside.[23]

Originally the stereochemistry assigned at C-3 to vincoside and strictosidine[5, 6] was opposite to those shown in [10] and [9] respectively. The work of Smith, however, indicated that the original assignments were incorrect and needed to be reversed at C-3. The controversy was resolved through the independent efforts of Battersby and Smith. Strictosidine [9] was correlated via a derivative [23] of vallesiachotamine [2] with dihydroantirhine acetate [25],[24] a substance of known absolute stereochemistry[25] (Scheme 3.1).

Brown and co-workers have isolated a number of alkaloids from *Adina rubescens*, and these are undoubtedly formed in the earlier stages of the biosynthetic pathway. They include vincosamide [27],[26] rubescine [28],[27] 10-β-D-glucosyloxyvincosamide [29],[28] deoxycordifoline [6],[29] 5α-carboxystrictosamide [30], and 5α-carboxyvincosamide [31].[29] Degradation of vincosamide tetraacetate [32] afforded the triols [33] and [34].[26] The enantiomeric triol [35] was obtained by degradation of corynantheine [36]

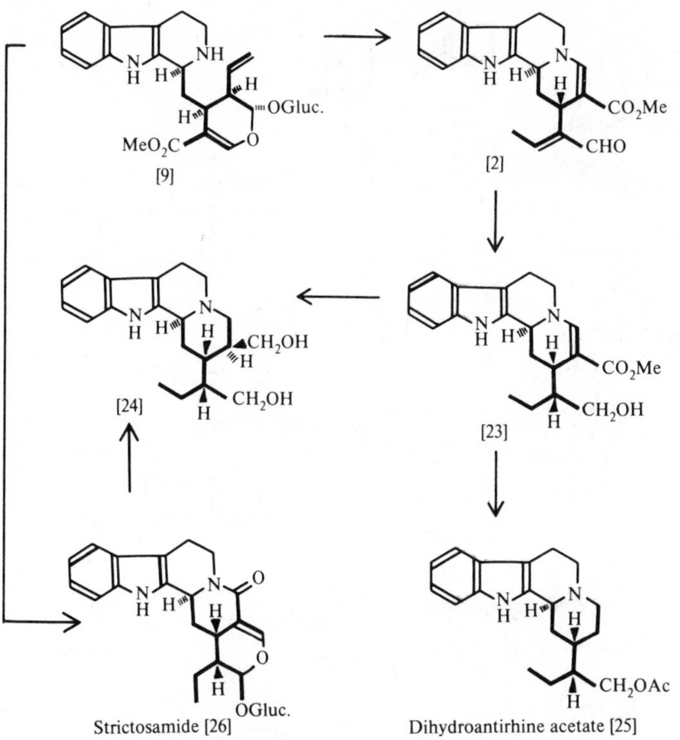

Strictosamide [26]   Dihydroantirhine acetate [25]   SCHEME 3.1

of known absolute stereochemistry.[30, 31] It was shown that epimerization could not have occurred during the conversions, thus establishing that strictosidine [9] had the same (α) stereochemistry at C-3 as compared to corynantheine [36]. A conformational analysis study of ipecoside [37][32] has indicated that the C-1 H β-configuration is the more stable one.

Smith and co-workers[24] have also obtained results establishing the stereochemistry of strictosidine as shown in [9]. Kennard and co-workers[33] have published a revised structure of ipecoside [37] with a reversal of stereochemistry at C-1, based on X-ray structure determination of O,O-dimethylipecoside [38]. Since vincoside had been previously correlated with ipecoside [37], the stereochemistry of vincoside was also reversed and established as shown in [10].[34] The correlation of another alkaloid, alangiside [39] isolated from *Alangium lamarckii*, with desacetyl ipecoside [40] also established the C-3 H α stereochemistry of strictosidine [9] at C-3,[35] vincoside [10] having the opposite C-3 H β-configuration. Unambiguous confirmation of these assignments has come from a recent X-ray structure and absolute stereochemistry determination of (N-*p*-bromobenzyl)-vincoside by Clardy and co-workers.[36]

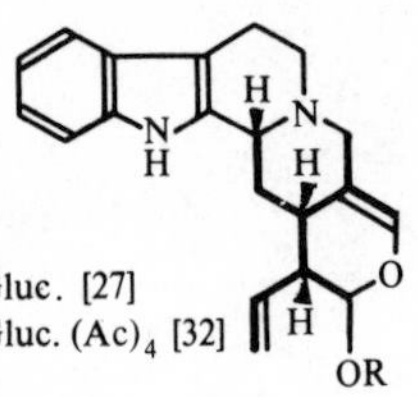

R=Gluc. [27]
R=Gluc. (Ac)$_4$ [32]
OR

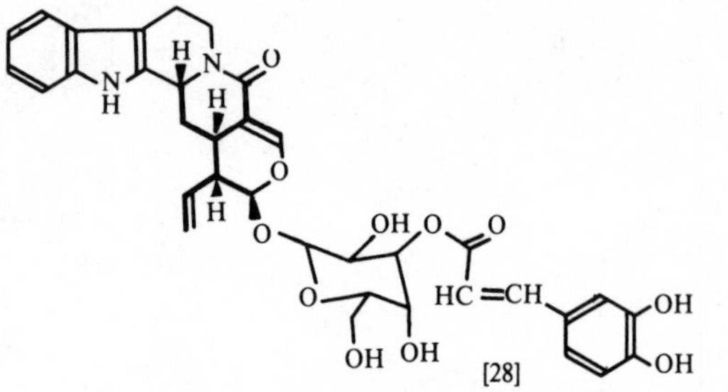

[28]

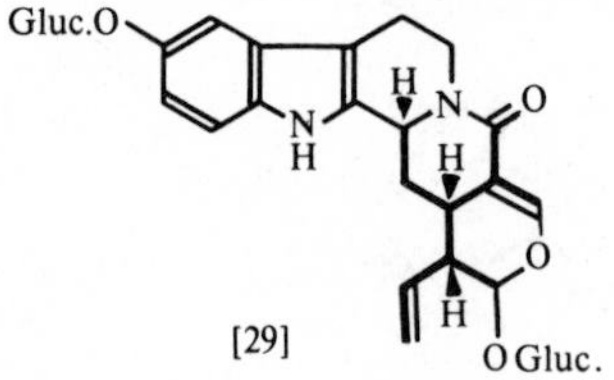

Gluc.O

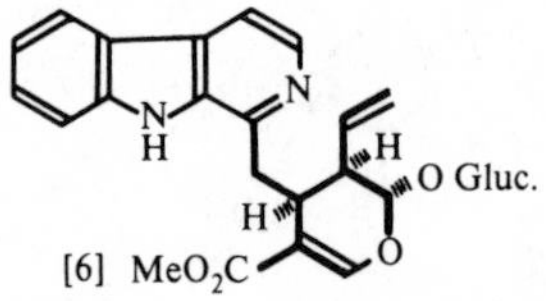

[29] OGluc.

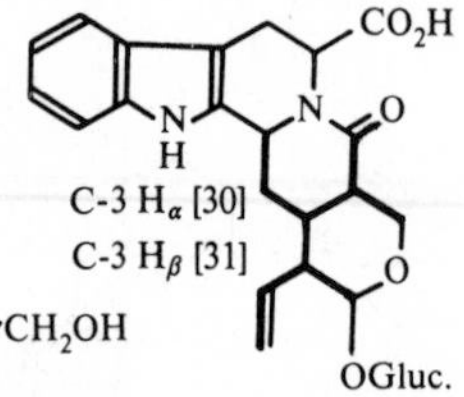

[6] MeO$_2$C O Gluc.

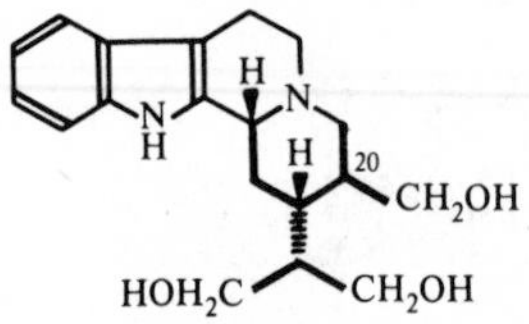

20 CH$_2$OH
HOH$_2$C CH$_2$OH
C-20 H$_\alpha$ [33]
C-20 H$_\beta$ [34]

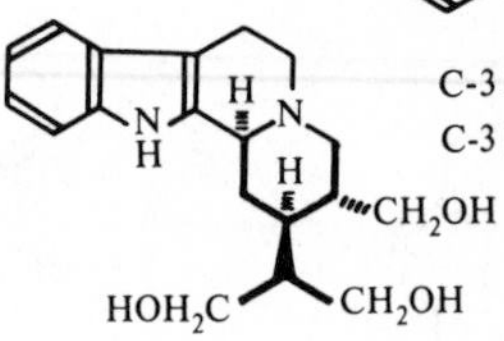

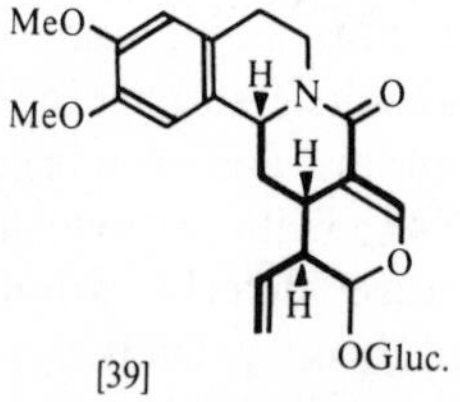

CH$_2$OH
HOH$_2$C CH$_2$OH
[35]

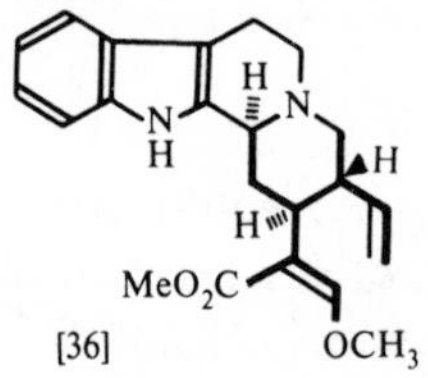

CO$_2$H
C-3 H$_\alpha$ [30]
C-3 H$_\beta$ [31]
OGluc.

[36] MeO$_2$C OCH$_3$

MeO
MeO
[39] OGluc.

R$_1$O
R$_1$O
OGluc.
MeO$_2$C
R$_1$=H, R$_2$=Ac [37]
R$_1$=CH$_3$, R$_2$=Ac [38]
R$_1$=H, R$_2$=H [40]

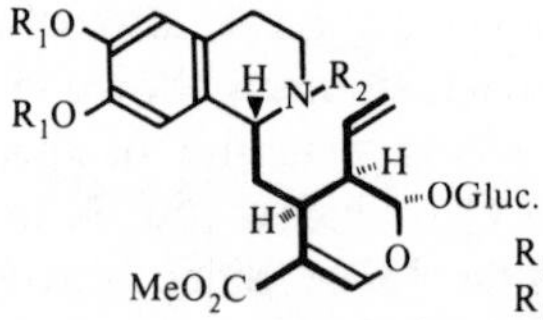

Feeding experiments have shown[8, 37-39] that stereoselective retention of tritium occurs from loganin [15] labelled at positions 1, 5, 7, and 8 and this label is probably incorporated at positions 21, 15, 3 and 19 of the alkaloids respectively. This lays down the limits of stereochemical requirements which any mechanism proposed to explain the formation and subsequent transformation of these alkaloids, must meet. For instance [1-³H]-loganin

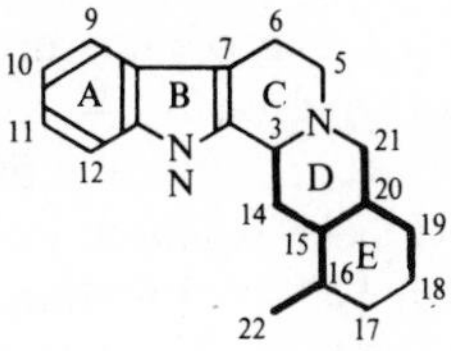
Yohimbane system [41]

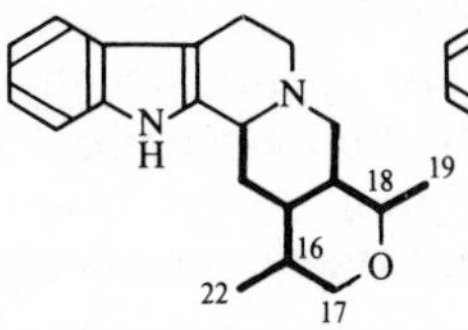
Heteroyohimbane system [42]

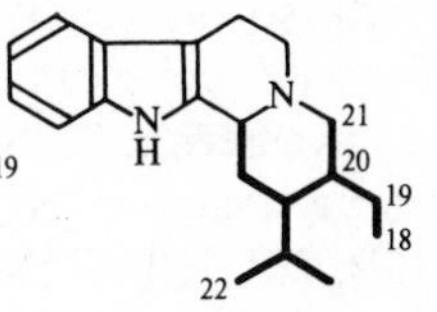
Secoyohimbane (corynane) system [43]

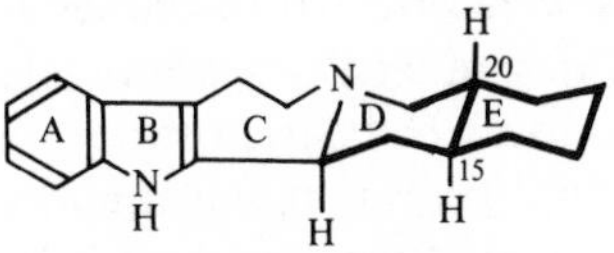
Yohimbane system [44]
D/E trans

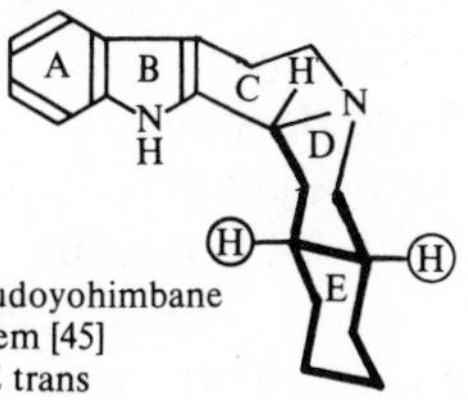
Pseudoyohimbane system [45]
D/E trans

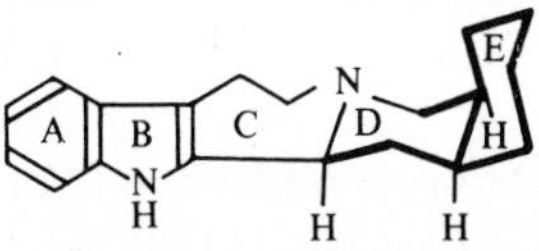
Alloyohimbane system [46]
D/E cis

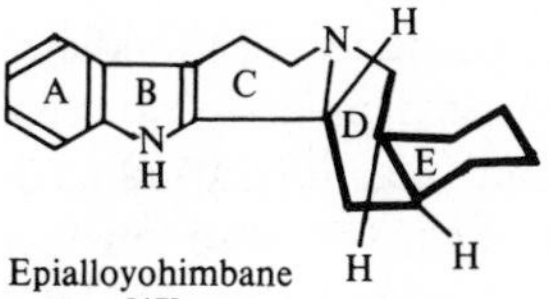
Epialloyohimbane system [47]
D/E cis

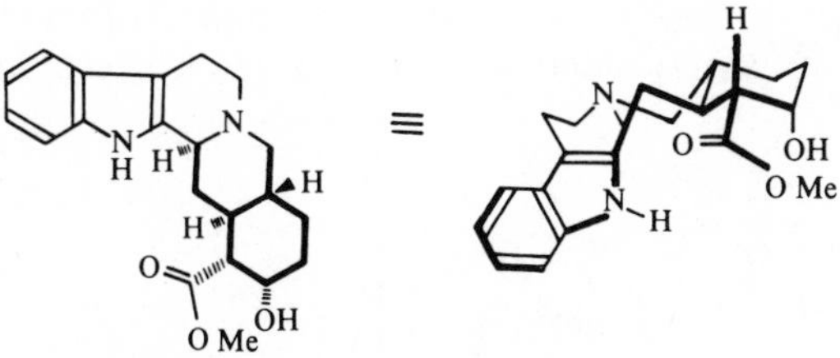
Yohimbine [59]

[15] is incorporated into ajmalicine [20][38] with retention of tritium at C-21. Thus any mechanism invoked to explain the formation of this alkaloid must explain the retention of tritium both at C-21 and C-3.

In addition to the glycoalkaloids mentioned above a number of tetrahydro-β-carboline alkaloids with an unrearranged secologanin skeleton are known. Three of the simplest systems of this type are based on yohimbane, heteroyohimbane, and secoyohimbane (corynane), structural systems differing in ring E which may be carbocyclic (yohimbane) [41], heterocyclic (heteroyohimbane) [42], or ring cleaved (secoyohimbane) [43].

The yohimbane system contains three asymmetric carbon atoms (C-3, C-15, and C-20) so that eight different diastereoisomeric compounds are possible. As only compounds containing the C-15 hydrogen in an α-disposition have been actually found in nature this reduces the number of possible isomeric structures to four. All four types have been isolated, i.e. yohimbane (3α, 20β) [44], pseudoyohimbane (3β, 20β) [45], alloyohimbane (3α, 20α) [46], and epialloyohimbane (3β, 20α) [47].

Once the key role played by strictosidine in the biosynthesis of indole alkaloids had been established[15-23] efforts were directed to unravelling the subsequent steps which lead to ajmalicine and other *Corynanthe* alkaloids. It has been shown by feeding of labelled 5α-carboxystrictosidine [3] and 5α-carboxyvincoside to several indole alkaloid-bearing plants that these two compounds are not involved in the biosynthesis of sarpagine- and ajmaline-type alkaloids.[22] Cathenamine [49] has been isolated from *Guettarda eximia* and it has been identified as 20,21-didehydroajmalicine.[40] Cathenamine accumulates when tryptamine and secologanin are incubated with an enzyme preparation from *C. roseus* cell suspension cultures and it was demonstrated to be a central intermediate in the enzymatic production of ajmalicine [20], 19-epi-ajmalicine [51], and tetrahydroalstonine [52].[41] Another novel intermediate isolated from *Guettarda eximia*, closely related to cathenamine, was identified as 20,21-didehydroheteroyohimbine [50][42] which could be converted to tetrahydroalstonine [52] and 19-epiajmalicine [51]. On mild dehydration [52] (19-R) was found to be converted almost quantitatively to cathenamine [49] (19-S) showing that the biogenetic intermediate [50] may be able to fragment to the conjugated immonium species [48] which can be equilibrated at C-19 via the corresponding

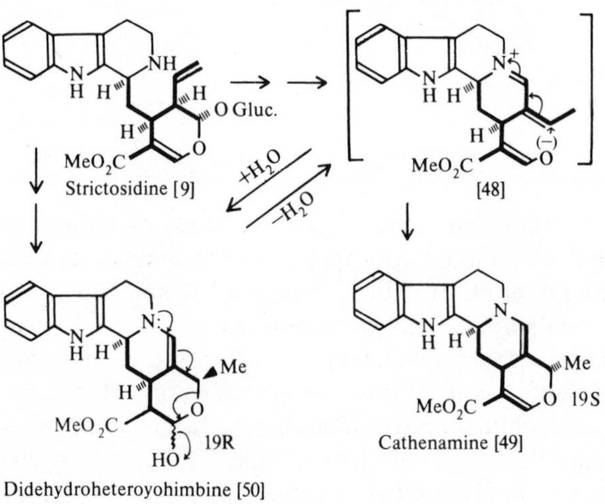

Strictosidine [9]

[48]

Didehydroheteroyohimbine [50]

Cathenamine [49]

SCHEME 3.2

dienamine into the 19S series. This suggests that the intermediate [50] may be involved in the formation of 19S and 19R series in the *in vivo* biogenetic transformation (Scheme 3.2).[122]

The major features of the probable biosynthetic pathway to the corynantheine group exemplified by ajmalicine [20], 19-epi-ajmalicine [51], and tetrahydroalstonine [52] are shown in Scheme 3.3. The condensation of tryptamine [53] with secologanin [1] is catalysed by the enzyme strictosidine synthetase which controls the C-ring closure mechanism, and results in the formation of strictosidine [9].[15, 19, 21] A second glycosidase enzyme converts strictosidine via a series of reactive intermediates [54]–[56] to 4, 21-dehydrocorynantheine aldehyde [57] by D-ring closure.[43] This aldehyde can then isomerize to [58] which cyclizes to

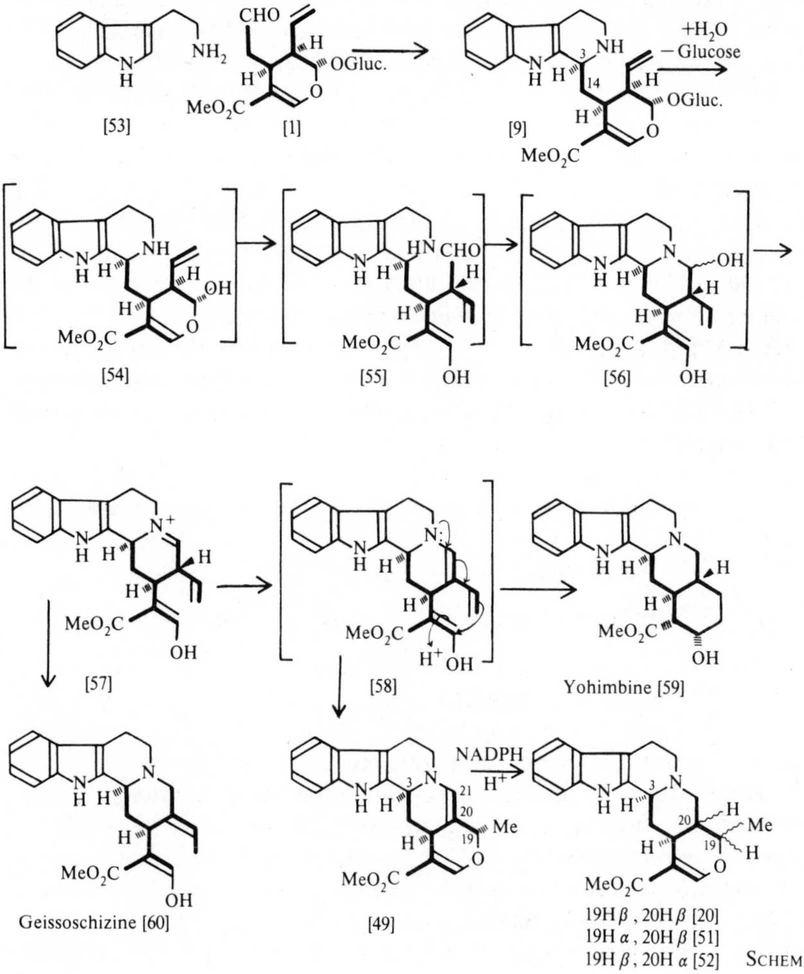

Scheme 3.3

cathenamine [49].[41] Ajmalicine [20], 19-epi-ajmalicine [51], and tetrahydroalstonine [52] can then be formed from cathenamine, the conversion being catalysed by the NADPH-dependent cathenamine reductase[44, 45] (Scheme 3.3).

The intermediate [58] could also be involved in the biosynthesis of the yohimbine group of alkaloids. An intramolecular attack of the vinylidene group (which is activated by conjugation with the basic nitrogen) on the acrylate system could, after reduction, afford yohimbine [59]. The biosynthetic origin of the yohimbine alkaloids has been a subject of much speculation in past[46-50] and Inouye has suggested that the *Yohimbe* group of alkaloids may arise by combination of tryptamine with an appropriate seco-iridoid.[63]

Most of the current studies are directed at the unravelling of the intra-and inter-class pathways. The search for 'link' alkaloids is being carried out by feeding various suitably-labelled indole alkaloids, and looking for incorporation at the expected positions in the corresponding alkaloids of other classes. Another approach is to try to repeat the predicted transformation pathways in the laboratory to demonstrate the mechanistic feasibility of such conversions.

Many papers have appeared in the literature attempting to rationalize the formation and interconversions of various alkaloid types.[5-7, 37, 38, 51-59] As mentioned earlier, the first class of alkaloids to be formed are those of the *Corynanthe* family, and there is overwhelming evidence[5, 6, 37, 38, 52, 53, 55, 58-60] that certain 'link alkaloids' of the *Corynanthe* group of the correct oxidation level and reactivity are then sequentially transformed as summarized below. The mechanistic aspects of these transformations will be dealt with in later chapters.

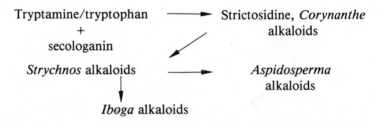

Feeding experiments with corynantheine aldehyde [49][38, 56] failed to show any significant incorporation into the later alkaloids except, understandably, the corresponding methyl ether. However [Ar-³H]-[61, 14], [OMe-³H]-[7], and [OMe-³H, Ar-³H]-[14]geissoschizine [60] were found to be incorporated into the *Corynanthe* alkaloid ajmalicine [20],[7, 14] serpentine [19],[14] the *Strychnos* alkaloid akuammicine [60],[7, 14] the *Aspidosperma* alkaloid vindoline [18],[7, 14] and coronaridine [62].[62] Geissoschizine

[60] was found to have a very dynamic radioprofile. On administration of
DL-[2-^{14}C]-tryptophan [16] to young seedlings of *Vinca rosea*, the
radioactivity of geissoschizine [60] was found to reach a maximum after
1.5–2 h (4 per cent of total radioactivity) and then declined slowly over 8
days to 1 per cent. There is therefore strong evidence that geissoschizine is
one of the main 'link' alkaloids, connecting the *Corynanthe* group with
other classes. In contrast, the radioprofile of ajmalicine [20] showed no
such effect. It has been suggested that two separate pathways may be
operative, one leading to geissoschizine [60] and the other to corynantheine
aldehyde with a possible interconnection in the seedlings,[56] the bifurcation
occurring when the cyclopentane ring is cleaved at the secoiridoid stage.[63]
More recently, it has been suggested that geissoschizine is not a central intermediate in the biosynthesis of ajmalicine and other related *Corynanthe*
alkaloids, but enters the main pathway by a NADP[+]-dependent reaction.[65]
On the other hand, recent work by Scott supports the view that
geissoschizine is a key intermediate in the formation of ajmalicine[66] and
16(R)-isositsirikine.[118] The interconversion of the enamine and immonium
ions from cathenamine[119] and the role of 4,21-dehydrogeissoschizine[120–121]
have also been studied. Further studies are required to clarify the precise
roles of cathenamine and geissoschizine in the biosynthetic process.

[60]

Akuammicine [61]

Coronaridine [62]

The biosynthesis of the heteroyohimbine alkaloids (ajmalicine) has been
demonstrated to be a multi-enzyme process in a cell-free system from
C. roseus.[67] It was found that the enzymic coupling of tryptamine with
secologanin afforded the 3-S series exclusively. It has been suggested[67] that
ajmalicine may be formed either via geissoschizine or cathenamine. The
conversion of geissoschizine to ajmalicine appeared to be linked to a
NADP-metabolizing enzyme.

A number of biogenetically-oriented syntheses of such alkaloid types have been reported by Brown and co-workers.[68] Thus condensation of methyl elenolate [63], prepared from secologanin with tryptamine and treatment of the resulting lactams with phosphorus oxychloride afforded ajmalicine [20], tetrahydroalstonine [52], and 19-epiajmalicine [51] (Scheme 3.4).

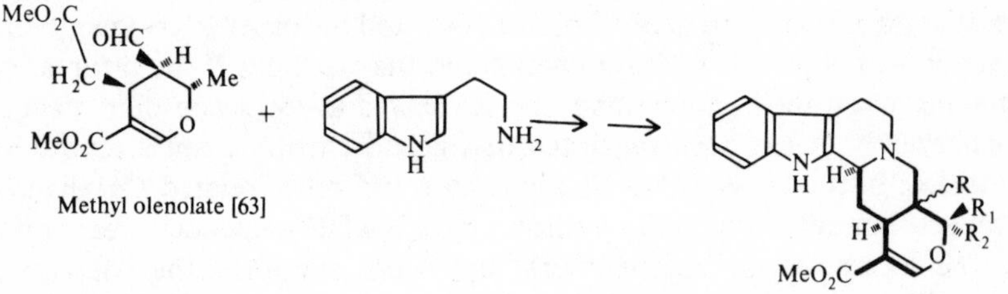

Methyl olenolate [63]

Ajmalicine: $R = \beta H$, $R_1 = H$, $R_2 = Me$ [20]
Tetrahydroalstonine: $R = \alpha H$, $R_1 = H$, $R_2 = Me$ [52]

SCHEME 3.4   19-Epiajmalicine: $R = \beta H$, $R_1 = Me$, $R_2 = H$ [51]

In a similar approach, condensation of secologanin with tryptamine at pH 4 followed by treatment with β-glucosidase at pH 5 and reduction with sodium cyanoborohydride afforded a 3:3:1 mixture of akuammigine [64], 2,3-secoakuammigine, and tetrahydroalstonine [52] in 70 per cent overall yield. Small amounts of ajmalicine were also formed in the reaction.[69] Similarly condensation of dihydrosecologanin with tryptamine in the presence of β-glucosidase at pH 5 afforded dihydromancunine [65].[70] Biogenetically-oriented syntheses of methyladirubine,[71] cathenamine and 19-epicathenamine have also been reported.[72]

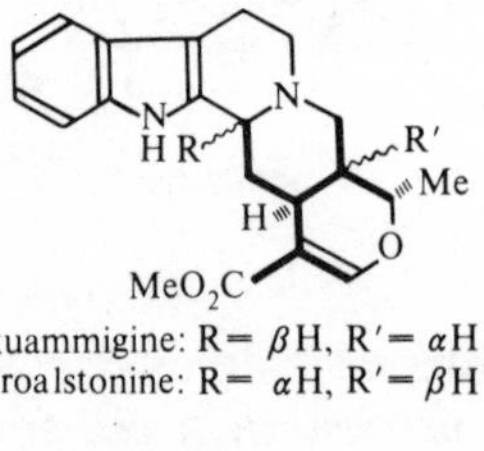

Akuammigine: $R = \beta H$, $R' = \alpha H$ [64]
Tetrahydroalstonine: $R = \alpha H$, $R' = \beta H$ [52]   Dihydromancunine: $R = \beta Et$ [65]

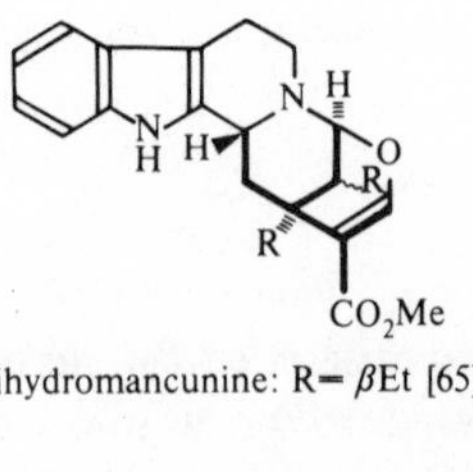

The biogenetic origin of the vincoside group of *Corynanthe* alkaloids was discussed earlier. The macrolidine group of alkaloids is exemplified by macrolidine [66] in which the carboxyl group has undergone intramolecular lactonization with the hydroxyl group of the glucose moiety.

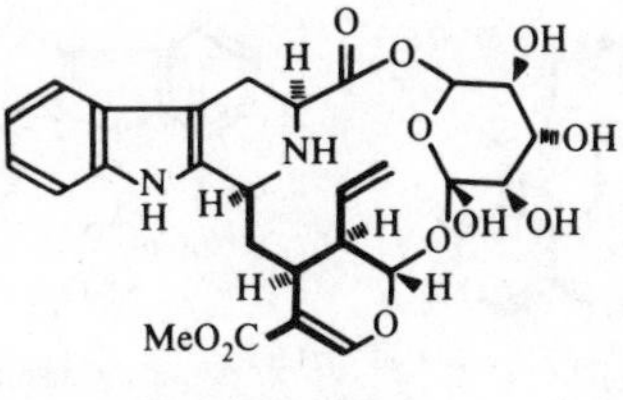

Macrolidine [66]

In the rubenine group, tryptophan is seen to have condensed with secologanin. The resulting α-carboxystrictosidine could have been converted to rubenine [69] by epoxidation of the olefinic double-bond, attack of the basic nitrogen on the epoxide to afford a new seven-membered ring, and the intramolecular lactonization of the resulting hydroxyl with the carboxyl group (Scheme 3.5).

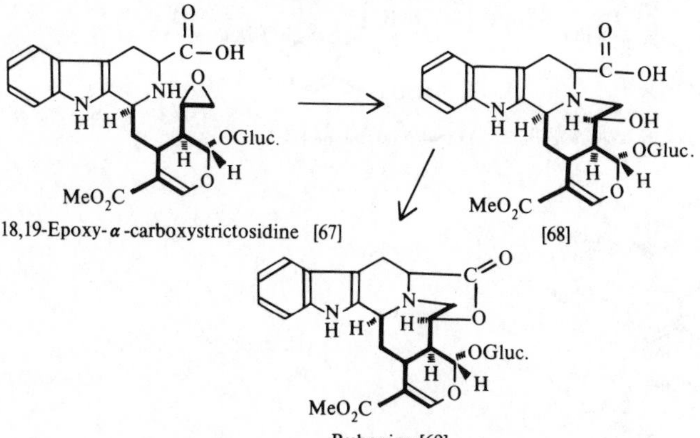

18,19-Epoxy-α-carboxystrictosidine [67]

[68]

Rubenine [69]

SCHEME 3.5

The same seven-membered ring found in rubenine is present in the cadambine group of bases, except that the tryptophan carboxyl group is lacking (e.g. 3α-dihydrocadambine) [71]. Dihydrocadambine could arise in nature by the intramolecular nucleophilic attack at C-18 by the basic nitrogen of 18,19-epoxystrictosidine [70] (Scheme 3.6). Isodihydrocadambine [72] which has been isolated from *Anthocephalus cadamba*,[59] could arise by the alternative attack on the same epoxide.

The cordifoline group of alkaloids is also derived from tryptophan [14]. In this group, ring C is aromatized and ring A is hydroxylated (e.g. cordifoline [5]).

In lyalidine [73] the oxygen atom of the secologanin moiety in cordifoline [5] has been replaced by a nitrogen atom and decarboxylation of the carboxyl group has occurred.

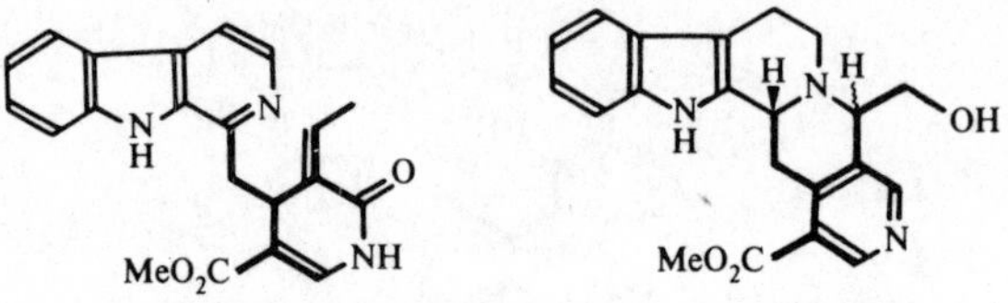

Lyalidine [73]   Cadamine [74]

A similar nitrogen-containing ring E is found in cadamine [74]. Cyclization of the basic tryptamine nitrogen atom with the ethylidine group has afforded a new hexacyclic system in [54] (c.f. isodihydrocadambine formation in Scheme 3.6).

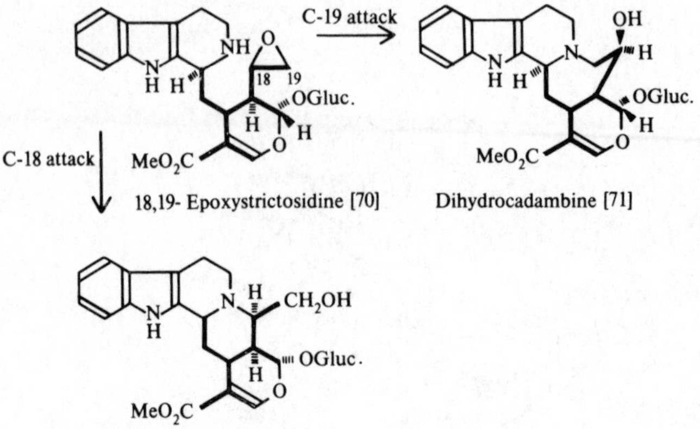

18,19- Epoxystrictosidine [70]   Dihydrocadambine [71]

Isodihydrocadambine [72]   SCHEME 3.6

In the adifoline group the condensation product of tryptophan and secologanin has undergone an intramolecular attack of the indole nitrogen on the vinyl group to afford a new seven-membered ring and ring C has been aromatized (Scheme 3.7).

The talbotine group can be formed from the hydrolytic cleavage product [75] of vincoside [10] which can undergo hydroxylation at C-16 and the corresponding phosphate ester [76] could then be displaced by the indole nitrogen atom. This would lead to talbotine [79] by intramolecular cyclization of the alcohol [78] with the C-17 aldehyde (Scheme 3.8). The bond between the indole nitrogen and C-16 could also be formed by a radical coupling mechanism. The intermediate [76] could give rise to pleiocarpamine [80] by cyclization after the loss of formyl group, or to C-fluorocurine via the quaternary alcohol [81] (Scheme 3.8). The mechanism of formation for such indoxyls is presented later.

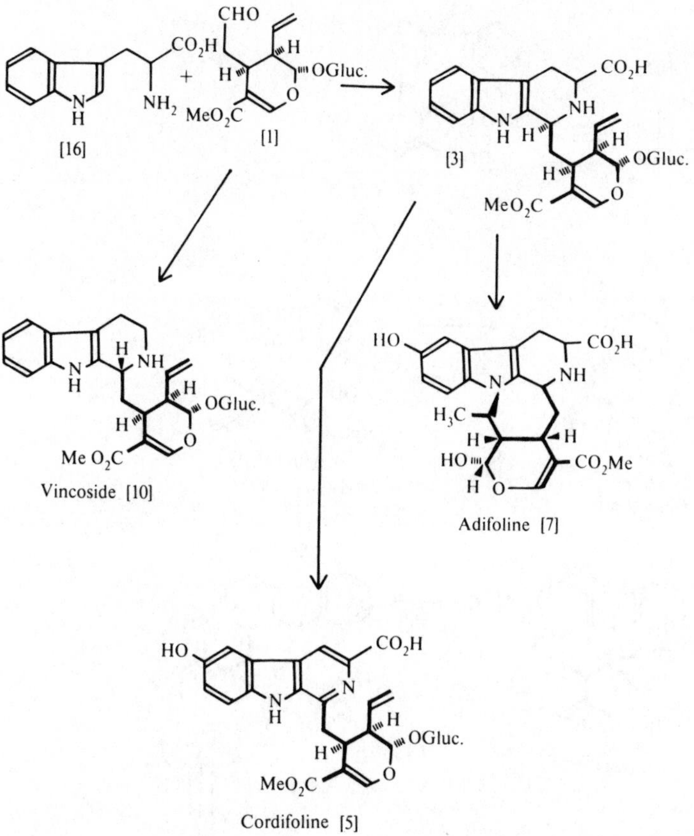

Vincoside [10]

Adifoline [7]

Cordifoline [5]

SCHEME 3.7

The vallesiachotamine group of alkaloids can be formed from strictosidine [9] by hydrolytic cleavage to [55], rotation at C_{14}—C_{15} bond followed by attack of the basic nitrogen on the acrylate system to afford vallesiachotamine [2] (Scheme 3.9).

The elegantine group of alkaloids [83] may arise from the ajmalicine system via the corresponding hydroxyindolenine (Scheme 3.10).

The alkaloids of the picraphylline group may arise from ajmalicine [20] or a common biogenetic precursor by N-methylation followed by hydrolytic cleavage of the C—N^+ bond and oxidation to afford picraphylline [84] on tautomerization (Scheme 3.11).

The oxogambirtannine group of alkaloids can arise from the intermediate [58] by the process shown in Scheme 3.12. Cyclization of the dienamine [58] followed by tautomerization and dehydration could afford oxogambirtannine [87], containing an aromatized ring E (Scheme 3.12).

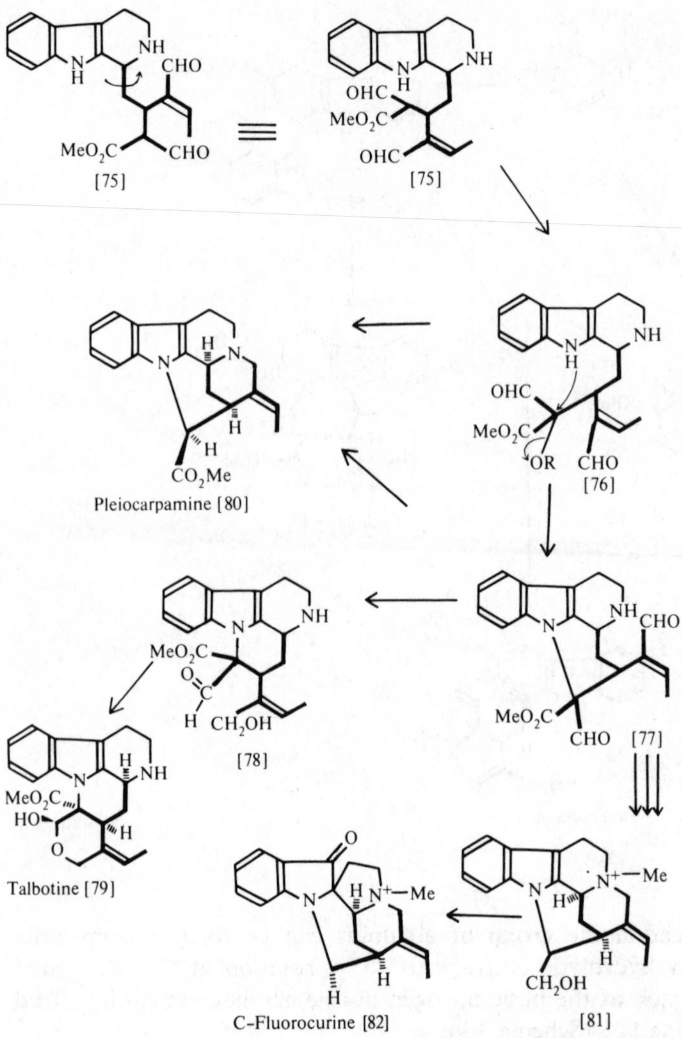

Pleiocarpamine [80]

Talbotine [79]

[78]

C-Fluorocurine [82]   [81]   SCHEME 3.8

The picraline group of bases, e.g. akuammiline [90] can result from the intermediate [89] where the indole-3-position is involved in cyclization with C-16. This may occur either by a radical coupling mechanism or by the displacement of a suitable leaving group (e.g. phosphate) at C-16 in [88] (Scheme 3.13). The related alkaloid aspidodasycarpine [91] can be envisaged to be formed from the vincoside precursor [3] by decarboxylative elimination and reduction of the corresponding aldehyde to the alcohol, and cyclization. The aldehyde function at C_{21} has also undergone cyclization with the basic nitrogen, followed by reduction (Scheme 3.13). In the

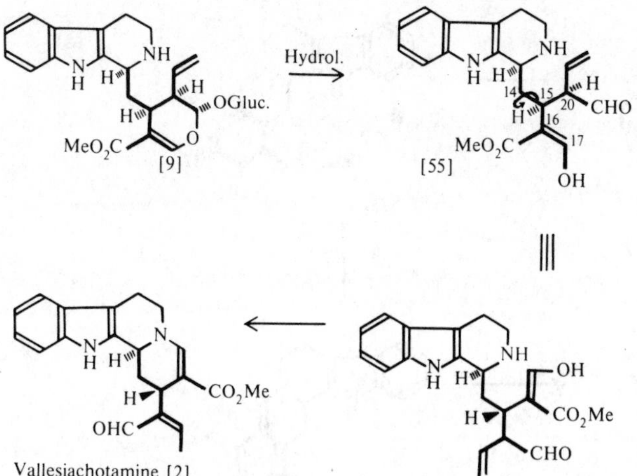

Vallesiachotamine [2]

[55]

SCHEME 3.9

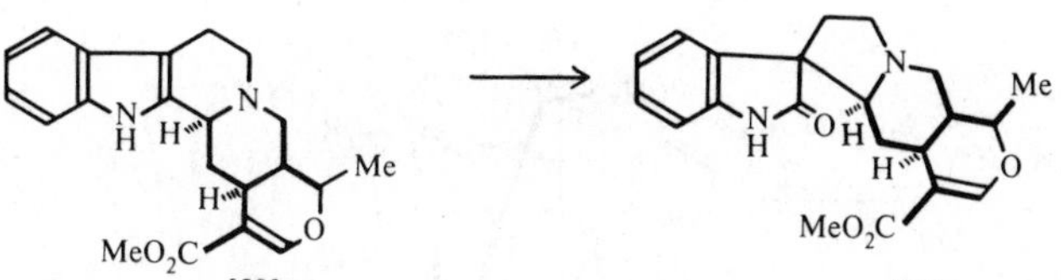

[20]

[83]

SCHEME 3.10

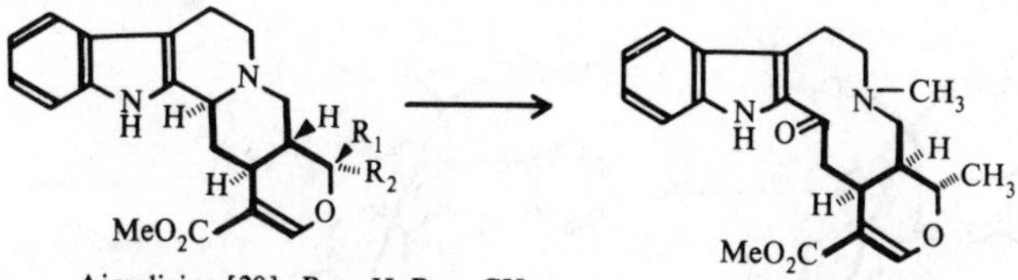

Ajmalicine [20] : $R_1 = H$, $R_2 = CH_3$
19-Epiajmalicine [51] : $R_1 = CH_3$, $R_2 = H$

Picraphylline [84]

SCHEME 3.11

closely related picrinine group, an additional ring is formed by an ether linkage across ring C [92]. This could occur by hydration of the immonium group, followed by attack of the hydroxyl group on the indolenine double bond (Scheme 3.13).

A number of routes to the mavacurine group of alkaloids can be considered. Clearly it is C-16 that is becoming attached to the indole nitrogen. This may occur by the direct attack of the indole nitrogen at C-16 with the displacement of a suitable leaving group in the intermediate [75] as illustrated in Scheme 3.8. A plausible alternative is the possibility of the in-

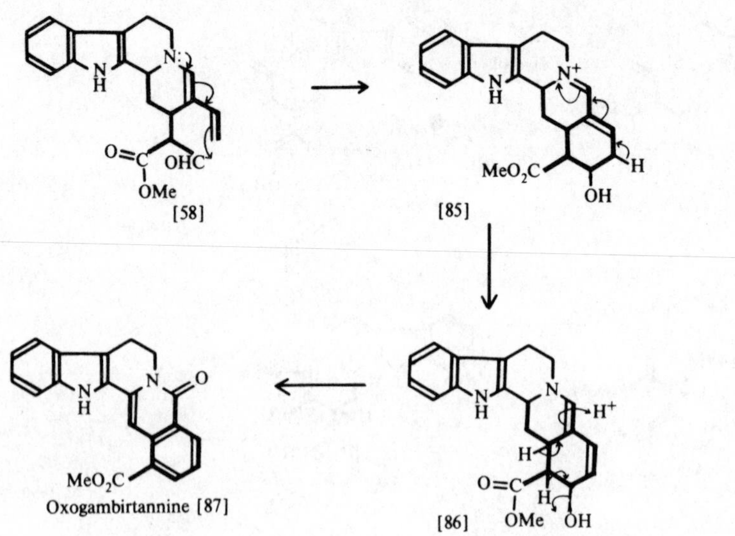

[58]   [85]

Oxogambirtannine [87]   [86]   SCHEME 3.12

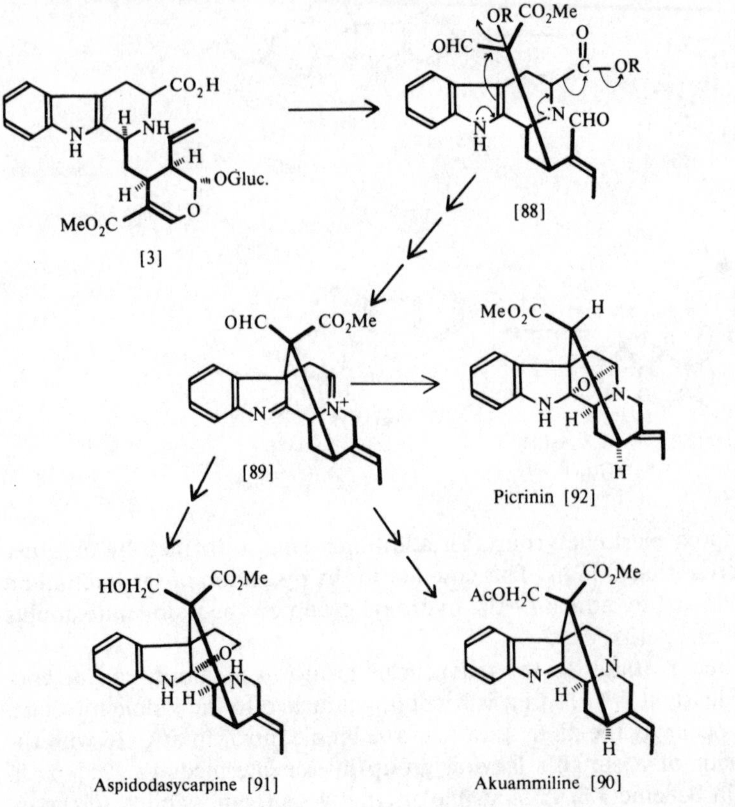

[3]   [88]

[89]   Picrinin [92]

Aspidodasycarpine [91]   Akuammiline [90]   SCHEME 3.13

termediacy of formylstrictamine [93] and [94]. These could, by a 1,3-shift and deformylation, afford pleiocarpamine [80] (Scheme 3.14). The co-occurrence of formylstrictamines [93] and [94] and pleiocarpamine [80] in *R. stricta*[72] may point to the viability of such a route.

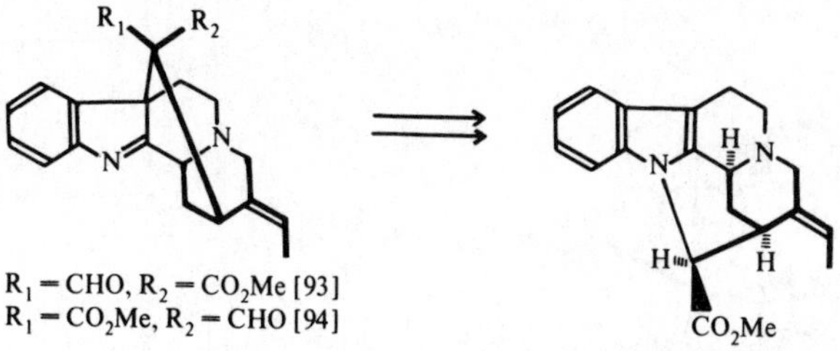

$R_1 = CHO, R_2 = CO_2Me$ [93]
$R_1 = CO_2Me, R_2 = CHO$ [94]

Pleiocarpamine [80]

SCHEME 3.14

A third possibility is the intermediacy of a substance such as [96] formed from geissoschizine [60]. This system may be too unstable to be isolated, but an *in vitro* conversion to [95] has been accomplished from akuam-micine [61],[73] (Scheme 3.15). Routes from the hypothetical intermediate [97] to stemmadenine [99], preakuammicine [100], formylstrictamines [93] and [94], and pleiocarpamine [80] can be envisaged (Scheme 3.16). Elaborate feeding experiments are required to establish whether such pathways between the *Corynanthe* and the *Strychnos* alkaloids actually exist.

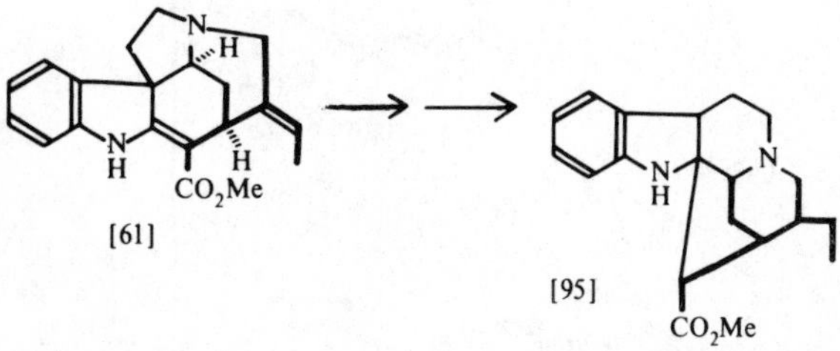

[61]

[95]

CO_2Me

SCHEME 3.15

The alkaloids of the aspidospermatin group probably arise from the intermediate [98] which can undergo a 1,3-hydrogen shift to afford [101] (Scheme 3.16). Cyclization of [101] with the β-position of the indole nucleus can then afford condylocarpine [102], a typical member of this group.

The vinoxine group of bases can be considered to be formed from geissoschizine [60] in a manner resembling the formation of the mavacurine system. The intermediate indoleninium ion [103] can undergo an attack by a hydroxide ion at C-6 accompanied by cleavage of the C—C bond to afford vinoxine [104] (Scheme 3.17).

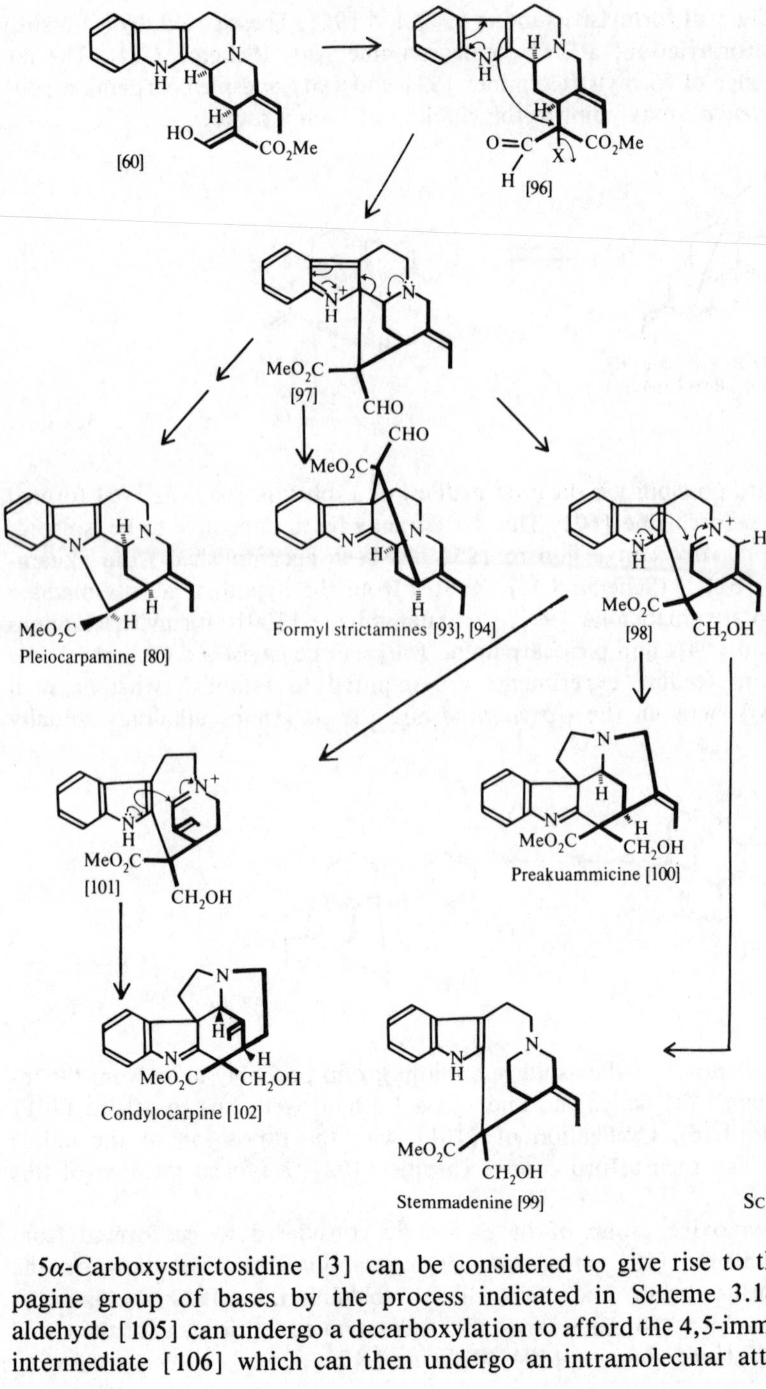

Scheme 3.16

5α-Carboxystrictosidine [3] can be considered to give rise to the sarpagine group of bases by the process indicated in Scheme 3.18. The aldehyde [105] can undergo a decarboxylation to afford the 4,5-immonium intermediate [106] which can then undergo an intramolecular attack by

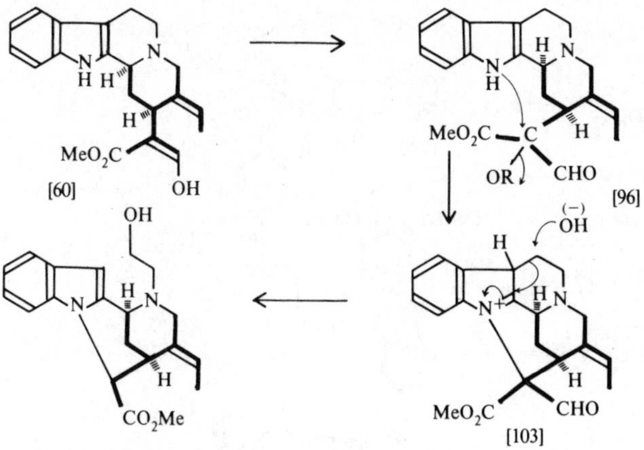

Vinoxine [104]

SCHEME 3.17

C-16, the carbon α-to the ester and aldyhyde functions, to afford [107]. Reduction of this would afford polyneuridine [108], a typical member of the sarpagine group.

The ajmaline group of bases may arise directly from the same intermediate [107] which is thought to be a precursor of the sarpagine group of bases (Scheme 3.18). An intramolecular attack of the β-position of the indole nucleus on the aldehyde function would afford the ajmaline [109] group of alkaloids (Scheme 3.18). An elegant synthesis of ajmaline has been accomplished by van Tamelen[74] by a similar route, in line with his earlier biogenetic proposals.[75]

The gardnutine [113] and eburnaphylline [114] groups may arise from the sarpagine group by cyclization of the primary alcoholic function on either ring C or D to afford a new five-membered tetrahydrofuran ring in each case. In gardnutine [113], this cyclization is accompanied by the loss of the carbomethoxyl group.

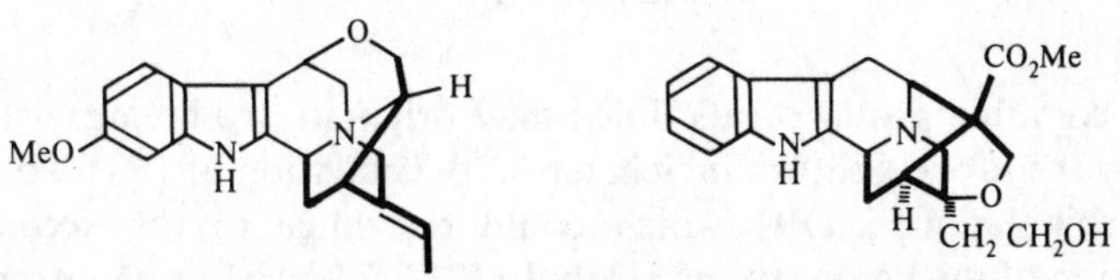

Gardnutine [113]

Eburnaphylline [114]

The cinchonamine group of alkaloids may be formed from the intermediate [105] by an oxidative decarboxylation followed by hydrolytic cleavage to afford [116]. The basic nitrogen can then attack the acrylate system. Hydrolysis, decarboxylation, and reduction would thus lead to cinchonamine [117] (Scheme 3.19).

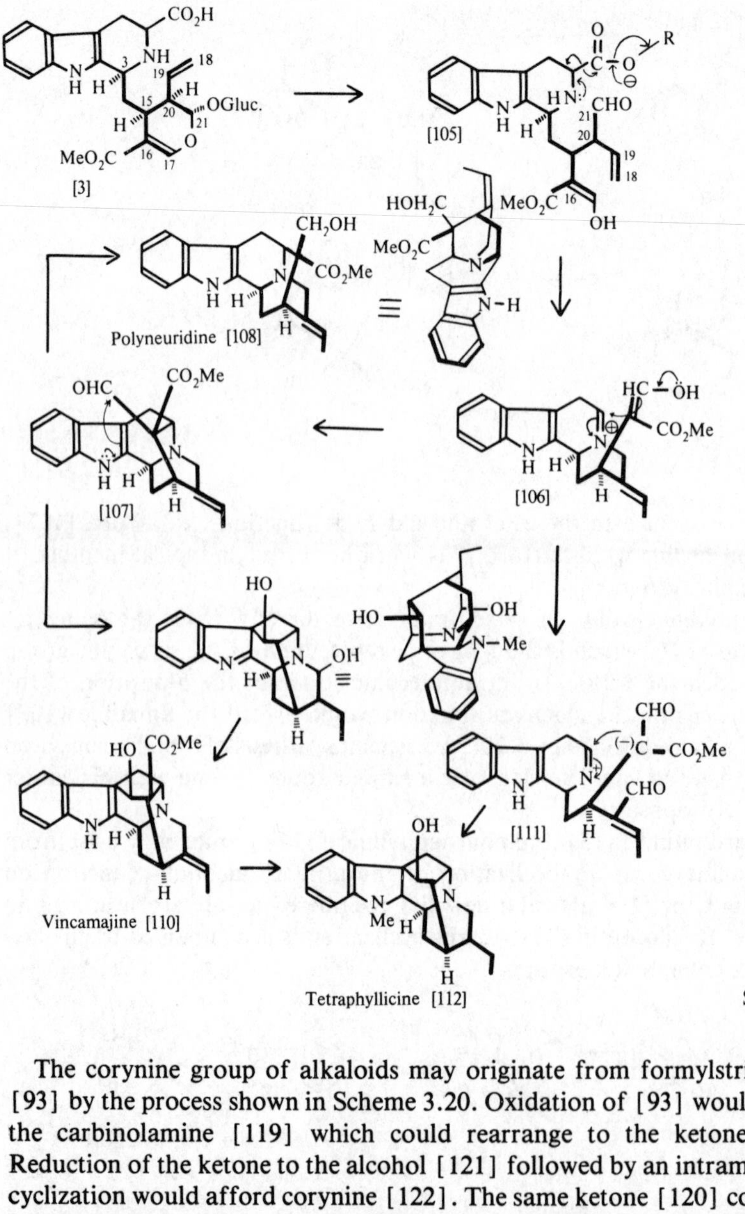

Tetraphyllicine [112]   SCHEME 3.18

The corynine group of alkaloids may originate from formylstrictamine [93] by the process shown in Scheme 3.20. Oxidation of [93] would afford the carbinolamine [119] which could rearrange to the ketone [120]. Reduction of the ketone to the alcohol [121] followed by an intramolecular cyclization would afford corynine [122]. The same ketone [120] could also give rise to vincoridine [123] by deformylation. Reduction of [93] can lead to vincaridine [124] (Scheme 3.20).

The erinine group of alkaloids may arise from the same intermediate [120] by reduction, quaternization of the nitrogen, oxidative cleavage, and cyclization (Scheme 3.21). An alternative route can be envisaged from the

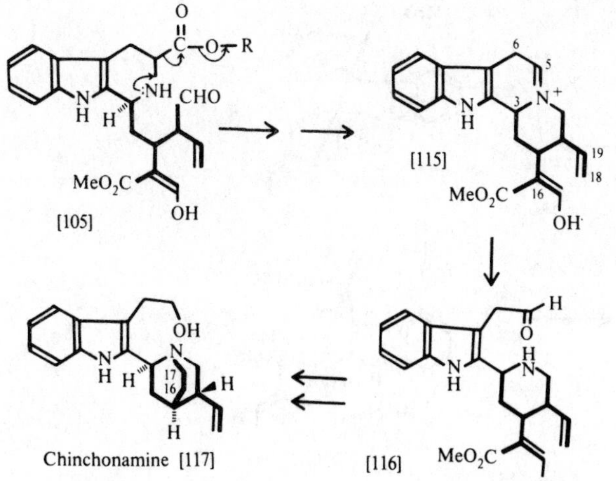

Chinchonamine [117]   [116]

SCHEME 3.19

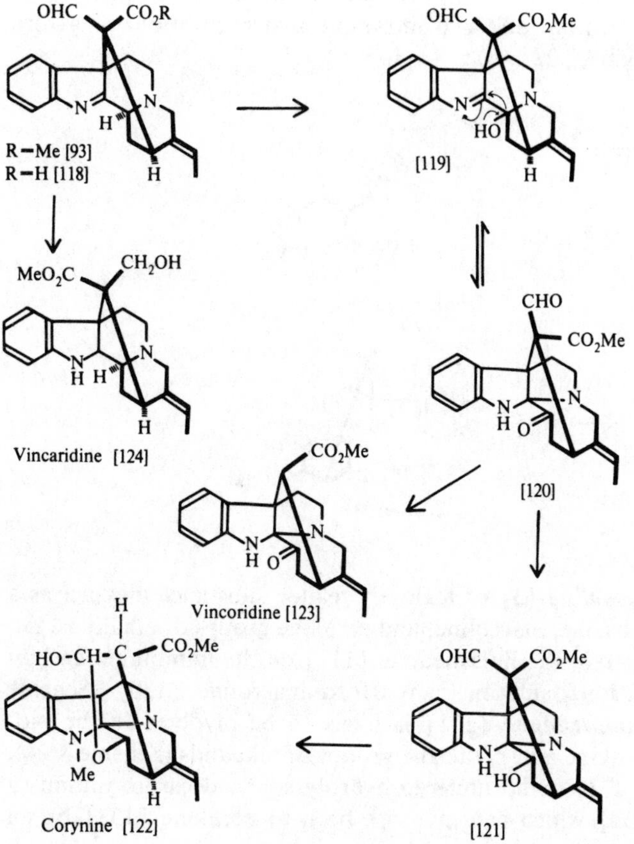

R—Me [93]
R—H [118]

Vincaridine [124]

Vincoridine [123]

[119]

[120]

[121]

Corynine [122]

SCHEME 3.20

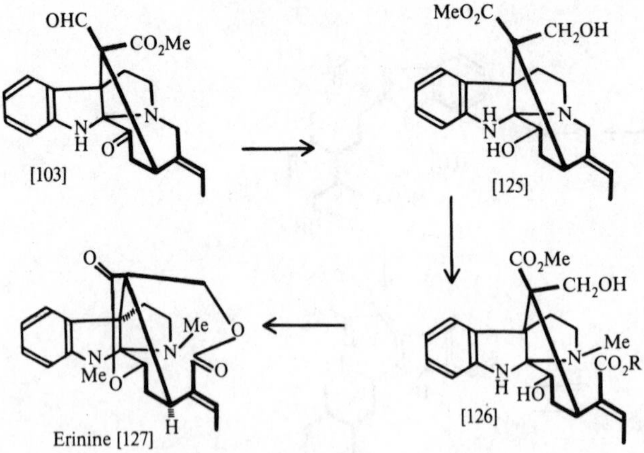

[103]

[125]

[126]

Erinine [127]

SCHEME 3.21

hydrolytic product of strictosidine [75]. This involves ring C contraction of the oxidized and methylated indolenine [128], radical coupling at the indole 3-position, reduction, and two intramolecular cyclizations to afford erinine [127] (Scheme 3.22).

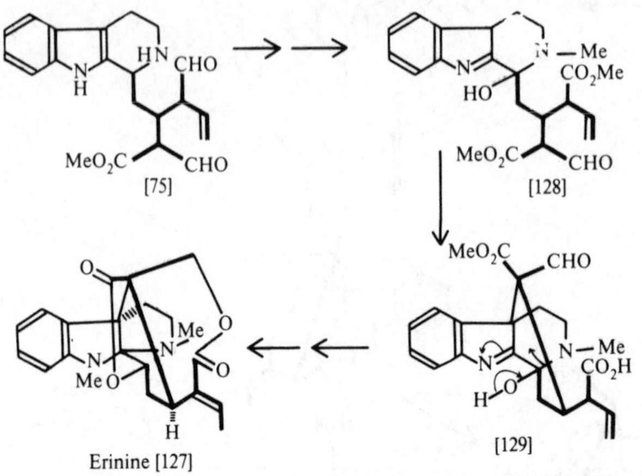

[75]

[128]

[129]

Erinine [127]

SCHEME 3.22

5α-Carboxystrictosidine [3] or a closely related substance may act as a precursor to the vobasine, macroline, and perakine groups of alkaloids. An intramolecular attack in the intermediate [111] on the immonium carbon followed by reduction/oxidation can afford macroline [130] (Scheme 3.23). The same intermediate [111] is likely to be involved in the biosynthesis of the perakine and peraksine groups of alkaloids (Scheme 3.24). The intermediate [131] could undergo hydrolysis and decarboxylation to the dialdehyde [132] which can give rise both to perakine [133] by an

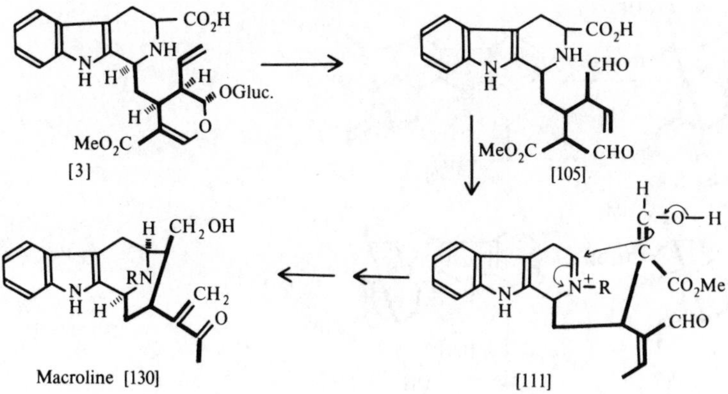

Macroline [130]   [111]   Scheme 3.23

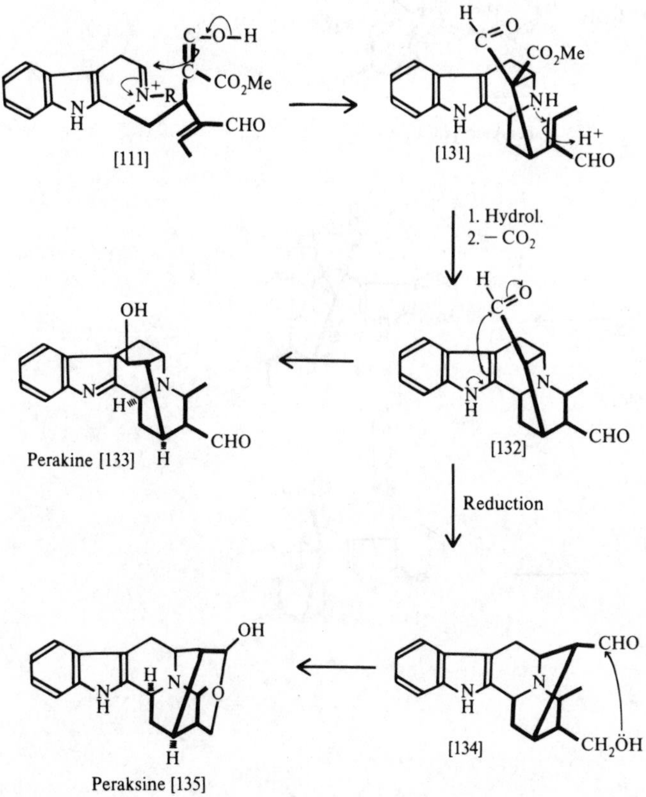

Perakine [133]   [132]

Peraksine [135]   [134]   Scheme 3.24

intramolecular cyclization as well as to peraksine [135] by reduction and cyclization. Alternatively, perakine and peraksine systems may be generated from the sarpagine/ajmaline groups by rearrangement as shown in Scheme 3.25.

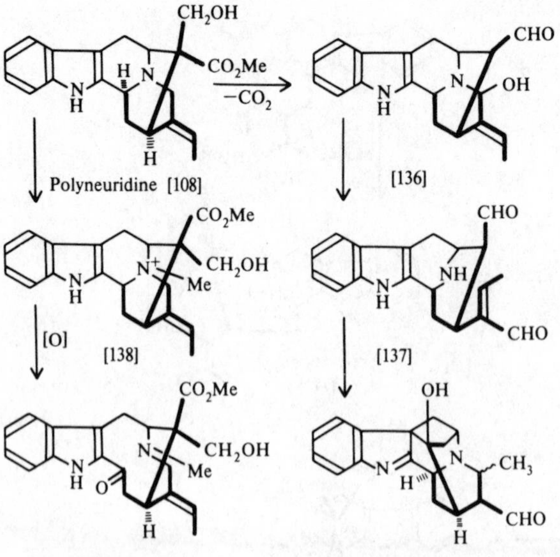

Polyneuridine [108]

[136]

[138]

[137]

Vincadiffine [139]

Perakine [133]

SCHEME 3.25

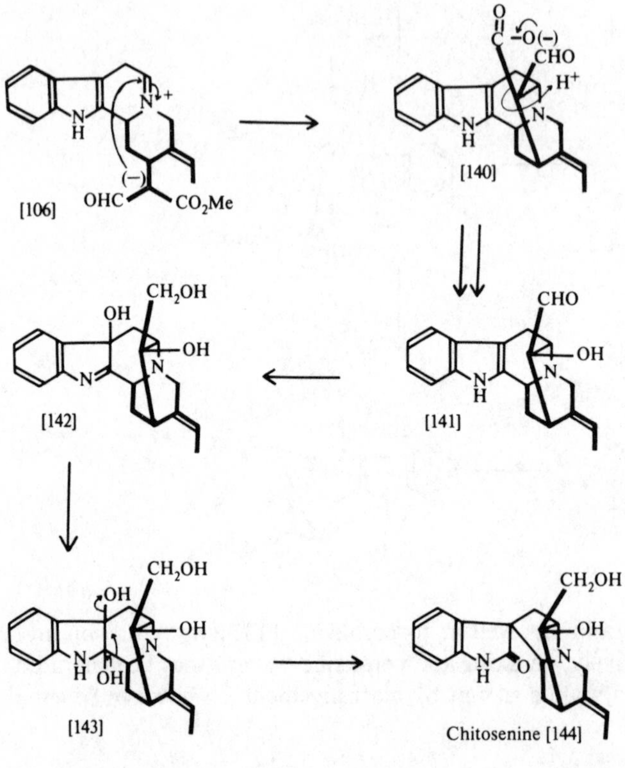

[106]

[140]

[142]

[141]

[143]

Chitosenine [144]

SCHEME 3.26

A sarpagine alkaloid, e.g. polyneuridine [108] can also act as a precursor for the vobasine group of 2-acyl indole alkaloids. Methylation of the basic nitrogen and oxidative cleavage of C—N⁺ bond would thus afford vincadiffine [139], a typical representative of this group (Scheme 3.25). The co-occurrence of vobasine, sarpagine, ajmaline, macroline, perakine, and peraksine groups of alkaloids supports such a closely linked biogenetic scheme.

The chitosenine group of alkaloids can arise from the intermediate [106] by attack of the carbanion in the side-chain on to the immonium carbon followed by a decarboxylation/oxidation sequence as shown in Scheme 3.26. The resulting intermediate [141] could then be transformed in several steps to the corresponding oxindole [144] (Scheme 3.26).

In the gardneramine group of alkaloids, the secologanin unit can be recognized intact in an unrearranged form. These alkaloids can arise from the intermediate [106] by the initial attack of the carbanion on the side chain on the immonium carbon atom. Subsequent formation of the corresponding oxindole, reduction, cyclization, and methoxylation would lead to gardneramine [147] (Scheme 3.27).

Gardneramine [147]

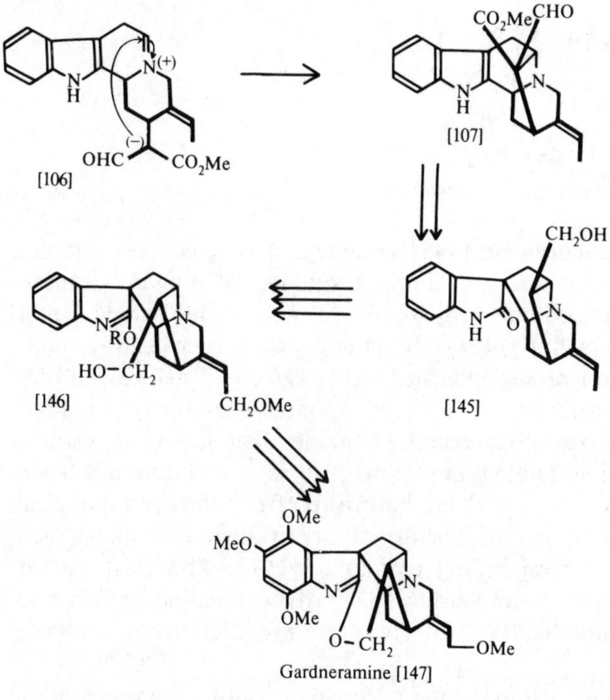

SCHEME 3.27

A related group of oxindole alkaloids, also bearing an unrearranged secologanin skeleton, is exemplified by rhyncophylline [151]. This can

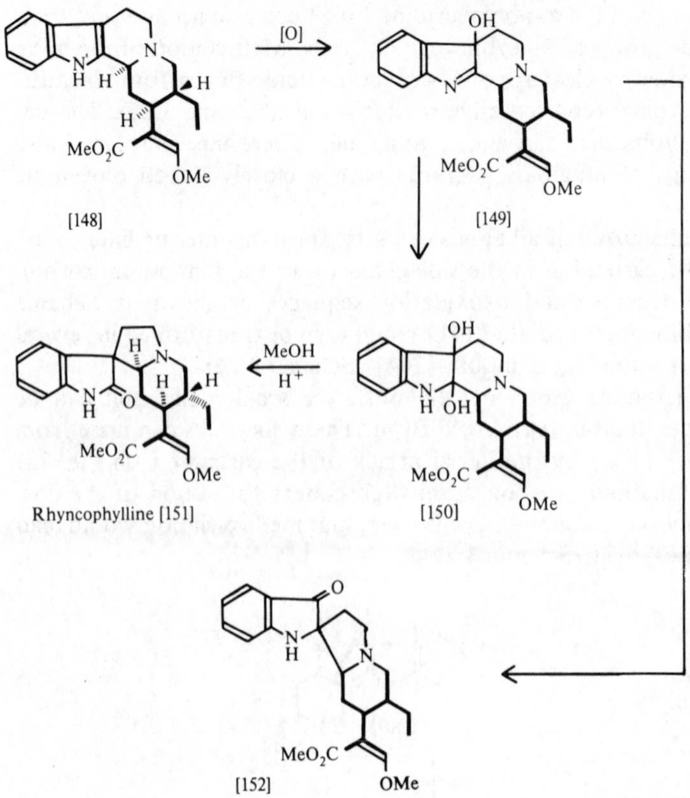

[148]

[149]

Rhyncophylline [151]

[150]

[152]

SCHEME 3.28

arise from dihydrocorynantheine [148] as shown in Scheme 3.28. Indoles are prone to oxidation at the β-position. Oxidation of dihydrocorynantheine [148] could afford the β-hydroxyindolenine [149] which could rearrange to rhyncophylline [151] in analogy with such known conversions.[77, 78] Hydration of the intermediate [149] could afford the diol [150] which could rearrange to the corresponding ψ-indoxyl alkaloid [152]. Many such *in vitro* conversions of indole alkaloids to the corresponding oxindoles and ψ-indoxyls are known,[77-88] as generalized in Scheme 3.29. β-Hydroxyindolenines [155] are known to afford the corresponding ψ-indoxyls [156] on treatment with acids or bases, while intermediates such as [158] afford the corresponding oxindoles [159]. The existence of C-profluorocurine [160], C-fluocurine [82], rhyncophylline [151], and coronaridine hydroxyindolenine [161] in nature provides strong evidence for such conversions.

Further indication of such oxidation occurring in nature comes from the ease of oxidation of ibogaine [162] to iboquine [170] and iboluteine [165]. These oxidation processes involve the intermediate generation of the

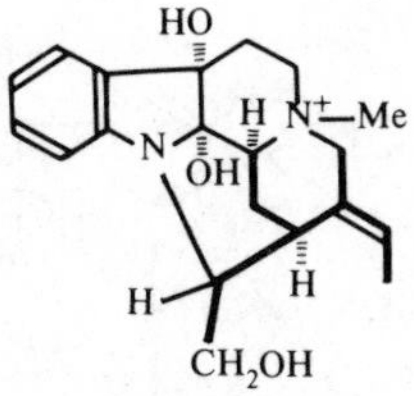

C-Profluorocurine [160]

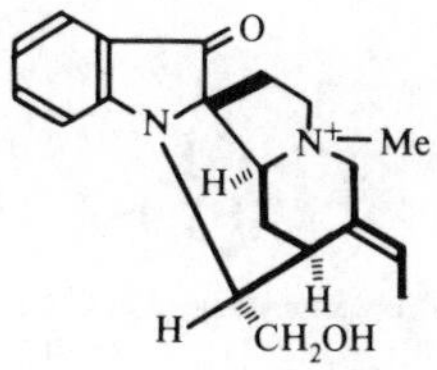

C-Flurocurine [82]

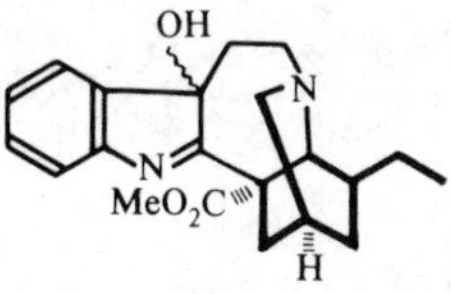

Coronaridine hydroxyindolenine [161]

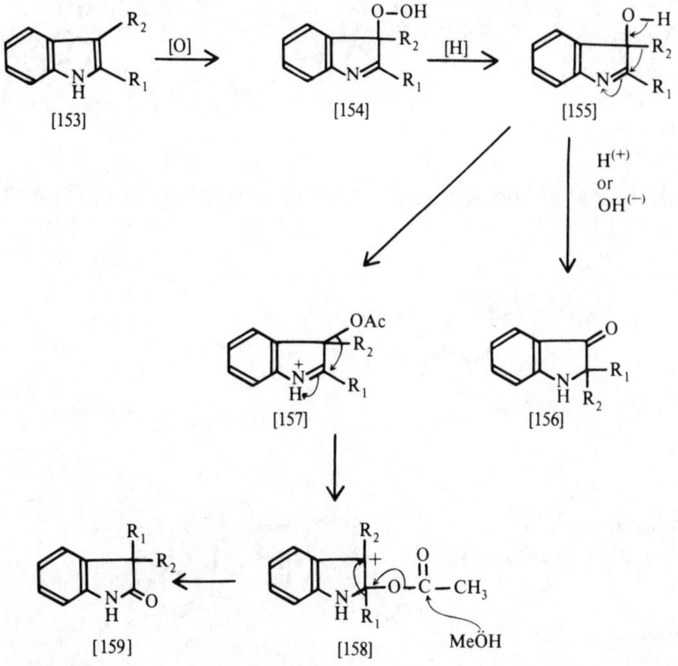

SCHEME 3.29

corresponding β-peroxy- and β-hydroxyindolenines [163] and [164], respectively, which then rearrange as shown in Scheme 3.30.

It has been shown that oxindoles are formed from corresponding indoles in *Mitragyna parvifolia*.[81, 82] Thus ajmalicine [20] and 3-isoajmalicine [171] were found to be incorporated into the oxindoles mitraphylline [172] and isomitraphylline [173], respectively. Ajmalicine has been oxidized with tert. butyl hypochlorite and the resulting 7-chloroindolenine,

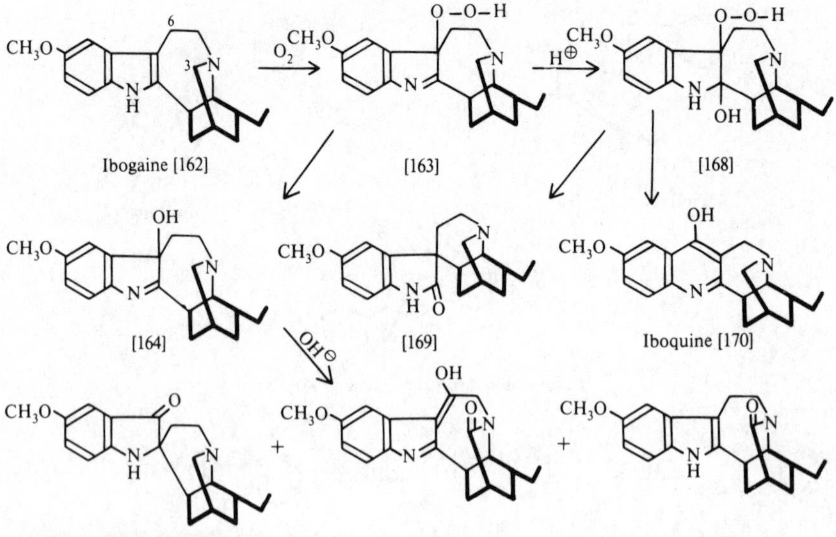

SCHEME 3.30   Iboluteine [165]   [166]   [167]

when refluxed in methanol and acidified, afforded mitraphylline [172] and isomitraphylline [173].[77, 78]

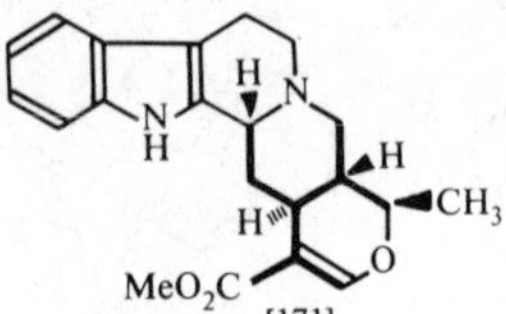

[171]

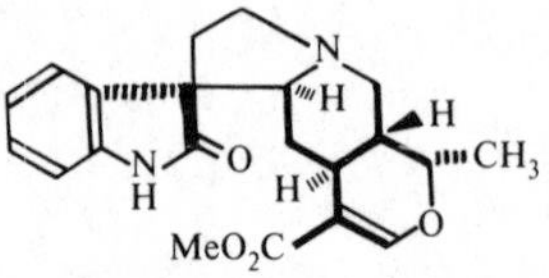

Mitraphylline [172] 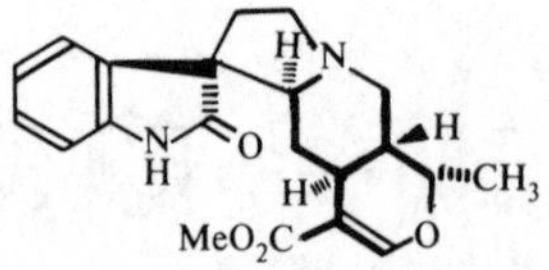   Isomitraphylline [173]

A number of oxindoles alkaloids belonging to the *Gelsemium* species have been isolated. In these, the C_{10} unit of the *Corynanthe* alkaloid system can be recognized intact, except for the loss of the methoxycarbonyl group. A scheme for the biogenesis of the major alkaloid gelsemine [182] was proposed[83] in 1959, but it remains to be verified by labelling experiments. It involves the generation of the oxindole [175] which undergoes an oxidative cleavage after quaternization of the basic nitrogen. The carbonium ion formed from the tertiary hydroxyl group is attacked by the enamine (formed on decarboxylation) to afford the spiro compound [178]. Subsequent attack of the carbanion on the immonium carbon, ether formation and decarboxylation can afford gelsemine [182] (Scheme 3.31).

Gelsemine [182]

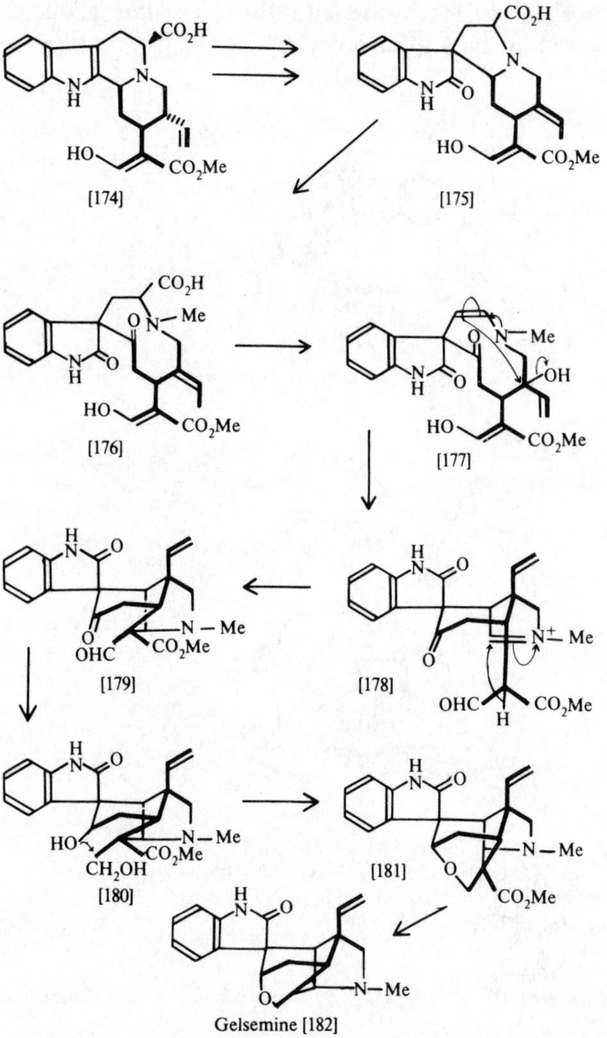

SCHEME 3.31

It has been mentioned above that the *Corynanthe* alkaloid geissoschizine was found to have a particularly dynamic radioprofile, and it was found to be incorporated not only into serpentine and ajmalicine but also into the *Strychnos* alkaloid akuammicine, the *Aspidosperma* alkaloid vindoline, and the *Iboga* alkaloid catharanthine.[84] Further evidence came from studies following administration of DL-[2-^{14}C]-tryptophan to young *V. rosea* seedlings. The radioactivity of geissoschizine was found to reach a maximum after about 2 h (4 per cent of the total activity of the alkaloid fractions) and then declined slowly to a steady 1 per cent over an 8-day period.[85] The conversion of labelled geissoschizine into the *Strychnos* alkaloid akuammicine was followed by mass spectroscopy.[61]

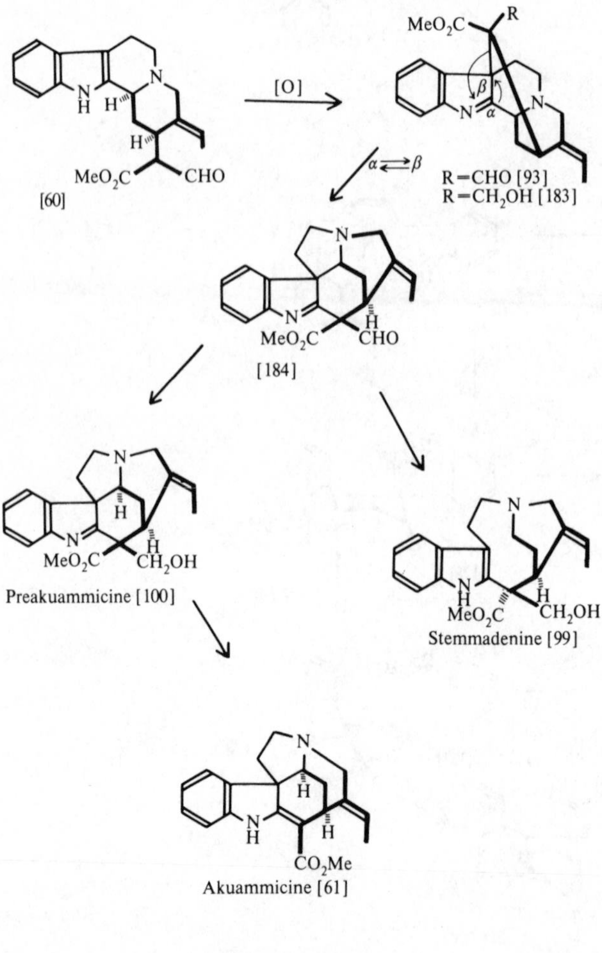

[60]

[O]

$\alpha \rightleftharpoons \beta$

R=CHO [93]
R=CH$_2$OH [183]

[184]

Preakuammicine [100]

Stemmadenine [99]

Akuammicine [61]

SCHEME 3.32

Several mechanisms have been advanced to interconnect the *Corynanthe* alkaloids with the *Strychnos* type. Since the *Strychnos* alkaloids are at a higher oxidation level than those of the *Corynanthe* series, it has been suggested[54, 86] that geissoschizine [60] could be first converted to formylstrictamine [93] (R = CHO) by a one-electron oxidative coupling. Rearrangement of compounds such as [183] to akuammicine [61] is known[87] so that the indolenine [184] could be formed by such a route (Scheme 3.32). It is notable that desacetyldesformoakuammiline (strictamine) [185] and the corresponding alcohol have been converted *in vitro* to akuammicine [61] and norfluorocurarine [188] under Oppenauer oxidation conditions,[88] thus linking the picraline group with akuammicine-type alkaloids[89, 90] (Scheme 3.33).

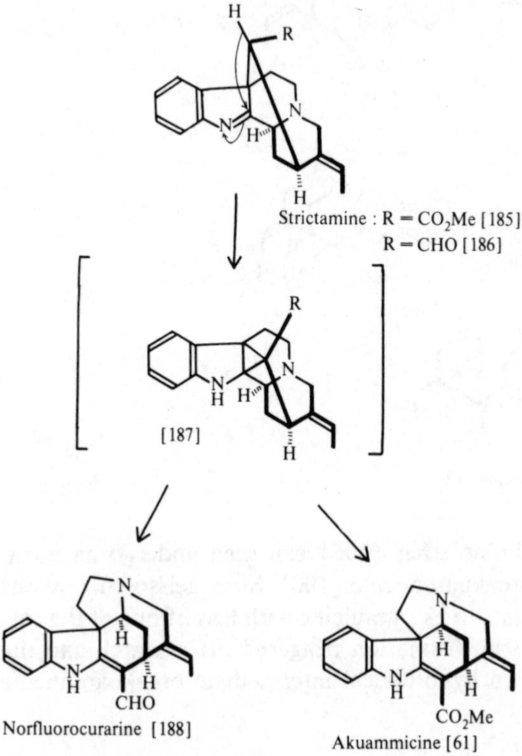

Strictamine : R = CO$_2$Me [185]
R = CHO [186]

[187]

Norfluorocurarine [188]

Akuammicine [61]

SCHEME 3.33

An alternative mechanism which can be advanced to connect the *Corynanthe* and *Strychnos* alkaloids is based on an *in vitro* analogy to the work of Harley-Mason.[91, 92] This involves α-protonation of the indole nucleus followed by an α–β rearrangement (Scheme 3.34). Another more direct mechanism which can be advanced is based on the spiroindolenine derivative [191] formed from tryptamine and secologanin to afford the akuammicine type of bases (Scheme 3.35).

Presently the most favoured mechanism[93] connecting the *Corynanthe* alkaloids geissoschizine with the *Strychnos* series involves the intermediate formation of geissoschizine oxindole [195] via the β-hydroxyindolenine

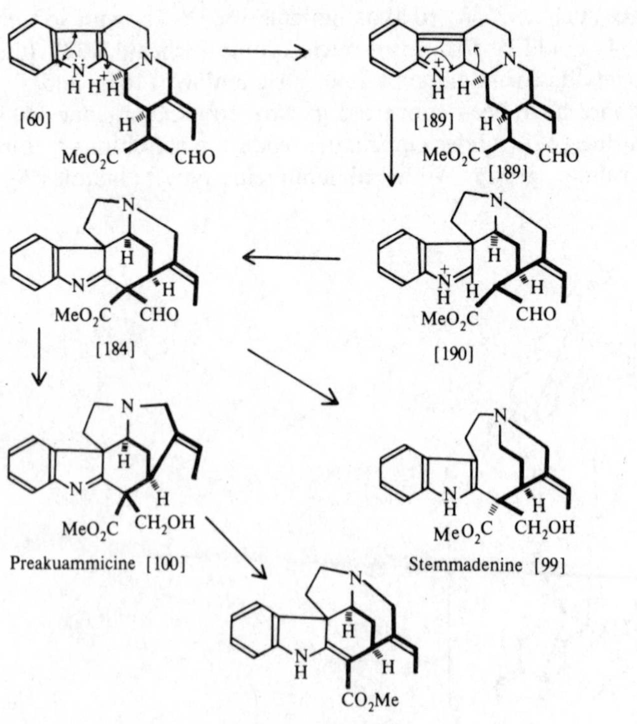

[60]
MeO2C
CHO
[189]
MeO2C
CHO
[189]
MeO2C
CHO
[184]
MeO2C
CHO
[190]
MeO2C
CH2OH
Preakuammicine [100]
MeO2C
CH2OH
Stemmadenine [99]
CO2Me

Akuammicine [61] SCHEME 3.34

[193]. The corresponding imino-ether [196] can then undergo an intramolecular attack to afford preakuammicine [100]. Since geissoschizine was shown to be converted in plants to akuammicine with loss of one of the carbons of the 'C_{10}' unit, the transformation triggered off a search and the ultimate discovery of the then hypothetical intermediate preakuammicine

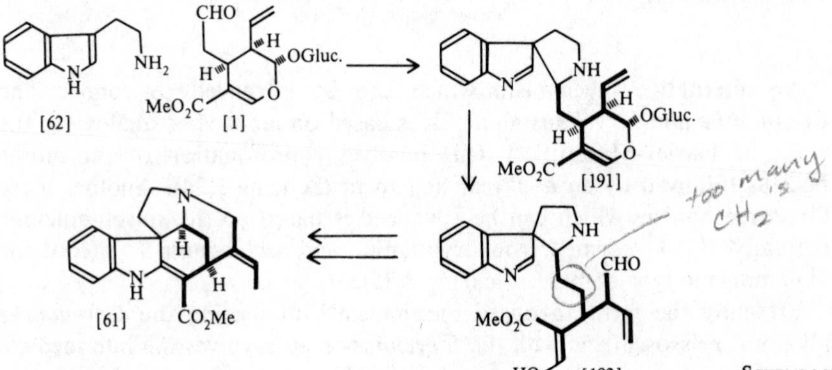

CHO
NH2
OGluc.
MeO2C
[62]
[1]
NH
OGluc.
MeO2C
[191]
NH
CHO
[61]
CO2Me
MeO2C
HO
[192]
too many CH2's

SCHEME 3.35

[100], which was found to undergo a ready loss of formaldehyde to give akuammicine [61] (Scheme 3.36). The isolation of geissoschizine oxindole from *V. rosea* and the incorporation of the labelled compound into akuammicine (and vindoline)[94] has added strength to such a proposal. An indole alkaloid with an imino-ether functionality has also been reported[95, 96] and it remains to be seen whether it is incorporated into akuammicine.

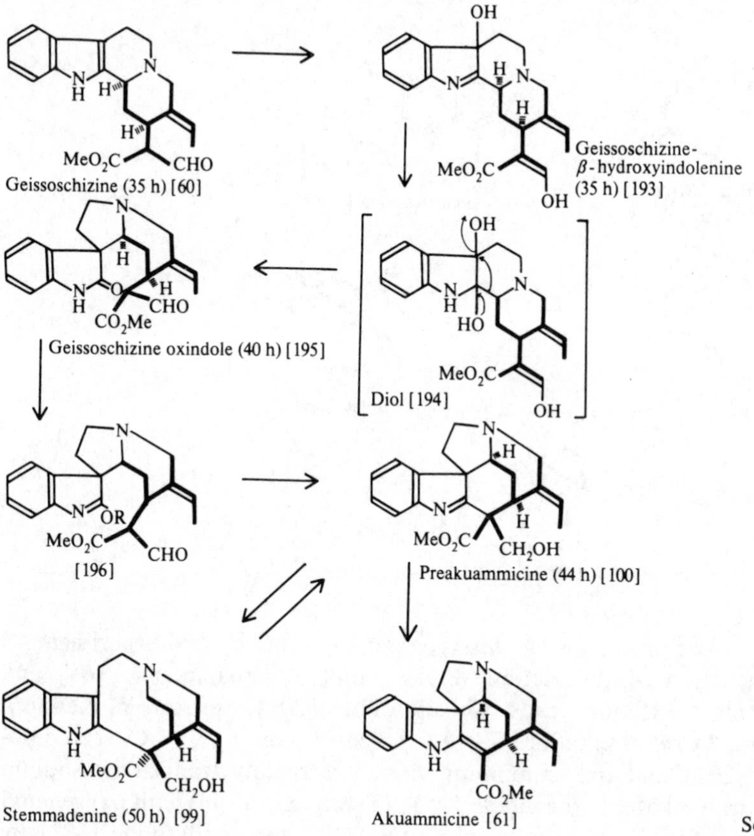

SCHEME 3.36

The stemmadenine group of alkaloids, which differ from the alkaloids of the akuammicine group by the absence of the C_7—C_3 bond, can arise from preakuammicine [100] by a ring-opening/reduction process as shown in Scheme 3.37. The same intermediate [197] can undergo a 1,3-hydrogen shift to afford the conjugated immonium compound [198] which may undergo attack at C-21 by the β-position of the indole nucleus to afford precondylocarpine [199]. This can loose formaldehyde by a retro-aldol process to afford the condylocarpine group of alkaloids (Scheme 3.37).

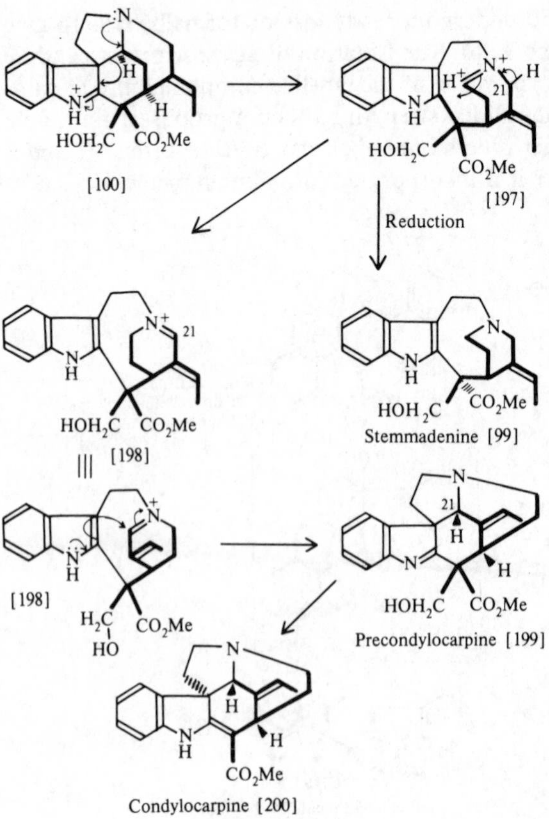

Condylocarpine [200]   SCHEME 3.37

Support for conversions of this type comes from *in vitro* experiments.[97] Hydrogenation of the olefinic double bond of akuammicine [61] and decarbomethoxylation leads to tubifoline [201]. In protic medium, tubifoline exists in equilibrium with the corresponding C_3—C_7 seco compound [202] and this immonium species is readily trapped by sodium borohydride to afford the indole [203]. Treatment of this with oxygen and platinum affords a mixture of tubifoline [201] and condyfoline [207] by the process shown in Scheme 3.38. The reaction probably proceeds via the intermediate peroxyindolenine [204] which may be attacked intramolecularly to afford the N-oxide [205]. This could readily be converted to either of the two immonium ions [202] or [206] respectively. Subsequent intramolecular ring closures could lead to the tubifoline group of alkaloids. The conversion of tubifoline [201] to condyfoline [207] by heating[98] can also be explained by the initial generation of the immonium ion [202] followed by a 1,3-hydrogen shift to afford [206] which can undergo ring closure to give condyfoline [207] (Scheme 3.38).

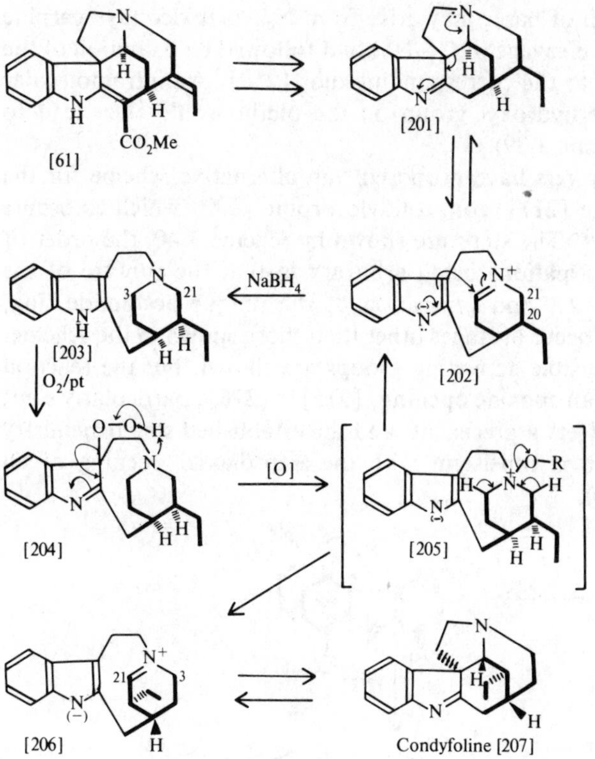

SCHEME 3.38

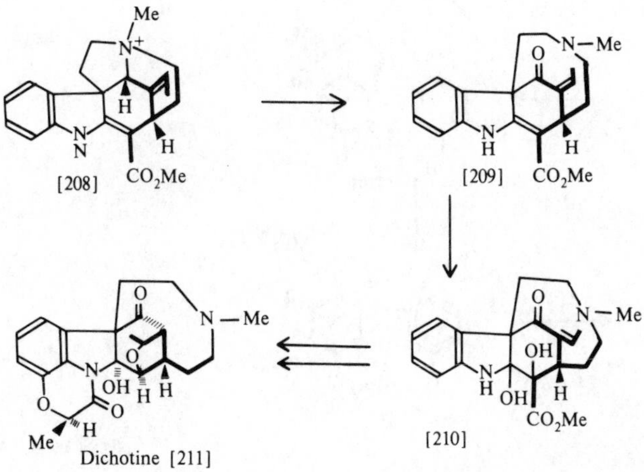

SCHEME 3.39

The dichotine group of bases may arise from $N_{(b)}$-methylcondylocarpine [208] by an oxidative cleavage of C—N$^+$ bond followed by oxidation of the acrylate double bond to the corresponding diol [210]. An intramolecular attack of the tertiary hydroxyl group on the olefin would then lead to dichotine [211] (Scheme 3.39).

Djerassi and co-workers have proposed[99] an alternative scheme for the biogenesis of dichotine [211] from condylocarpine [200] which co-occurs in *Vallesia dichotma*.[100] The steps are shown in Scheme 3.40, the order of cleavage and bond formations being arbitrary in that the rupture of the 3,4-bond, [200] → [212], and generation of the unique acetamide ring, [216] → [217], could occur at stages other than those shown in the scheme. No biochemically plausible activating groups are shown, but the reaction may proceed through an epoxide opening, [215] → [216], particularly since squalene epoxide[101] offers a precedent, and the established stereochemistry at positions 2 and 16 is consistent with the transdiaxial opening of an epoxide (Scheme 3.40).

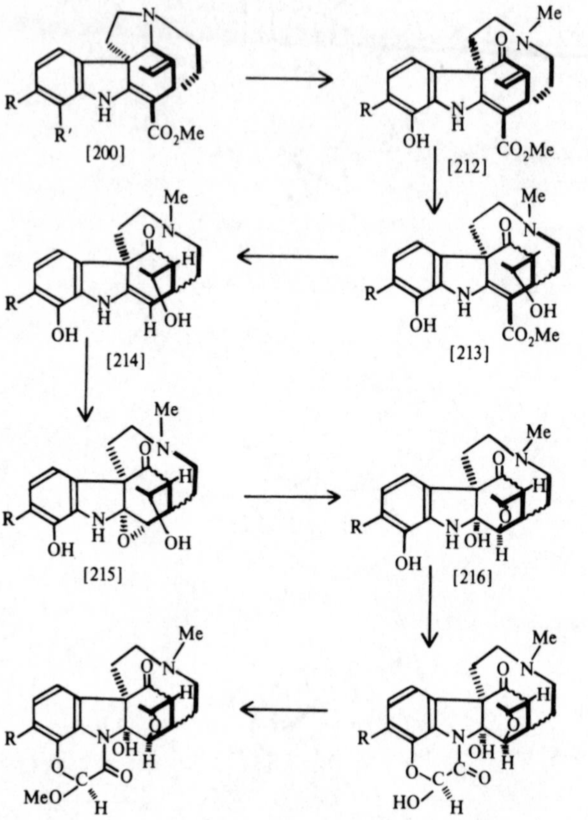

SCHEME 3.40

The discovery of enantiomeric forms of the *Strychnos* alkaloids akuammicine [61] and the diastereoisomers (−)-lochneridine [218] and (+)-20-epilochneridine [224] has led to the explanation that a non-stereospecific step must be involved in the pathway to account for the inversion at C_3, C_7, and C_{15} by rupture of C_3—C_7 and C_{15}—C_{16} bonds.[102] Experiments have also shown that geissoschizine [60] is a good precursor of both akuammicine [61] and strychnine [250]. It has been suggested[63] that the *Strychnos* alkaloids can undergo inversion at C_3, C_7, and C_{15} via the intermediacy of a non-stereodiscerning dihydropyridine intermediate [221] which can undergo intramolecular ring closures to afford either (+)-lochneridine [222] or (−)-lochneridine [218] and thence by dehydration lead to either of the two enantiomeric forms of akuammicine (Scheme 3.41).

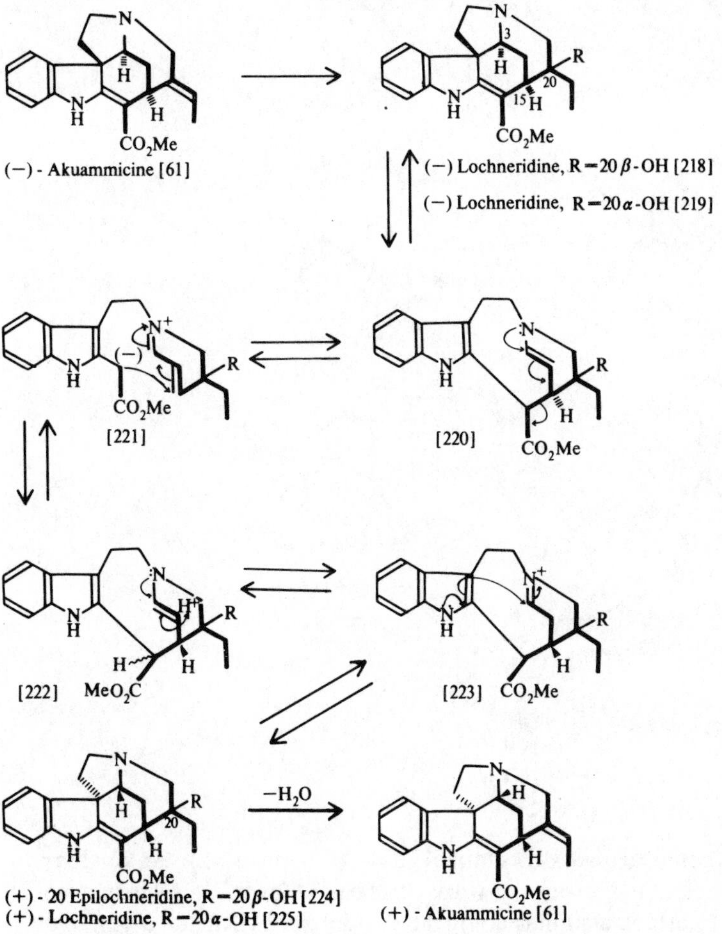

(−) - Akuammicine [61]

(−) Lochneridine, R = 20 β-OH [218]
(−) Lochneridine, R = 20 α-OH [219]

[221]

[220]

[222]   MeO$_2$C

[223]   CO$_2$Me

(+) - 20 Epilochneridine, R = 20 β-OH [224]
(+) - Lochneridine, R = 20 α-OH [225]

−H$_2$O

(+) - Akuammicine [61]

SCHEME 3.41

It has been demonstrated by feeding experiments[104] that stemmadenine [99] can also act as a precursor for the vallesamine group of bases. The conversion of stemmadenine [99] to vallesamine [229] and apparicine [231] involves loss of C_5 adjacent to the nitrogen. The formation of these alkaloids is rationalized in Scheme 3.42. The hydroxyindolenine [226] from stemmadenine may undergo a cleavage of C_5—C_6 bond with the departure of a suitably functionalized group to afford the seco compound [227]. Hydrolytic cleavage of the immonium methylene group followed by

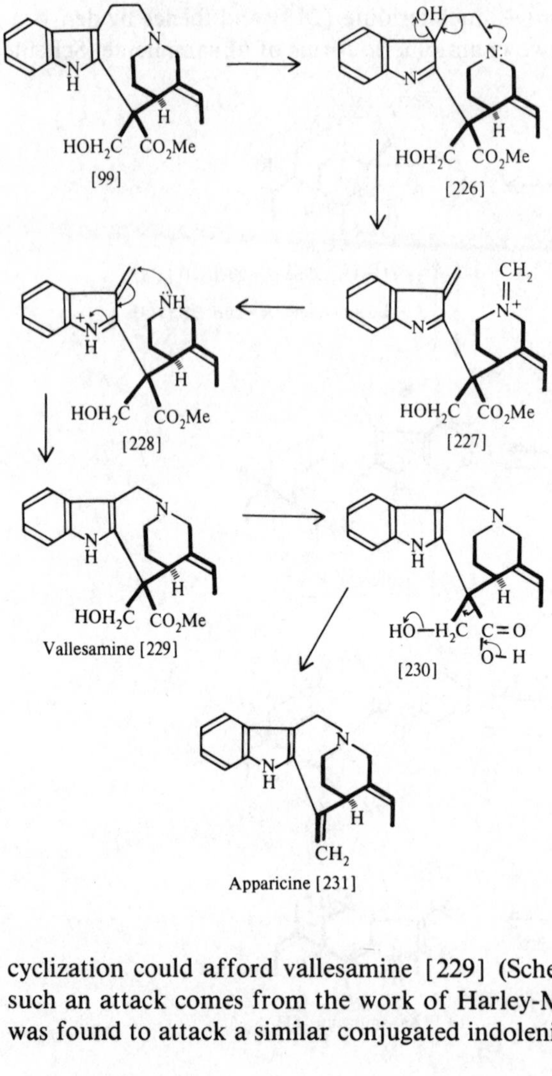

HOH$_2$C CO$_2$Me
[99]

HOH$_2$C CO$_2$Me
[226]

HOH$_2$C CO$_2$Me
[228]

HOH$_2$C CO$_2$Me
[227]

HOH$_2$C CO$_2$Me
Vallesamine [229]

HO—H$_2$C C=O
O—H
[230]

Apparicine [231]

Scheme 3.42

cyclization could afford vallesamine [229] (Scheme 3.42). An analogy to such an attack comes from the work of Harley-Mason[105] in which cyanide was found to attack a similar conjugated indolenine (Scheme 3.43).

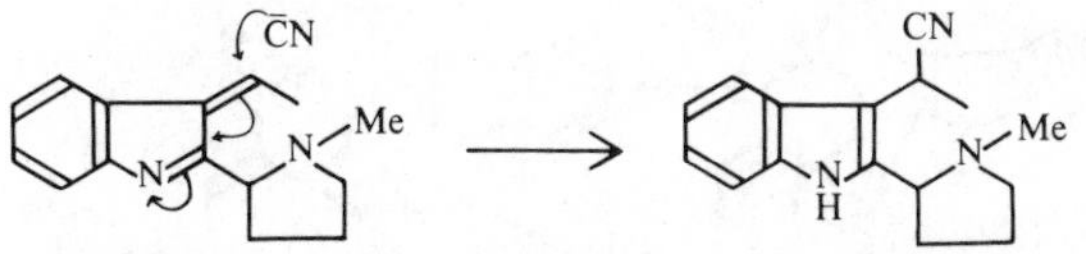

[232]   [233]   SCHEME 3.43

Vallesamine [229] can be converted to apparicine [231] by a concerted decarboxylative-dehydration reaction. Feeding experiments have shown that tryptophan is incorporated with retention of C_3 in apparicine [231].[104] Labelled stemmadenine, and to lesser extent vallesamine, were also found to be incorporated into apparicine,[106] and it was suggested that the loss of C_5 from stemmadenine occurs at a late stage.[106] An alternative to the mechanism shown in Scheme 3.42 would be the formation of an intermediate N-oxide followed by rearrangement (Scheme 3.44).[107]

It has been proposed[108] that the formation of the uleine group of alkaloids from stemmadenine [99] may involve the initial generation of the corresponding N-oxide [236] which could undergo a decarboxylative fragmentation of the C_5—C_6 bond. The resulting immonium compound [237] can, by a 1,3-H shift, afford the conjugated enimmonium substance [238]. This process could also be accompanied by the attack of a hydroxyl group at C_6, resulting in the allylic-type displacement of the exocyclic hydroxyl function by a concerted process. The β-position of the indole nucleus can then attack C_{21} to afford uleine [240] in two steps as shown in Scheme 3.44. Apparicine [231] can also be formed from the same C_5—C_6 seco intermediate [237] by an intramolecular cyclization reaction.

An alternative possibility to the biogenesis of uleine [240] could be by oxidative removal of the tryptamine bridge from condylocarpine [200]. It has, however, been shown[110] that the two carbon substituents located at the β-position of an α-methylene indoline system cannot be cleaved, in contrast to β C-1 units which can be cleaved readily. It has therefore been proposed that the formation of uleine [240] involves an initial C_5—C_6 bond cleavage followed by an oxidative degradation of the β C-1 unit (Scheme 3.44). It is possible that while both uleine and apparicine are derived from stemmadenine, two different pathways are followed after fission of the C_5—C_6 bond. In the case of uleine the carbon atom β to $N_{(b)}$ is lost, while in apparicine the carbon α to $N_{(b)}$ is degraded, as outlined in Scheme 3.44.

A surprising observation by Kutney and co-workers[111-113] was the incorporation of secodine [242] into apparicine [231] with specific retention of the ester carbonyl group. This would imply that the ester carbonyl is reduced and dehydrated to form the exocyclic methylene of apparicine [231] while the acrylate methylene in secodine is lost. This seems to be unlikely and these results need to be repeated. Even more surprising results by the same group was the non-incorporation of labelled [ar-^{3}H, 2-^{14}C]- and

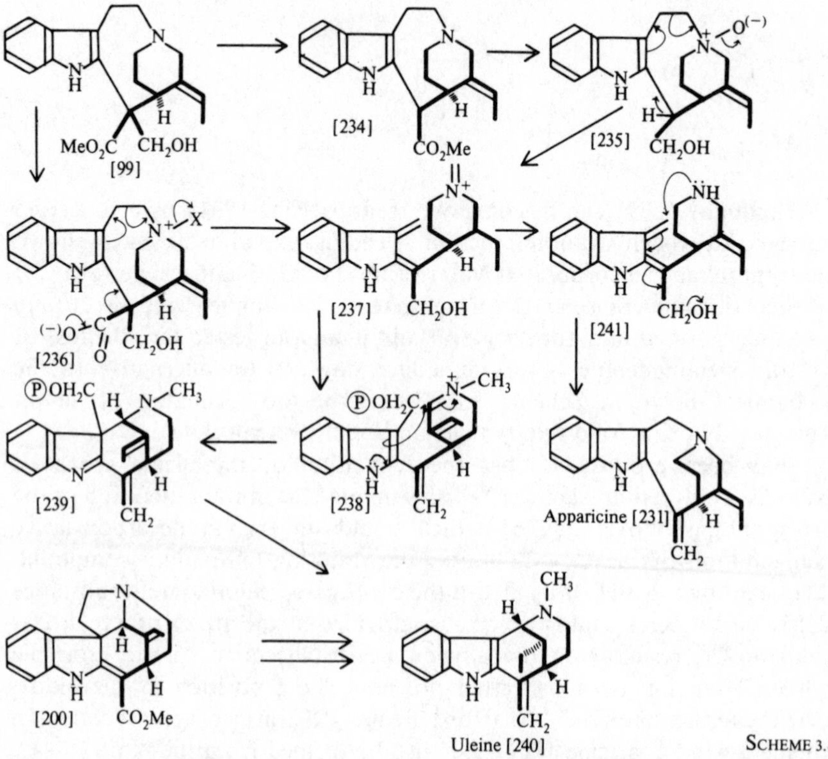

Uleine [240]   SCHEME 3.44

[ar-^{3}H, 3-^{14}C]-tryptophan, [ar-^{3}H]-stemmadenine, [ar-^{3}H]-vallesamine, and [ar-^{3}H]-3-aminomethyl indole into uleine.[78, 80] The non-tryptophan origin of uleine, particularly when secodine has been shown to be incorporated into uleine,[104, 106] is questionable (Scheme 3.45). More recently,

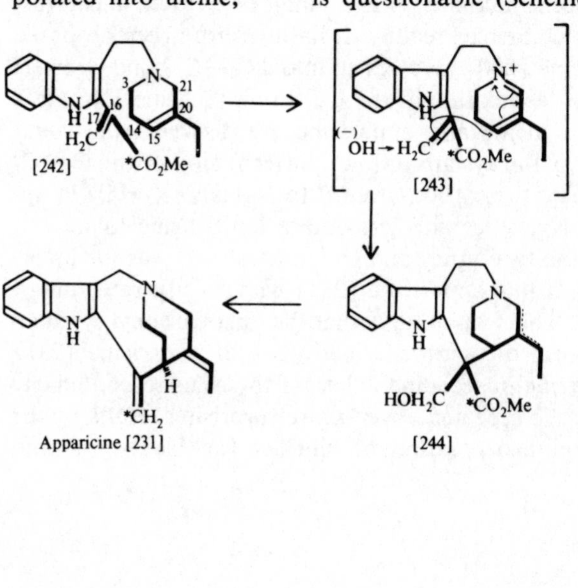

Apparicine [231]   [244]   SCHEME 3.45

Scott and co-workers have shown that stemmadinine-N-oxide can, by a Polonovski reaction followed by hydrolysis and cyclization, afford vallesamine.[114] It is likely that a similar process is operational in nature.

Many of the *Strychnos* alkaloids, e.g. those of the strychnine and vomicine groups can arise from the akuammicine system. These alkaloids possess an additional two-carbon unit between the indole nitrogen and C-17. Previously it was generally accepted that the Wieland–Gumlich aldehyde [245] or its N-acetyl derivative diaboline [246], served as precursors for such alkaloids, but feeding experiments[115] have not supported this hypothesis and it has been shown that the additional two-carbon unit comes from acetate at some stage of the biosynthesis. An intermediate such as [192] formed by the ring opening of the spiroindolenine [191] can serve as a precursor for these *Strychnos* alkaloids as shown in Scheme 3.46. It is pro-

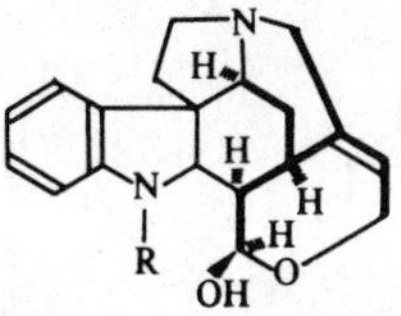

Wieland–Gumlich aldehyde R←H [245]
Diaboline R —COCH$_3$ [246]

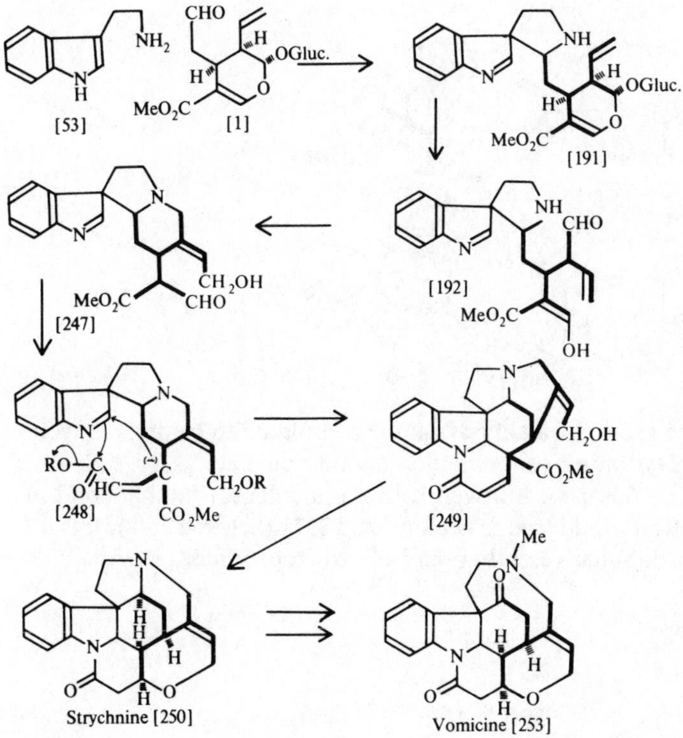

[53]   [1]

[191]

[192]

[247]

[248]

[249]

Strychnine [250]

Vomicine [253]

SCHEME 3.46

posed that the acetate unit can be incorporated at the indolenine stage [247] to afford [248]. This could lead to strychnine [250] by cyclization and decarbomethoxylation. An *in vitro* analogy of such an intramolecular cyclization to the α, β-unsaturated ketone system is provided in the final step of the synthesis of strychnine by R. B. Woodward and co-workers.[116]

An alternative route to strychnine [250] may be visualized from preakuammicine [100] which may be converted by a retro-aldol process to akuammicine [61]. Addition of a two-carbon unit, probably derived from acetate, to akuammicine, oxidation at the allylic carbon, and reductive cyclization could then lead to strychnine (Scheme 3.47). The related vomicine group of alkaloids can arise by methylation at the basic nitrogen in strychnine followed by oxidative cleavage of C—N$^+$ bond as shown in Scheme 3.46.

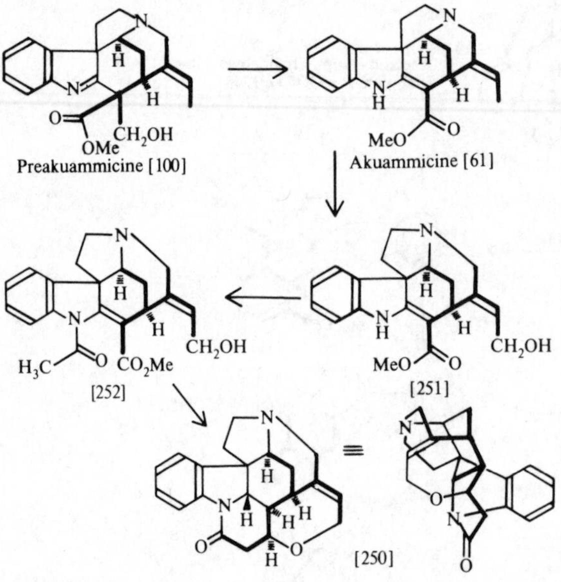

SCHEME 3.47

The echitamine group of alkaloids can be considered to be derived from rhazinol [254]. Hydration of the immonium intermediate [255] followed by recyclization of the basic nitrogen with the indolenine double bond in [256] and reduction would lead to echitamine [257] (Scheme 3.48). It is interesting that rhazinol has recently been isolated from *Rhazya stricta*.[117]

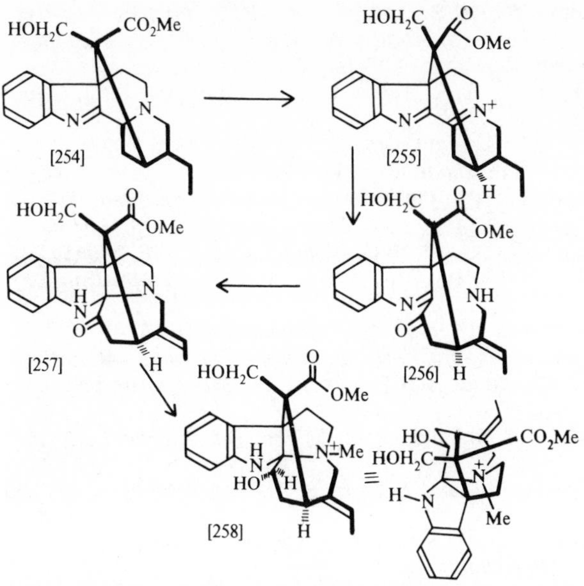

SCHEME 3.48

Summary

The condensation of secologanin with tryptophan leads to strictosidine [9] and thence to the various *Corynanthe*, *Yohimbe*, and *Strychnos* alkaloids through the intermediacy of certain key 'link' alkaloids. There is growing evidence that cathenamine [49] and geissoschizine [60], either directly or through appropriately functionalized derivatives, play an important role in the biosynthesis of tetrahydroalstonine [52], yohimbine [59], and other related alkaloids (Scheme 3.3). A number of other alkaloidal groups arise by intramolecular rearrangements and manifestations of enamine and dienamine chemistry. Routes connecting the *Corynanthe* alkaloids to those of the *Strychnos* type are discussed (Schemes 3.33–3.36).

References

1. DESILVA, K. T. D., KING, D., and SMITH, G. N. *Chem. Commun.* 908 (1971).
2. BROWN, R. T. and ROW, L. R. *Chem. Commun.* 453 (1967).
3. BROWN, R. T., RAO, K. V. J., RAO, P. V. S., and ROW, L. R. *Chem. Commun.* 350 (1968).
4. SMITH, G. N. *Chem. Commun.* 912 (1968).
5. BATTERSBY, A. R., BURNETT, A. R., and PARSONS, P. G. *Chem. Commun.* 1282 (1968).

6. BATTERSBY, A. R., BURNETT, A. R., and PARSONS, P. G. *J. Chem. Soc.* 1193 (1969).
7. SCOTT, A. I. *Acc. Chem. Res.* **3**, 151 (1970).
8. BATTERSBY, A. R., BURNETT, A. R., HALL, E. S., and PARSONS, P. G. *Chem. Commun.* 1582 (1968).
9. LEVESQUE, J., POUSSET, J. L., and CAVA, A. *Fitoterapia* **48**, 5 (1977).
10. COUNE, C. and ANGENOT, L. *Planta Med.* **34**, 53 (1978).
11. HOTELLIER, F., DELAVEAU, P., and POUSSET, J. L. *Planta Med.* **11**, 106 (1977).
12. ALI, E., GIRI, V. S., and PAKRASHI, S. C. *Indian J. Chem.* **14B**, 306 (1976).
13. BROWN, R. T., SMITH, G. N., and STAPLEFORD, K. S. J. *Tetrahedron Lett.* 4349 (1968).
14. BATTERSBY, A. R. and HALL, E. S. *Chem. Commun.* 793 (1969).
15. STOCKIGT, J. and ZENK, M. H. *J. Chem. Soc. Chem. Commun.* 646 (1977).
16. BATTERSBY, A. R., LEWIS, N. G., and TIPPETT, J. M. *Tetrahedron Lett.* 4849 (1978).
17. RUEFFER, M., NAGAKURA, N., and ZENK, M. H. *Tetrahedron Lett.* 1593 (1978).
18. BROWN, R. T., LEONARD, J., and SLEIGH, S. K. *Phytochemistry* **17**, 899 (1978).
19. SCOTT, A. I., LEE, S. L., CAPITE, P. DE., CULVER, M. G., and HUTCHINSON, C. R. *Heterocycles* **7**, 979 (1977).
20. VAN TAMELEN, E. E., HAARSTADT, V. R., and ORVIS, R. L. *Tetrahedron* **25**, 687 (1968).
21. VAN TAMELEN, E. E. and OLIVER, L. K. *Bio-org. Chem.* **5**, 309 (1976).
22. STOCKIGT, J. *Tetrahedron Lett.* 2615 (1979).
23. STOCKIGT, J. and ZENK, M. H. *FEBS Lett.* **79**, 233 (1977).
24. DESILVA, K. T. D., SMITH, G. N., and WARREN, K. E. H. *Chem. Commun.* 905 (1971).
25. JOHNS, S. R., LAMBERTON, J. A., and OCCOLOWITZ, J. C. *Chem. Commun.* 229 (1967); *Aust. J. Chem.* **21**, 1399 (1968).
26. BLACKSTOCK, W. P., BROWN, R. T., and LEE, G. K. *Chem. Commun.* 910 (1971).
27. BLACKSTOCK, W. P. and BROWN, R. T. *Tetradedron Lett.* 3727 (1971).
28. BROWN, R. T. and BLACKSTOCK, W. P. *Tetrahedron Lett.* 3063 (1972).
29. BLACKSTOCK, W. P., BROWN, R. T., CHAPPLE, C. L., and FRASER, S. B. *Chem. Commun.* 1006 (1972).
30. WENKERT, E. and BRINGI, N. V. *J. Am. Chem. Soc.* **81**, 1474 (1959).
31. OCHIAI, E. and ISHIKAWA, M. *Chem. Pharm. Bull.* **7**, 256 (1959).
32. WARREN, K. E. H. Ph.D. Thesis, University of Manchester (1970).
33. KENNARD, O. P., ROBERTS, J., ISSACS, N. W., ALLEN, F. H., MOTHERWELL, W. D. S., GIBSON, K. H., and BATTERSBY, A. R. *Chem. Commun.* 899 (1971).
34. INOUYE, H. *Pharmacognosy and phytochemistry* (eds. H. Wagner and L. Hornhammer). Springer, New York.
35. KAPIL, R. S., SHOEB, A., POPLI, S. P., BURNETT, A. R., KNOWLES, G. D., and BATTERSBY, A. R. *Chem. Commun.* 904 (1971).
36. MATTES, K. C., HUTCHINSON, C. R., SPRINGER, J. P., and CLARDY, J. *J. Am. Chem. Soc.* **97**, 6270 (1975).
37. BATTERSBY, A. R. and GIBSON, K. H. *Chem. Commun.* 902 (1971).
38. BATTERSBY, A. R., BYRNE, T. C., KAPIL, R. S., MARTIN, J. A., PAYNE, T. G., ARIGONI, D., and LOEW, P. *Chem. Commun.* 951 (1968).

39. BATTERSBY, A. R., HUTCHINSON, C. R., and LARSON, R. A. *Abstracts (Organic Section No. 11), A. C.S. National Meeting*, Boston, MA, 1972.
40. HUSSON, H-P. and KAN-FAN, C. *Tetrahedron Lett.* 1889 (1977).
41. STOCKIGT, H., HUSSON, H-P., KAN-FAN, C., and ZENK, M. H. *J. Chem. Soc. Chem. Commun.* 164 (1977).
42. HUSSON, H-P. and KAN-FAN, C. *Chem. Soc. Chem. Commun.* 618 (1978).
43. STOCKIGT, J., RUEFFER, M., ZENK, M. H., and HOYER, G. A. *Planta Med.* **33**, 188 (1978).
44. TREIMER, J. F. and ZENK, M. H. *Phytochemistry* **17**, 227 (1978).
45. STOCKIGT, J. *Phytochemistry* **18**, 965 (1979).
46. BARGER, G. and SCHOLZ, C. *Helv. Chim. Acta.* **16**, 1343 (1933).
47. BARGER, G. *IX Congresso Internacional de Quimica Pura Applicabda*, Madrid, Conferencies de Introduccion 177 (1934).
48. HAHN, G. and WERNER, H. *Justus Liebigs Ann. Chem.* **520**, 123 (1935).
49. WOODWARD, R. B. *Nature (Lond.)* **162**, 155 (1948).
50. WOODWARD, R. B. *Angew. Chem.* **68**, 13 (1956).
51. BEGUE, J. P. *Bull. Soc. Chim. Fr.* 2545 (1969).
52. LEETE, E. *Advances in enzymology and related areas, Molecular biology*, Vol. 32, p. 373. Interscience, New York (1969).
53. SUNDBERG, R. J. *The chemistry of indoles*, Academic Press, New York (1970).
54. WENKERT, E. *J. Am. Chem. Soc.* **84**, 98 (1962).
55. BATTERSBY, A. R. *Chimia* **22**, 310 (1968).
56. QURESHI, A. A. and SCOTT, A. I. *Chem. Commun.* 948 (1968).
57. KUTNEY, J. P., CRETNEY, W. J., HADFIELD, J. R., HALL, E. S., NELSON, V. R., and WIGFIELD, D. C. *J. Am. Chem. Soc.* **90**, 336 (1968).
58. QURESHI, A. A. and SCOTT, A. I. *Chem. Commun.* 945 (1968).
59. KUTNEY, J. P., EBRET, C., NELSON, V. R., and WIGFIELD, D. C. *J. Am. Chem. Soc.* **90**, 5929 (1968).
60. BOJTHE-HORVATH, K., VARGA-BALAZS, M., and CLAUDER, O. *Planta Med.* **17**, 328 (1969); *Chem. Abs.* **71**, 109801.
61. SCOTT, A. I., CHERRY, P. C., and QURESHI, A. A. *J. Am. Chem. Soc.* **91**, 4932 (1969).
62. LANGLOIS, N. and POTIER, P. *C. R. Acad. Sci. Paris, Ser., C* **273**, 994 (1971).
63. INOUYE, H., UEDA, S., and TAKEDA, Y. *Tetrahedron Lett.* 3453 (1968).
64. BROWN, R. T., FRASER, S. B., and BANERJI, J. *Tetrahedron Lett.* 3335 (1974).
65. STOCKIGT, J. *J. Chem. Soc. Chem. Commun.* 1097 (1978).
66. LEE, S. L., HIRATA, T., and SCOTT, A. I., *Tetrahedron Lett.* 691 (1979).
67. SCOTT, A. I., LEE, S. L., HIRATA, T., and CULVER, M. G. *Rev. Latinoam. Quium.* **9**, 131 (1978).
68. BROWN, R. T., CHAPPLE, C. L., DUCKWORTH, D. M., and PLATT, R. *J. Chem. Soc. Perkin Trans. I* 2, 160 (1976).
69. BROWN, R. T., LEONARD, J., and SLEIGH, S. K. *J. Chem. Soc. Chem. Commun.* 636 (1977).
70. BROWN, R. T., CHAPPLE, C. L., PLATT, R., and SLEIGH, S. K. *Tetrahedron Lett.* 203 (1976).
71. VAN TAMELEN, E. E. and DORSCHEL, C. *Bio-org. Chem.* **5**, 203 (1976).
72. BROWN, R. T. and LEONARD, J. *J. Chem. Soc. Chem. Commun.* 877 (1979).
73. HINSHAW, W. B., LEVY, J., and LEMEN, J. *Tetrahedron Lett.* 955 (1971).
74. VAN TAMELEN, E. E. and OLIVER, L. K. *J. Am. Chem. Soc.* **92**, 2136 (1970).
75. VAN TAMELEN, E. E., HARSTAAD, V. R., and ORVIS, R. L. *Tetrahedron* **24**, 687 (1968).

76. TAYLOR, W. I., FREY, A. J., and HOFMANN, A. *Helv. Chim. Acta.* **45**, 611 (1962).
77. FINCH, N. and TAYLOR, W. I. *J. Am. Chem. Soc.* **84**, 1318, 3871 (1962).
78. SHAVEL J., JR. and ZINNES, H. *J. Am. Chem. Soc.* **84**, 1320 (1962).
79. FINCH, N., GEMENDEN, C. W., HSIU-CHU HSU, I., and TAYLOR, W. I. *J. Am. Chem. Soc.* **85**, 1520 (1963).
80. SAXTON, J. E. *The alkaloids* (ed. R. H. F. Manske) Vol. 10. Academic Press, New York (1968).
81. SHELLARD, E. J. and HOUGHTON, P. J. *Planta Med.* **21**, 16 (1972).
82. SHELLARD, E. J., SARPONG, K., and HOUGHTON, P. J. *J. Pharm. Pharmacol.* **23**, 244 (1971).
83. CONROY, H. and CHAKRABARTI, J. K. *Tetrahedron Lett.* 6 (1959).
84. BATTERSBY, A. R. and HALL, E. S. *Chem. Commun.* 793 (1969).
85. SCOTT, A. I., SLAYTOR, M. D., REICHARDT, P. B., and SWEENY, J. G. *Bio-org. Chem.* **1**, 157 (1971).
86. WENKERT, E. and WICKBERG, B. *J. Am. Chem. Soc.* **81**, 1580 (1965).
87. SAXTON, J. E. *The alkaloids* (ed. R. H. F. Manske) Vol. 10. Academic Press, New York (1968).
88. OLIVIER, L., LEVY, J., LEMEN, J., JANOT, M-M., BUDZIKIEWICZ, H., and DJERASSI, C. *Bull. Soc. Chim. Fr.* 868 (1965).
89. POUSSET, J.-L., POISSON, J., OLIVIER, L., LEMEN, L., and JANOT, M.-M. *C. R. Acad. Sci., Paris* **261**, 5538 (1965).
90. AHMAD, Y., FATIMA, K., RAHMAN, ATTA-UR-, OCCOLOWITZ, J. L., SOLHEIM, B., CLARDY, J., GARNICK, R. L., and LE QUESNE, P. W. *J. Am. Chem. Soc.* **99**, 1943 (1977).
91. HARLEY-MASON, J. and WATERFIELD, W. *Tetrahedron* **19**, 65 (1963).
92. GASKELLE, A. J. and JOULE, J. A. *Tetrahedron* **23**, 4053 (1967).
93. SCOTT, A. I. and QURESHI, A. A. *J. Am. Chem. Soc.* **91**, 5874 (1969).
94. SCOTT, A. I. *International reviews of science (alkaloids)* Vol. 9, p. 105. M.T.P. Press, Lancaster (1973).
95. SAKAI, S., AIMI, N., KUBO, A., KITAGAWA, M., SHIRATORI, M., and HAGINIWA, J. *Tetrahedron Lett.* 2057 (1971).
96. AIMI, N., SAKAI, S., LITAKA, Y., and ITAI, A. *Tetrahedron Lett.* 2061 (1971).
97. SCHUMANN, D. and SCHMID, H. *Helv. Chim. Acta.* **46**, 1997 (1963).
98. KOMPIS, I., HESSE, M., and SCHMID, H. *Lloydia* **34**, 269 (1971).
99. LING, N. C. and DJERASSI, C. *J. Am. Chem. Soc.* **92**, 6019 (1970).
100. WALSER, A. and DJERASSI, C. *Helv. Chim. Acta.* **48**, 391 (1965).
101. CLAYTON, R. B. *Q. Rev. Chem. Soc.* **19**, 68 (1965).
102. LING, N. C., DJERASSI, C., and SIMPSON, P. G. *J. Am. Chem. Soc.* **92**, 222 (1970).
103. SCOTT, A. I. and YEH, C. L. *J. Am. Chem. Soc.* **96**, 2273 (1974).
104. KUTNEY, J. P., NELSON, V. R., and WIGFIELD, D. C. *J. Am. Chem. Soc.* **91**, 4278 (1969).
105. FOSTER, G. H. and HARLEY-MASON, J. *Chem. Commun.* 1440 (1968).
106. KUTNEY, J. P., NELSON, V. R., and WIGFIELD, D. C. *J. Am. Chem. Soc.* **91**, 4279 (1969).
107. AHOND, A., CAVE, A., KAN-FAN, C., LANGLOIS, Y., and POTIER, P. *Chem. Commun.* 517 (1970).
108. POTIER, P. and JANOT, M. M. *C. R. Acad. Sci., Paris, Ser. C.* **276**, 1727 (1973).
109. HUSSON, A., LANGLOIS, Y., NICHE, C., HUSSON, H.-P., and POTIER, P. *Tetrahedron* **29**, 3095 (1973).

110. KUN, I. K. and ERICKSON, K. L. *Tetrahedron* **27**, 3979 (1971).
111. KUTNEY, J. P., BECK, J. F., EHRET, C., POULTON, G., SOOD, R. S., and WESTCOTT, N. D. *Bio-org. Chem.* **1**, 194 (1971).
112. KUTNEY, J. P., NELSON, V. N., and WIGFIELD, D. C. *J. Am. Chem. Soc.* **91**, 4278, 4179 (1969).
113. KUTNEY, J. P. *Heterocycles* **4**, 429 (1976).
114. SCOTT, A. I., YEH, C.-L., and GREENSLADE, D. *J. Chem. Soc. Chem. Commun.* 947 (1978).
115. SCHLATTLER, C. H., WALDNER, E. E., GROGER, D., MAIER, W., and SCHMID, H. *Helv. Chim. Acta.* **52**, 776 (1969).
116. WOODWARD, R. B., CAVA, M. P., OLLIS, W. D., HUNGER, A., DAENIKER, H. U., and SCHENKER, H. *J. Am. Chem. Soc.* **76**, 4749 (1954).
117. AHMAD, Y., FATIMA, K., LE QUESNE, P. W., and RAHMAN, ATTA-UR- *J. Chem. Soc. Pakistan* **1**(1), 69 (1979).
118. HIRATA, T., LEE, S-L., and SCOTT, A. I. *J. Chem. Soc. Chem. Commun.* 1081 (1979).
119. HEINSTEIN, P., STOCKIGT, J., and ZENK, M. H. *Tetrahedron Lett.* 141 (1980).
120. KAN-FAN, C. and HUSSON, H. P. *J. Chem. Soc. Chem. Commun.* 1015 (1979).
121. RUEFFER, M., KAN-FAN, C., HUSSON, H. P., STOCKIGT, J., and ZENK, M. H. *J. Chem. Soc. Chem. Commun.* 1016 (1979).
122. SCOTT, A. I., LEE, S.-L., CULVER, M. G., WAN, W., HIRATA, T., GUERITTE, F., BAXTER, R. L., NORDLÖV, H., DORSCHEL, C. A., MIZUKAMI, H., and MACKENZIE, N. E. *Heterocycles* **15**, 1257 (1981).

4

BIOSYNTHESIS OF CLASS II ALKALOIDS
(*Aspidosperma–Hunteria Type*)

The next major class of indole alkaloids to be encountered is that of *Aspidosperma–Hunteria* type, exemplified by aspidospermidine [1]. The non-tryptophan C-10 portion has undergone rearrangement in these alkaloids, so that they differ significantly in structure from the alkaloids of the *Corynanthe–Strychnos* class which are formed from tryptophan and an unrearranged secologanin unit.

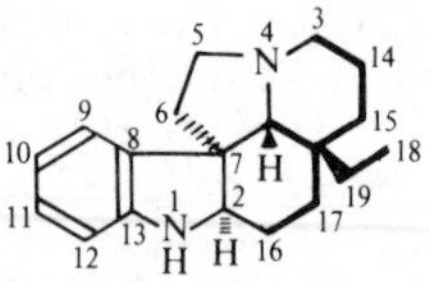

[1]

As shown in Scheme 4.1, this rearrangement involves the cleavage of the C_3–C_4 bond and attachment of the $C_{2,6}$ carbon atom at C_4 in accordance with Thomas and Wenkert's original hypothesis.[1, 2] For many years it was not clear whether this rearrangement occurs before or after combination with tryptophan, and it is only during the last decade that the broad features of the pathways to these alkaloids have been unravelled. Several reasonable possibilities suggested themselves for the formation of the *Aspidosperma* class of alkaloids. One involved the combination of an appropriately re-arranged iridoid (bearing the correct non-tryptophan skeleton) with tryptophan. Other possible alternative modes of generation considered included the initial formation of loganin or secologanin and its rearrangement to the *Aspidosperma*-type C-10 unit, prior to combination with tryptamine/tryptophan.

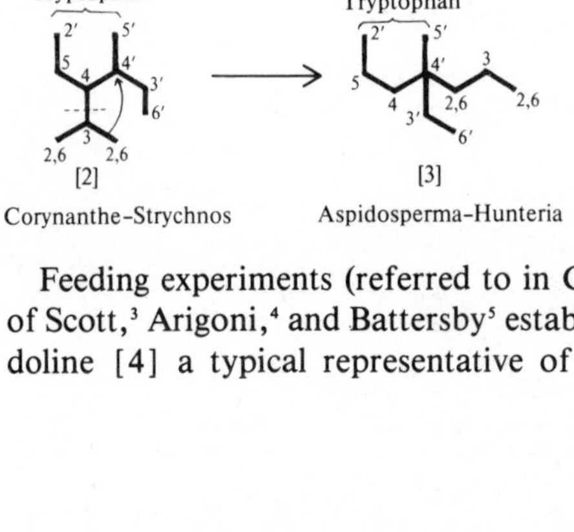

[2]   [3]

Corynanthe-Strychnos   Aspidosperma-Hunteria   SCHEME 4.1

Feeding experiments (referred to in Chapter 1) by three separate groups of Scott,[3] Arigoni,[4] and Battersby[5] established the mevalonoid origin of vin-doline [4] a typical representative of the *Aspidosperma* class. Further, feeding of sodium mevalonate labelled at various positions into *Vinca rosea*[5, 6] plants afforded vindoline labelled at the expected positions, and the extent of incorporation was in accordance with that predicted in Thomas and Wenkert's hypothesis.[1, 2] Degradation of dehydroaspidospermidine, another member of the *Aspidosperma* class, supported the head-to-tail combination of two C-5 units prior to rearrangement.

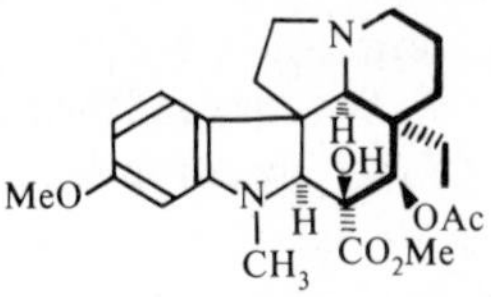

[4]

Subsequent feeding experiments with geraniol [5],[7-12] 10-hydroxygeraniol [6],[13, 14] deoxyloganin [7],[15] loganin [8],[16-21] and

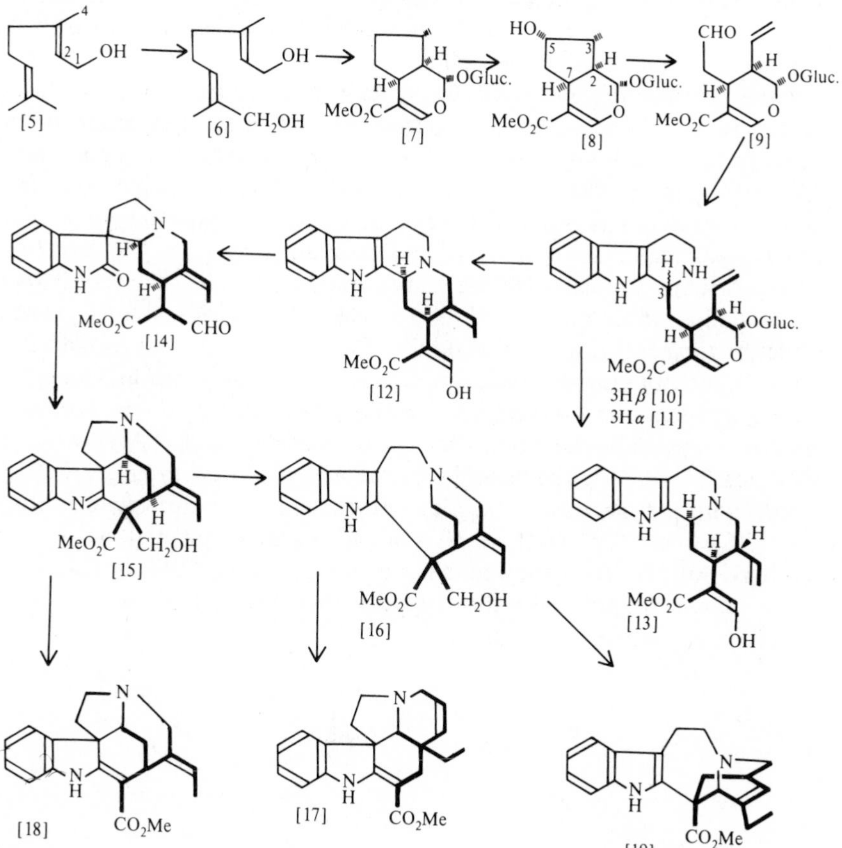

SCHEME 4.2

secologanin [9][22-24] labelled at various positions all provided firm evidence for the formation of vindoline [4] by the mevalonate → geraniol → deoxyloganin → secologanin pathway (Scheme 4.2).

Evidence for the rearrangement of the C-10 unit occurring at some stage after combination of secologanin [9] with tryptophan was provided by the incorporation of labelled strictosidine [11][20, 25-30] into vindoline [4]. It has been shown recently that vincoside [10] is not incorporated, contrary to earlier reports. Positive incorporation results were also obtained with labelled geissoschizine [12],[27, 31] corynantheine aldehyde [13],[31, 32] geissoschizine oxindole [14],[31] and the *Strychnos* alkaloid stemmadenine [16].[32] These studies pointed to the initial formation of a *Corynanthe* alkaloid which connects with the *Aspidosperma* skeleton via the *Strychnos* system. Labelled stemmadenine [16] was also observed to be incorporated into tabersonine [17], another member of the *Aspidosperma* class[32] (Scheme 4.2).

Sequential isolation work also supported the above results. There are only two plants, *V. rosea* and *Stemmadenia pubescens*, in which all the three main classes of indole alkaloids (*Corynanthe–Strychnos, Aspidosperma–Hunteria,* and *Iboga*) have been found to occur. Initial studies in *V. rosea* seedlings suggested the systematic conversion of the *Corynanthe* alkaloids into *Strychnos, Aspidosperma,* and *Iboga* alkaloids.[32] In another series of experiments, geissoschizine [12] was detected in seedlings within 28–40 h, along with geissoschizine oxindole [14], preakuammicine [15], and two new alkaloids.[31, 32] The *Strychnos* alkaloids stemmadenine [16] and akuammicine [18], and the *Aspidosperma* alkaloids, tabersonine [17], were detected after 40–50 h of growth, while the *Iboga* alkaloids catharanthine [19] was found to be formed after 100 h.[33] The isolation of preakuammicine [15][33, 33] and precondylocarpine [20][34] provided strong support for the intermediacy of the *Strychnos* alkaloid systems between the *Corynanthe* and *Aspidosperma* alkaloids.[26, 32, 35, 36] As the possible mechanistic pathways linking the *Corynanthe* with the *Strychnos* alkaloids have already been dealt with in some detail (Chapter 3), it suffices to state here that sequential, as well as labelling, studies clearly showed that one or more of the *Strychnos* alkaloids act as connecting links with the *Aspidosperma–Hunteria* series.

MeO₂C   CH₂OH
[20]

The first suggestion of the involvement of C_3–C_7 'seco' species connecting the *Strychnos* alkaloidal system with the *Aspidosperma–Hunteria* (and *Iboga*) alkaloids was made by Wenkert,[2] who proposed that the stemmadenine-type intermediate [21] could be converted to the α-formyl ester [22]. This could undergo reduction and dehydration to afford the acrylic ester [23] and [24]. These could then react by Michael and Mannich type reactions to afford the *Aspidosperma* and *Iboga* classes of alkaloids, respectively (Scheme 4.3). This hypothesis was based purely on mechanistic speculation and the fact that within a decade there is overwhelming evidence to support the bulk of these proposals is a reflection on the brilliant insight of Wenkert's rationalization.

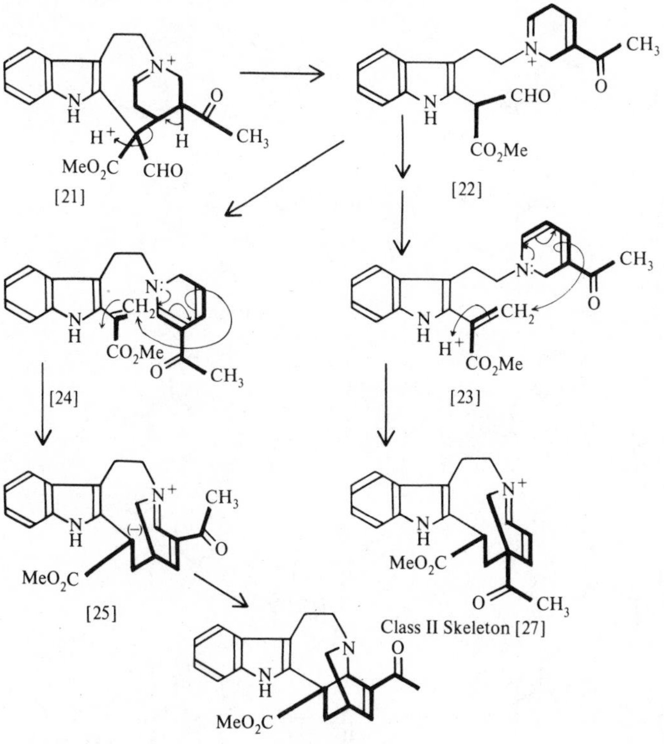

SCHEME 4.3

The first evidence for the existence of a 'seco' intermediate such as [23] in the biosynthetic scheme came from the work of the Manchester-based group who isolated the dimeric alkaloids, the secamines, from *Rhazya orientalis*.[37, 38] It was demonstrated that secamine has the structure [28] by degradative and synthetic studies.[37, 39] This was an important discovery as it can be clearly seen to be derived by the combination of two 'seco' units. In

1971, Japanese workers isolated tetrahydrosecamine [29] from *Amsonia elliptica*.[40, 41] Subsequent studies led to the isolation of several alkaloids bearing the seco unit, namely decarbomethoxytetrahydrosecodine [30][42] from *Tabernaemontana cumminsii*[43] and *Vinca minor*[44, 45] and the secodines [31], [32], and [33] were also shown to be present by dilution analysis.[46] Another group of alkaloids, the presecamines [37], have been isolated from *Rhazya* and they contain the secodine skeleton in a dimerized form.[47] The structures [37] for presecamine and [38] for tetrahydro-presecamine were confirmed by their rearrangement to secamines under mild acid conditions.[39] The presecamines are probably formed by the Diels–Alder dimerization of secodine [34] and 15,20-dihydrosecodine [36], and a biomimetic synthesis of presecamines is based on this proposal.[47]

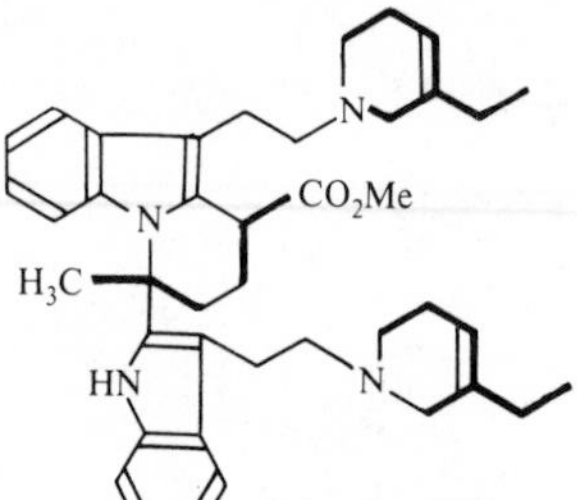

Secamine [28]

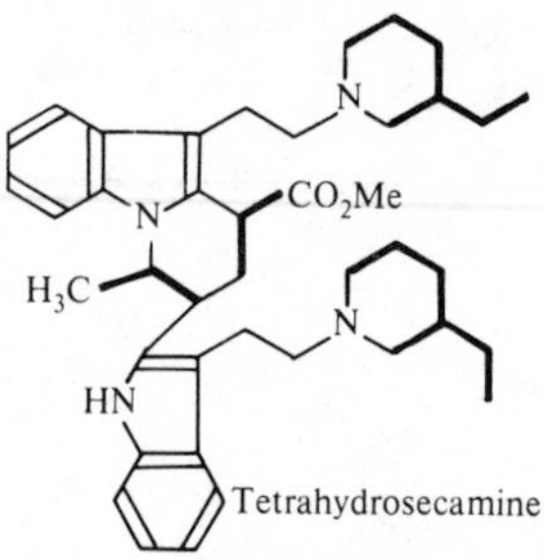

Tetrahydrosecamine [29]

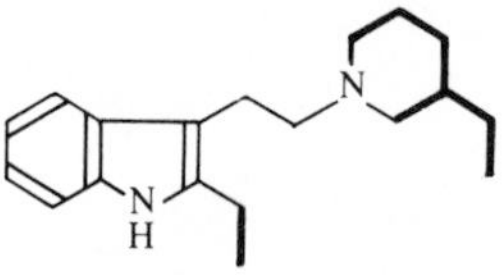

[30]

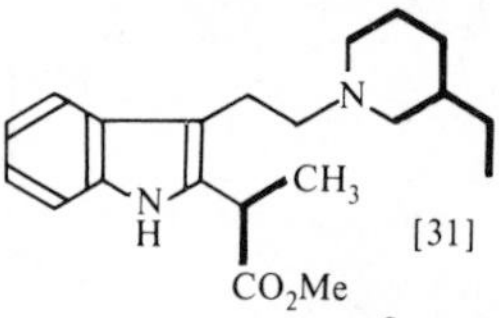

[31]

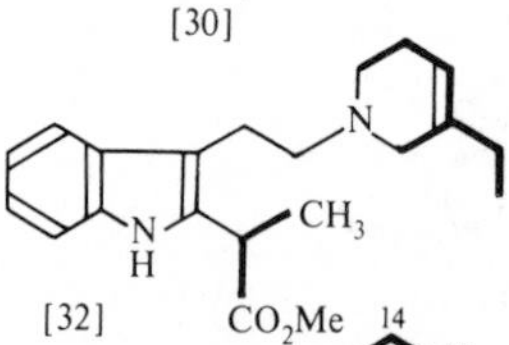

[32]

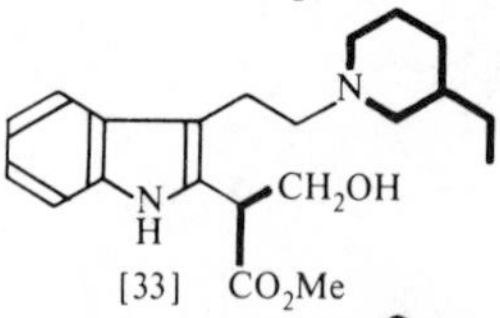

[33]

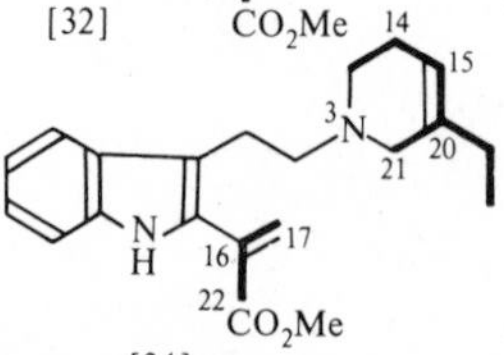

[34]

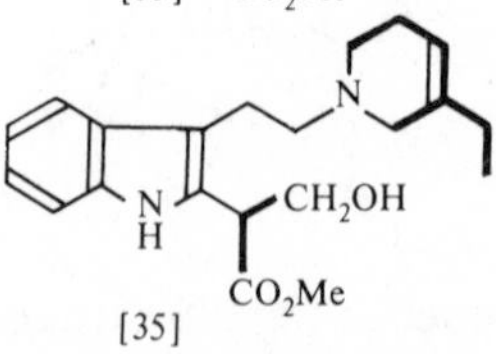

[35]

Initial incorporation studies[48] with [Ar–³H]–16,17-dihydrosecodine-17-ol [35] and [Ar–³H]-secodine [34] did not result in any significant incorporation into vincamine [39] or minovine [40] in *V. minor*. Kutney and co-workers claim that some incorporation has been observed into vindoline

[41] in *V. rosea*.[48, 49] Low incorporations of [Ar-³H]-16-,17-dihydro-secodine-17-ol [35] into the alkaloids of *C. roseus* and *V. minor* have also been claimed.[50] Doubly-labelled secodines were also found to be incorporated into vindoline [41], vincamine [39], and minovine [40]. The uncertainty of labelling of many of the compounds and the lack of rationalization of their results by Kutney and co-workers has led to them being dubbed as 'enigmatic' and 'ambiguous'.[39] Conclusive results for the high incorporation of secodine [34] into the *Aspidosperma* alkaloids are still awaited.

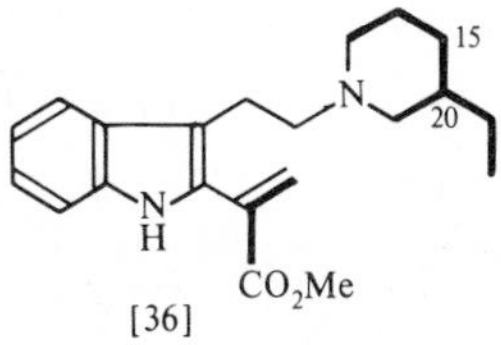

[36]

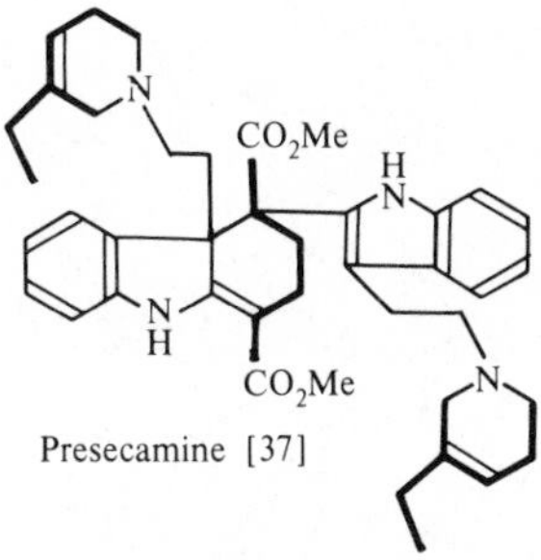

Presecamine [37]

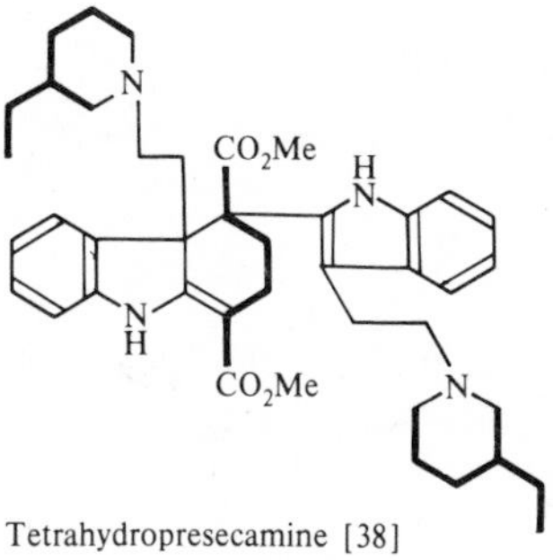

Tetrahydropresecamine [38]

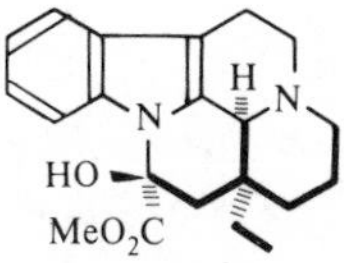

Vincamine [39]

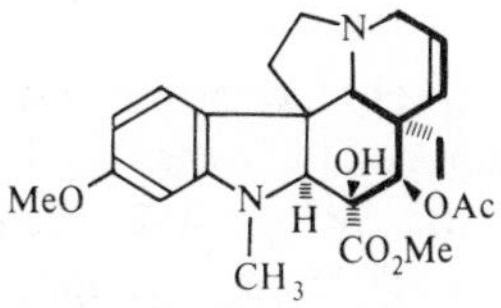

Vindoline [41]

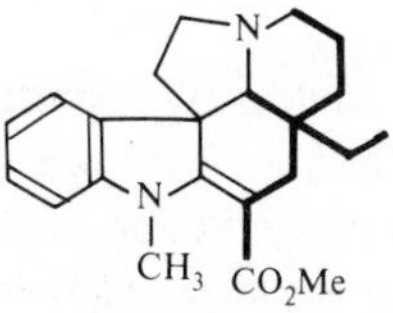

Minovine [40]

The first *in vitro* evidence for the formation of secodine [34] as an intermediate in the biosynthetic pathway was claimed by Scott and Qureshi in 1968.[51, 52] These workers reported that when stemmadenine [16] was refluxed in glacial acetic acid for 34 h, it was transformed to a mixture of (±)-tabersonine [17] (12 per cent), (±)-catharanthine [19] (9 per cent), and pseudocatharanthine [42] (16 per cent).[51] This appeared to offer powerful evidence for the conversion of *Strychnos* alkaloids to those of the *Aspidosperma* and *Iboga* classes. Similarly, it was claimed that tabersonine

[17], after reflux in glacial acetic acid for 16 h, was converted to catharanthine [19], (12 per cent) and pseudocatharanthine [42] (28 per cent), some unreacted tabersonine remaining.[51] It was suggested that the reaction proceeded through the intermediacy of the acrylic ester [43] (Scheme 4.4). In

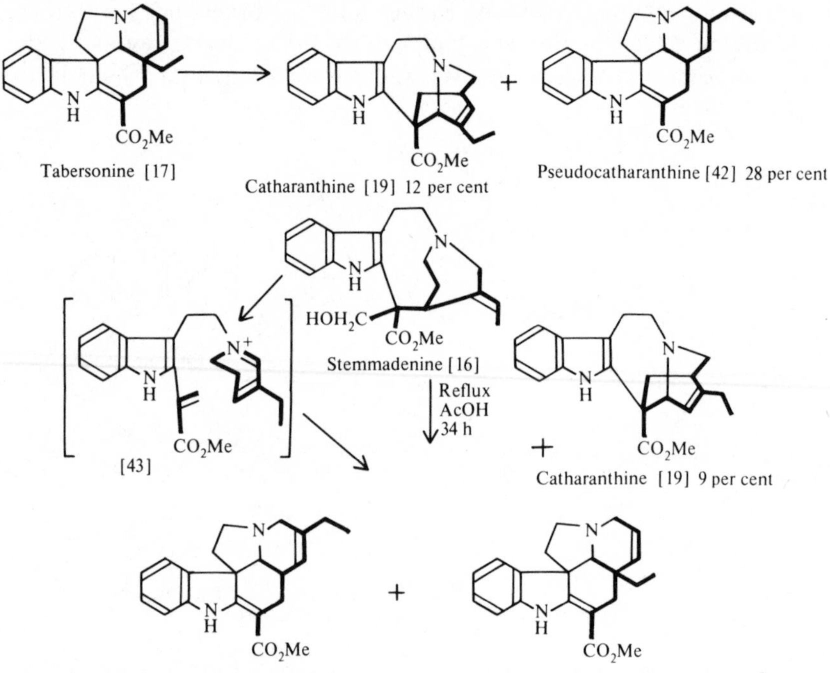

Tabersonine [17]

Catharanthine [19] 12 per cent

Pseudocatharanthine [42] 28 per cent

Stemmadenine [16]

Reflux AcOH 34 h

[43]

Catharanthine [19] 9 per cent

Pseudocatharanthine [42] 16 per cent

Tabersonine [17] 12 per cent

SCHEME 4.4

another series of experiments the same workers reported that when corynantheine aldehyde [13] and geissoschizine [12] were refluxed in glacial acetic acid for 72 h they were converted to catharanthine [19] (5 per cent) and pseudocatharanthine [42] (15 per cent). It was proposed that because of the equilibrium, the conversion to catharanthine [19] was of 'considerable synthetic utility'. Since it was found that some optical activity was retained in the isolated pseudocatharanthine [42], it was suggested that the reaction proceeded via the intermediacy of the dihydropyridine [46] (Scheme 4.5).

Scott's results appeared to provide concrete evidence towards the intermediacy of a secodine derivative in the biosynthesis of the *Aspidosperma* and *Iboga* classes of indole alkaloids, and two independent groups attempted to repeat them. Surprisingly, none of this work has been found to be reproducible and contradictory results have been published.[53-55]

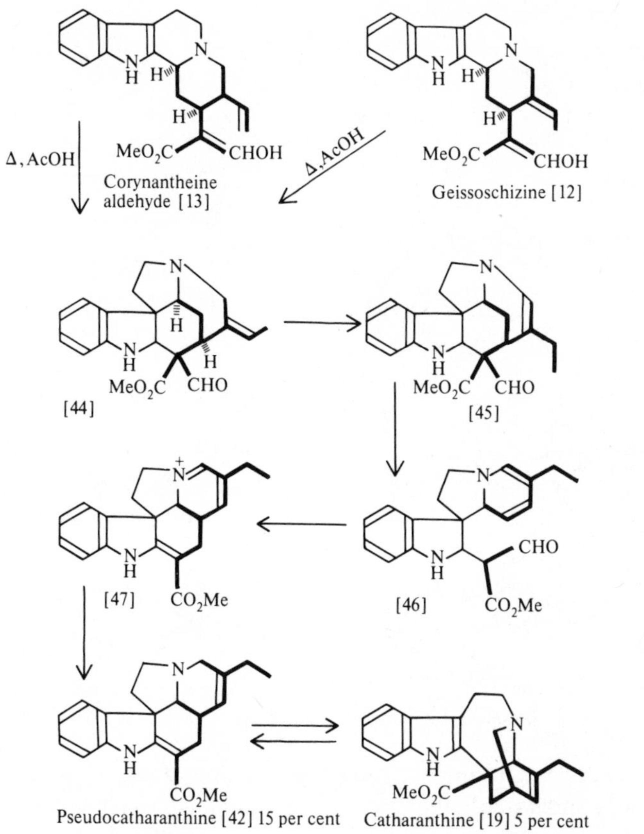

SCHEME 4.5

The British group, as well as the French workers, have reported that tabersonine [17] does not give catharanthine [19] or pseudocatharanthine [42] upon reflux in acetic acid.[53-55] Moreover, stemmadenine [16] does not give tabersonine [17], catharanthine [19], or pseudocatharanthine [42] as claimed by Scott. Instead, upon reflux in glacial acetic acid for 16 h at 130°C, tabersonine [17] was transformed to vincadifformine [48] (14 per cent), allocatharanthine [49] (17 per cent), dihydroallocatharanthine [50] (15 per cent), acetoallocatharanthine [51] (6 per cent), while 48 per cent was recovered unchanged (Scheme 4.6). When pseudocatharanthine [42] was refluxed in glacial acetic acid, less than 1 per cent yield of catharanthine [19] resulted, the reaction therefore having little synthetic value.

When stemmadenine [16] was similarly refluxed in glacial acetic acid for 15 h, 50 per cent yield of O-acetyl stemmadenine [53] was isolated along with 15 per cent unchanged starting material. When the reflux time was increased to 50 h, the yield of isolated O-acetyl stemmadenine [53] increased

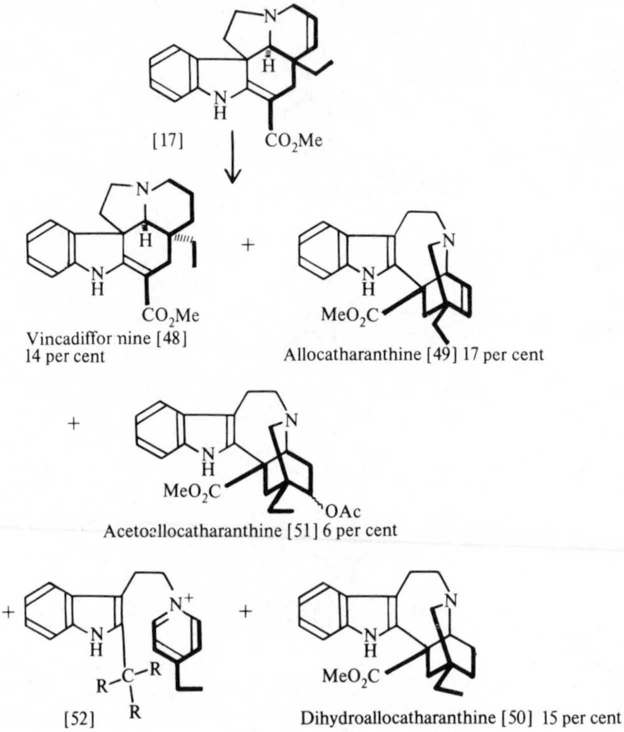

[17]   CO₂Me

Vincadifformine [48]
14 per cent

Allocatharanthine [49] 17 per cent

Acetoallocatharanthine [51] 6 per cent

[52]

Dihydroallocatharanthine [50] 15 per cent   SCHEME 4.6

to 55 per cent, while 5 per cent unchanged starting material was recovered (Scheme 4.7). It was suggested[53] that the C_{17}–C_{20} bond is not broken in

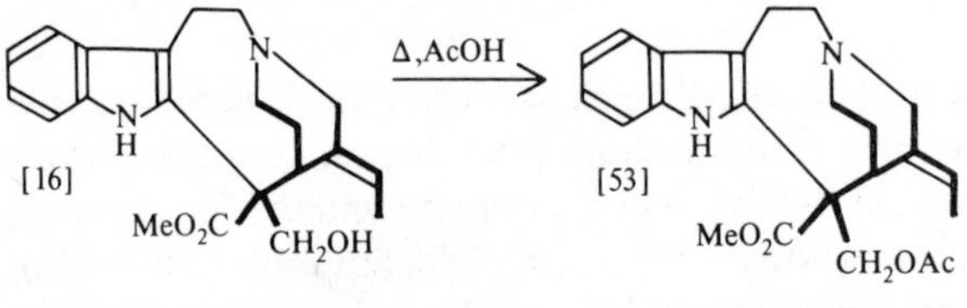

[16] $\xrightarrow{\Delta, AcOH}$ [53]

SCHEME 4.7

these conversions and that the intermediacy of secodine derivatives need not be involved. The mechanism suggested by G. F. Smith and co-workers is illustrated in Scheme 4.8.

Scott has subsequently published details of more experiments in this field in support of his earlier results. However, only one of the previous experiments has been repeated[56] and the results proved contradictory to those published earlier.[51] On heating tabersonine [17] in glacial acetic acid for 16 h, Scott did not obtain 12 per cent catharanthine [19] or 28 per cent pseudocatharanthine [42] as claimed previously, but isolated allocatharanthine [49] (11 per cent), in broad agreement with the earlier work of

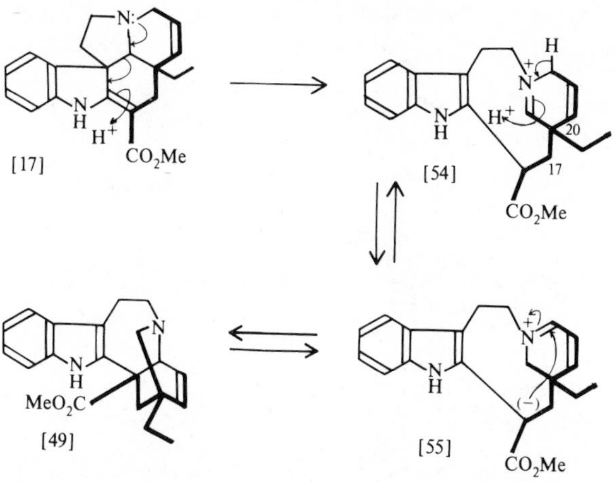

[17]   CO$_2$Me

[54]

[55]

[49]

MeO$_2$C

CO$_2$Me

CO$_2$Me

SCHEME 4.8

Smith[53] and Poisson.[55] As such, Scott's claim of 'complete vindication and extension of the original observations' needs further substantiation.

It is relevant to mention here that the earlier work of Neuss and co-workers[57] had established that on 2.5 h reflux of catharanthine [19] in glacial acetic acid, no starting material remained, the major product isolated being pseudocatharanthine [42]. Scott has claimed yields of 12 per cent catharanthine [19] after 16 h, 9 per cent after 34 h, and 5 per cent after 72 h in various experiments. While it is conceivable that the rates of conversion of tabersonine [17] or stemmadenine [16] to catharanthine [19] are such that it is detectable after long reaction times, the results are doubtful. This is particularly so in view of the subsequent studies of Smith and co-workers,[54] who have shown that on reflux in acetic acid for 2 h, catharanthine [19] afforded partially racemic pseudocatharanthine [42] (30 per cent), coronaridine [56] (10 per cent), dihydropseudocatharanthine (20-epipseudovincadifformine) [57] (5 per cent), pseudovincadifformine [58], 16S, 20R-carbomethoxydihydrocleavamine [59] (3 per cent), 16R, 20R-carbomethoxydihydrocleavamine [60] (2 per cent), and an oxygenated catharanthine C$_{21}$H$_{24}$N$_2$O$_4$, [61] (5 per cent), (Scheme 4.9). More recently, the life-time of catharanthine in refluxing glacial acetic acid has been re-investigated and it has been found that less than 4 per cent catharanthine remains after refluxing in glacial acetic acid for 12–16 h.[58]

It was observed that pseudocatharanthine [42] obtained by heating catharanthine [19] or 16S-carbomethoxycleavamine [62] was partially

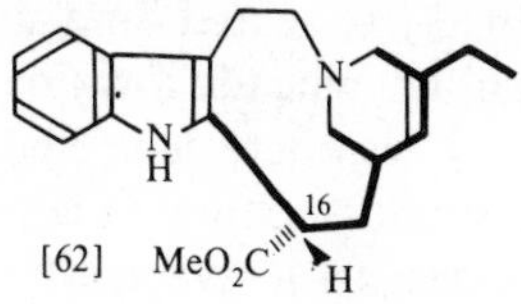

[62]   MeO$_2$C$^{\text{,,,}}$ H

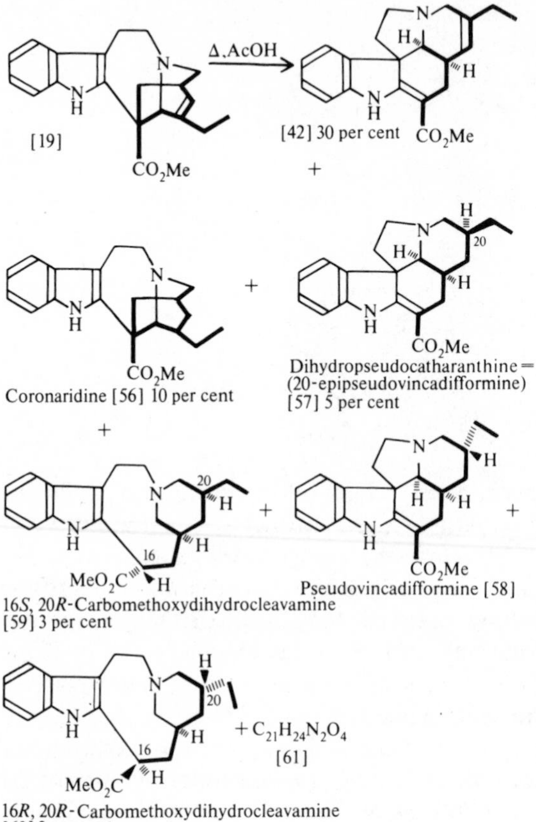

[19] → [42] 30 per cent CO_2Me

+

Coronaridine [56] 10 per cent

+

Dihydropseudocatharanthine = (20-epipseudovincadifformine) [57] 5 per cent

+

16S, 20R-Carbomethoxydihydrocleavamine [59] 3 per cent

+

Pseudovincadifformine [58]

+

16R, 20R-Carbomethoxydihydrocleavamine [60] 2 per cent

$+ C_{21}H_{24}N_2O_4$ [61]

SCHEME 4.9

racemic, but the dihydropseudocatharanthine [57] obtained from 16S, 20S-dihydrocarbomethoxycleavamine was optically pure. This suggests that when the double bond was present in the piperidine ring, partial racemization occurred. The explanation has been advanced that when there is no double bond in the ring, either K_1 or K_2 are too slow so that optically pure products are obtained. However, in the presence of a double bond, K_1 or K_2 are faster so that racemization is observed (Scheme 4.10).

In a series of experiments[56, 59-61] attempting to establish the intermediacy of secodine [34] in the conversions of the *Corynanthe–Strychnos* alkaloids into those of the *Aspidosperma* class, Scott has reported[60] that when stemmademine-O-acetate [53] is heated on a silica surface, it is converted to the *Aspidosperma* alkaloid vincadifformine [48] in 0.15–2 per cent yields. No tabersonine [17] was detected. The explanation advanced for this transformation is that stemmademine-O-acetate [53] first isomerizes to the enamine [66] which undergoes a reverse Mannich reaction to afford the

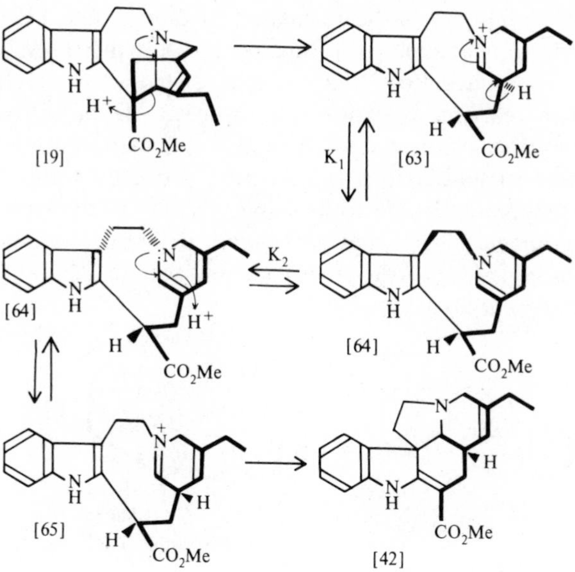

SCHEME 4.10

dehydrosecodine [43]. This undergoes reduction to the secodine [34] which cyclizes to vincadifformine [48] (Scheme 4.11).

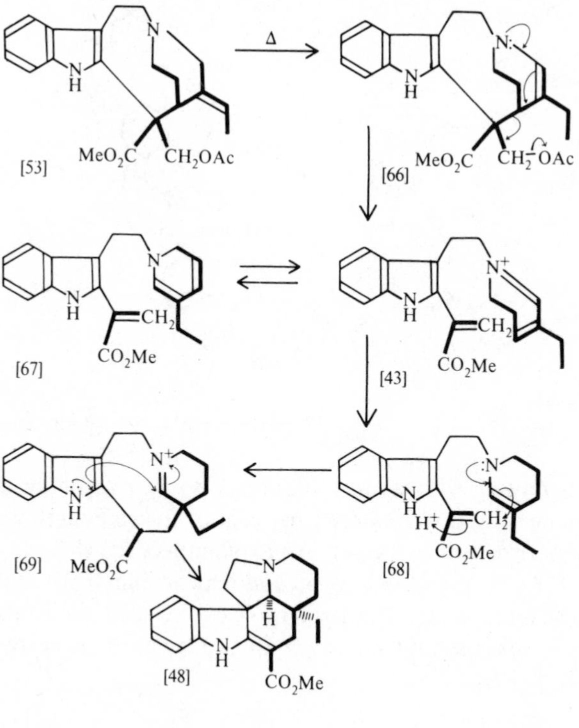

SCHEME 4.11

Another interesting observation was the formation of tabersonine [17] (0.2 per cent), and (±)-vincadifformine [48] (0.02 per cent) by thermolysis of dihydroprecondylocarpine acetate[60] [72] prepared from O-acetyl-stemmadenine [53]. The reaction is claimed to proceed via dehydro-secodine B [67]. Since no trace of the *Iboga*-alkaloid pseudocatharanthine could be detected, it was suggested that the reaction proceeds without isomerization of dehydrosecodine B [67] to its isomer A [75] as shown in Scheme 4.12. These experiments suggest that the isomeric dehydrosecodines [67] and [75] are produced *in-vitro* from stemmadenine acetate [53] and converted to the *Aspidosperma* alkaloids.

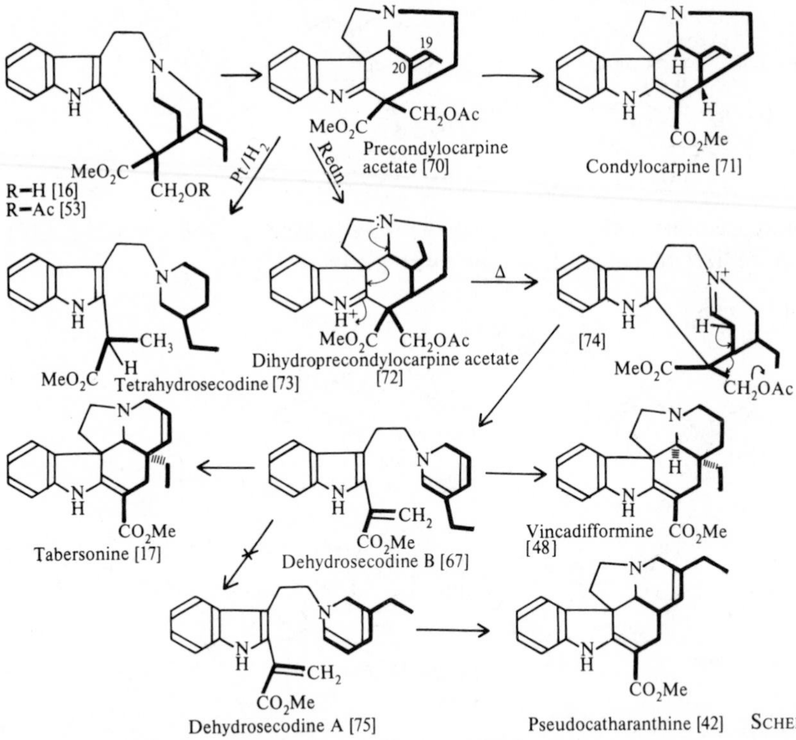

SCHEME 4.12

Surprisingly stemmadenine [16] was not found to possess a dynamic radioprofile, a steady value of approximately 1 per cent of the radioactivity from tryptophan being recorded over 8 days. An explanation for this may be that stemmadenine [16] is converted to precondylocarpine [20] and preakuammicine [15] (Scheme 4.13). The formation of the seco alkaloids and the subsequent *Aspidosperma* and *Hunteria* alkaloids can then be rationalized as indicated in Scheme 4.14.

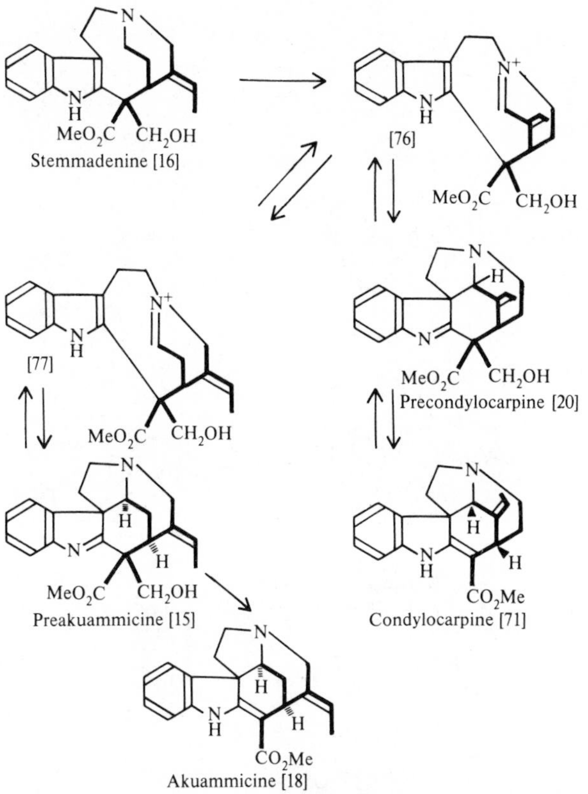

Stemmadenine [16]

[76]

[77]

Precondylocarpine [20]

Preakuammicine [15]

Condylocarpine [71]

Akuammicine [18]

SCHEME 4.13

Scott has suggested that the C_{15}–C_{16} bond in stemmadenine acetate [53] is cleaved very readily, even at room temperature, to provide the secodine system. Similarly precondylocarpine acetate [70] rearranges readily to [81] thus providing strong evidence for similar *in vivo* conversions[62] (Scheme 4.15).

Similarly, cleavage of 'a' and 'b' bonds has been achieved both in the *Aspidosperma* alkaloid tabersonine [17] as well as in the *Iboga* alkaloid catharanthine [19] to afford the secodine system as the pyridinium salt [82] in high yield.[62] Borohydride reductions of this salt afford dihydro-secodine [32] and tetrahydrosecodine [31], thus demonstrating that the acrylic-ester system can be generated from all three major classes of indole alkaloids (Scheme 4.16).

Scott has reported[63] that heating xylene solutions of (−)-tabersonine [17], (+)-catharanthine [19], and (±)-pseudocatharanthine [42] in sealed tubes affords 3-ethyl pyridine [84] and 1-methyl-2-hydroxycarbazole [85]. It was suggested[63] that these reactions take place by a retro-Diels–Alder reaction to afford the ester [67]. The dihydropyridine

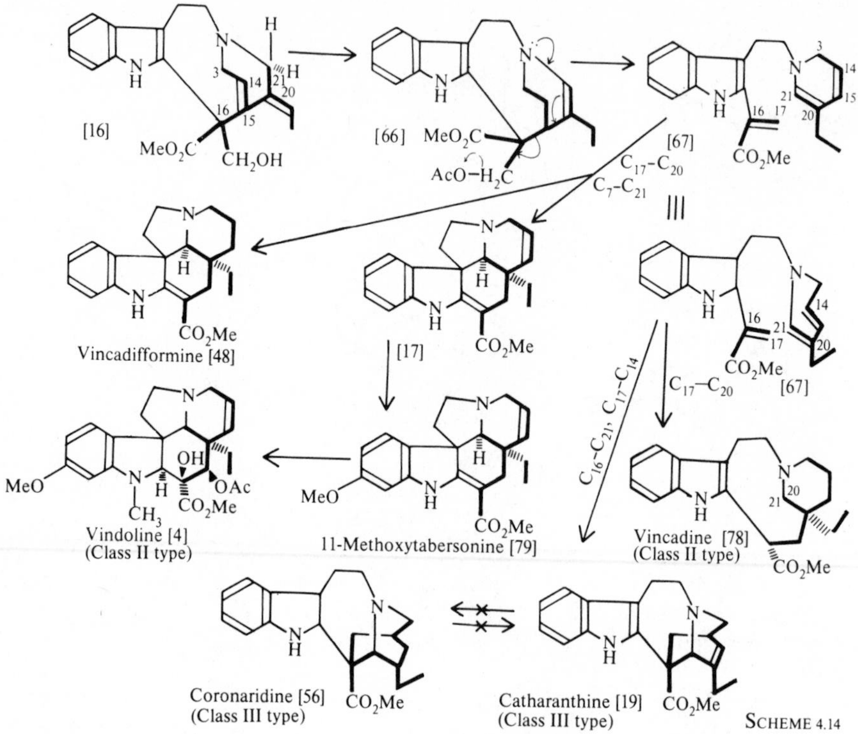

[16]

[66]   MeO$_2$C

AcO—H$_2$C

[67]

C$_{17}$-C$_{20}$

C$_7$-C$_{21}$

III

[67]

Vincadifformine [48]

[17]   CO$_2$Me

C$_{16}$-C$_{21}$, C$_{17}$-C$_{14}$

C$_{17}$-C$_{20}$

Vindoline [4]
(Class II type)

11-Methoxytabersonine [79]

Vincadine [78]
(Class II type)

Coronaridine [56]   CO$_2$Me
(Class III type)

Catharanthine [19]   CO$_2$Me
(Class III type)

SCHEME 4.14

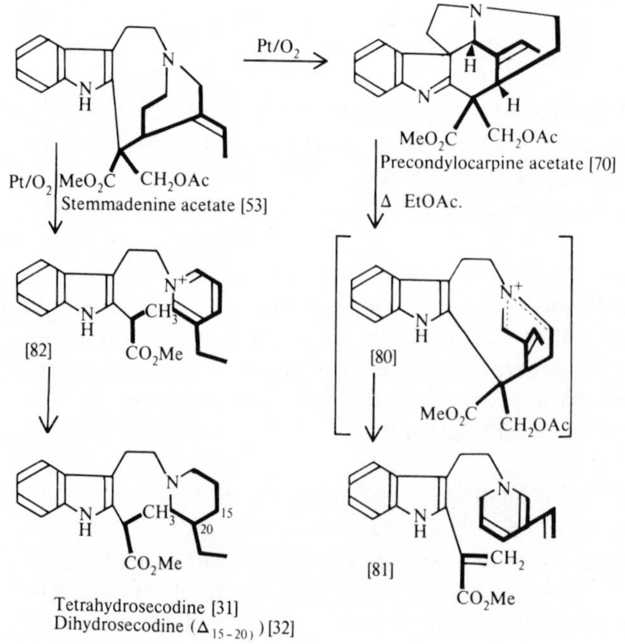

Pt/O$_2$

Precondylocarpine acetate [70]

Pt/O$_2$ MeO$_2$C   CH$_2$OAc
Stemmadenine acetate [53]

Δ EtOAc.

[82]

[80]

MeO$_2$C   CH$_2$OAc

Tetrahydrosecodine [31]
Dihydrosecodine (Δ$_{15-20}$) [32]

[81]

SCHEME 4.15

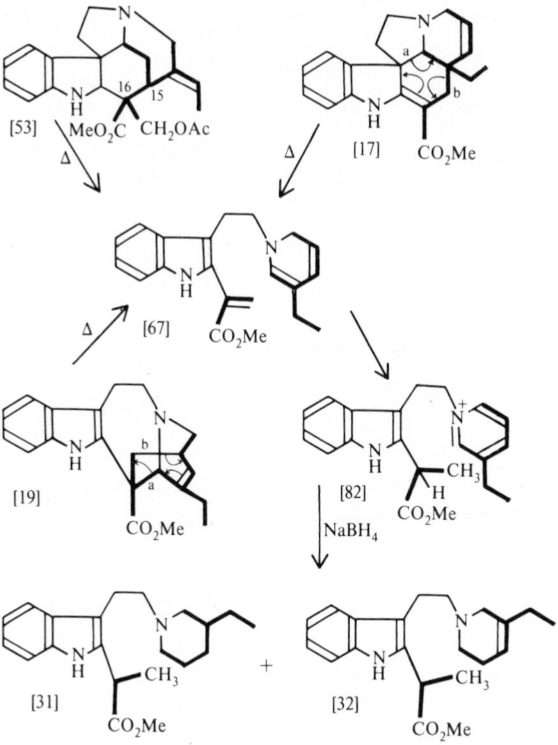

SCHEME 4.16

intermediate [67] then undergoes an intramolecular rearrangement and hydrogen transfer from the dihydropyridine to the ester function, yielding the hemi-ketal [83]. Elimination of methanol and further rearrangement affords the carbazole [85]. The isolation and characterization of 1-methyl-2-methoxycarbazole [86] supports the occurrence of such a process (Scheme 4.17). More direct evidence for the intermediacy of a secodine-type system was obtained when catharanthine [19] was heated in methanol for 2 h. It was found to be converted in 50 per cent yield to the racemic salt [82], which on heating eliminated 3-ethylpyridine to afford the 3-vinyl indole derivative [87] (Scheme 4.17).

The preparation of pro-5S tritiated and pro-5R tritiated mevalonic acids[64–67] and their feeding into *V. rosea* plants has produced results which indicate that the pro-5S-proton is lost stereospecifically in the formation of vindoline [4]. This loss most probably occurs from the intermediate [89] where the C-14 proton is lost during the cyclization step (Scheme 4.18).

Several of the *Aspidosperma* alkaloids exhibit both the antipodal series. Thus both (±)- and (−)-quebrachamine, [90] and [91] respectively, are known. Similarly (−)-tabersonine [17] and the closely related

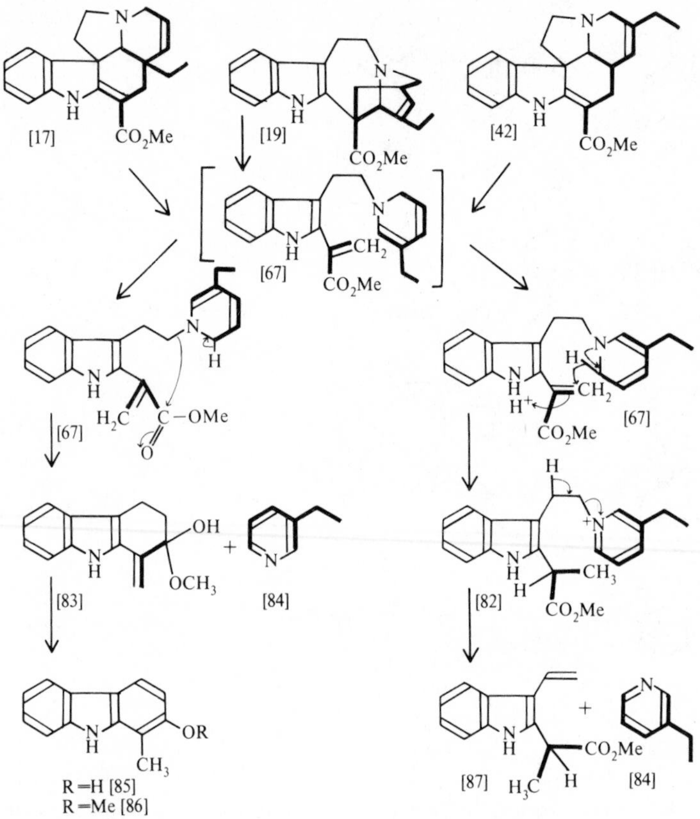

SCHEME 4.17

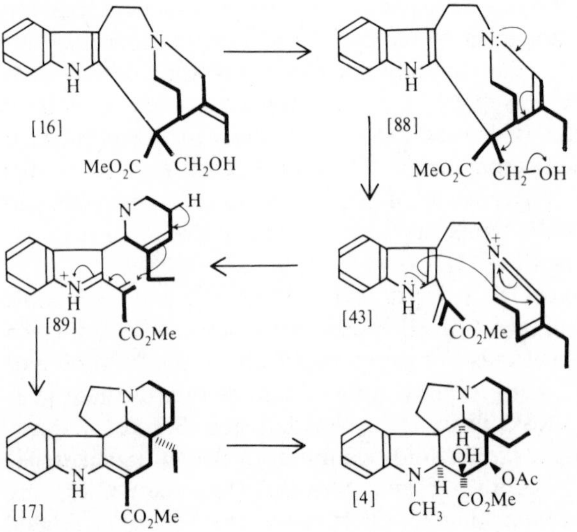

SCHEME 4.18

(+)-minovincine [92] are other examples. Generally, alkaloids of both antipodal series do not occur in the same plant. However, *Amsonia tabernaemontana* was found to contain (+)-1,2-dehydroaspidospermidine [93][68, 69] and (+)-vincadifformine [48],[69-71] as well as (−)-tabersonine [17].[72-74]

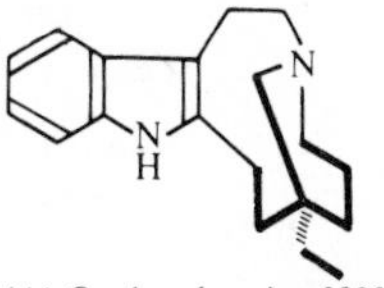

(+)-Quebrachamine [90] 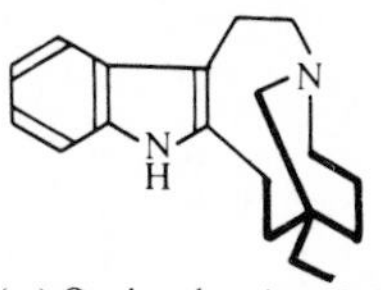   (−)-Quebrachamine [91]

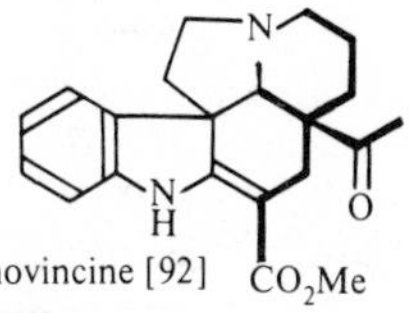

(+)-Minovincine [92] CO_2Me

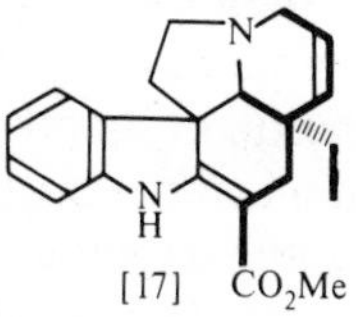

[17] CO_2Me [93]

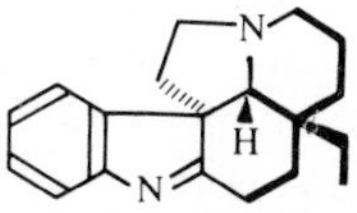

The closely-related alkaloids of the *Hunteria* type also exhibit both antipodal series. The co-occurrence of both the (+)- and (−)-forms of the *Aspidosperma* and *Hunteria* alkaloids in *V. minor* is strong indication of the closely-linked biogenetic relationship of these two types of alkaloids. Thus both (±)-vincadifformine[75] [47] and (±)-eburnamonine [94] and [95][45] were found in *V. minor*. (+)-Eburnamenine[45] [96] and (−)-eburnamonine [95][76] also co-occur in the same plant.

It would be interesting to learn how the antipodal series of alkaloids arise. The differentiation may occur at the preakuammicine [15], stemmadenine [16], or achiral secodine [34] stages. Experiments with various alkaloids labelled at C-21 are needed to establish if this is the case. Enzyme-controlled stereospecific attack of C-16 of stemmadenine [16] could also give rise to the antipodal series.

As mentioned earlier, the *Aspidosperma* alkaloids are considered to arise via the intermediacy of an acrylate ester 'seco' intermediate [34]. The simplest members of this class are those of the quebrachamine group. These can arise by an enamine ring closure of the seco intermediate [68] to afford vincadine [78]. Hydrolysis and decarboxylation of vincadine [78] would afford quebrachamine [90]. The achiral nature of the enamine [68] results in both the antipodal forms existing in quebrachamine (Scheme 4.19).

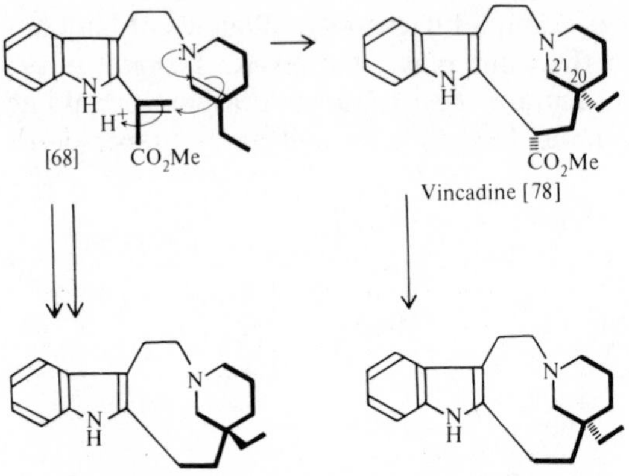

[68]   CO₂Me

Vincadine [78]

(−)-Quebrachamine [91]   (+)-Quebrachamine [90]   SCHEME 4.19

While quebrachamine [90] and vincadine [78] do not have a C—C bond linking the indole β-position to C-21, the majority of the *Aspidosperma* alkaloids show such a linkage and two possible routes to them can be considered. The first involves the *in vivo* oxidation of an alkaloid such as vincadine [78] to the corresponding immonium compound [69], followed by an intramolecular ring closure to afford the aspidospermine group of alkaloids exemplified by vincadifformine [48] (Scheme 4.20). It has been

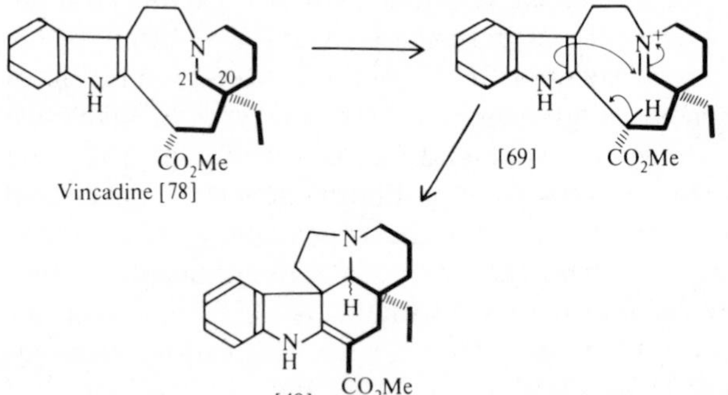

CO₂Me

Vincadine [78]   [69]   CO₂Me

[48]   CO₂Me   SCHEME 4.20

found that the quebrachamine and aspidospermine systems are interconvertible. This was first discovered by Smith and Wrobel[77] during chemical investigations on a degradation product from akuammicine [18]. Such interconversions depend on the equilibrium between the 1,2-dehydroaspidospermidine system [93] and the corresponding seco derivative [97]. Biemann utilized this finding by oxidizing desacetylaspidospermine [98] with iodine in alkali to the indolenine [93].[78] Since this is in equilibrium

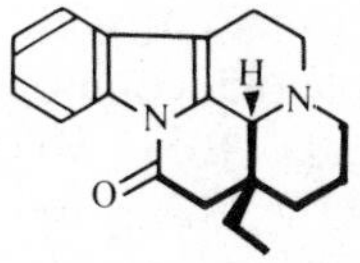

(+)-Eburnamonine [94]

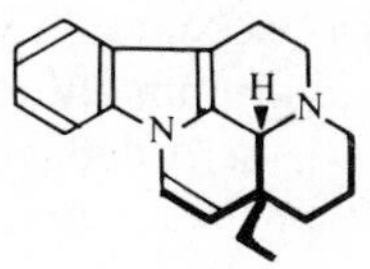

(+)-Eburnamenine [96]

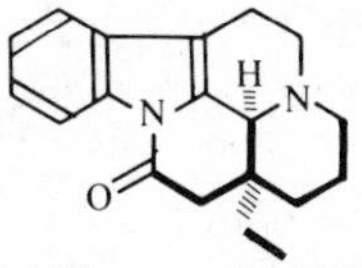

(−)-Eburnamonine [95]

with the tetracyclic compound [97], borohydride reduction of this afforded 12-methoxyquebrachamine [99] (Scheme 4.21).

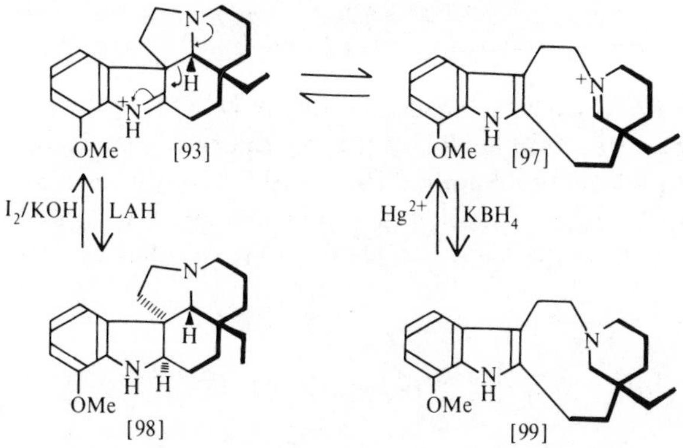

SCHEME 4.21

Wenkert had originally suggested[2] that intermediates such as [100] may be general precursors to the aspidospermine alkaloids, which would arise by transannular cyclization reactions (Scheme 4.22). It has been found that

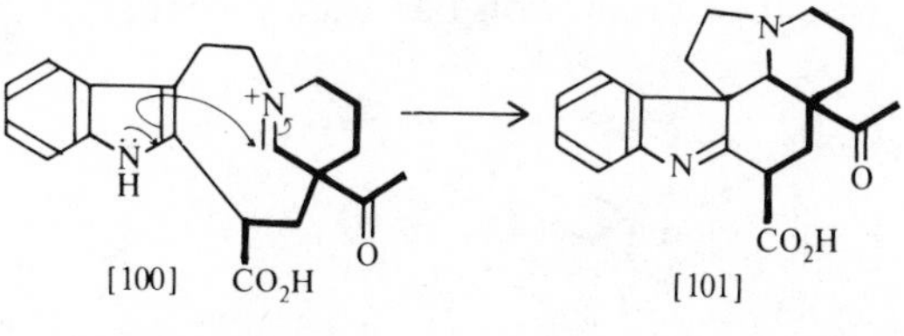

SCHEME 4.22

such transannular cyclizations could be carried out in the laboratory. This reaction was first attempted on dihydrocleavamine [102], a tetracyclic indole formed by degradation of the dimeric *Vinca* alkaloids vinblastine and vinleurosine, the synthesis of which was first reported by Harley-Mason and

Atta-ur-Rahman.[79] Oxidation of dihydrocleavamine with mercuric acetate in acetic acid and lithium aluminium hydride reduction of the intermediate indolenine [104] afforded the compound [105] which possessed an *Aspidosperma*-type skeleton (Scheme 4.23).[60] It is interesting that the

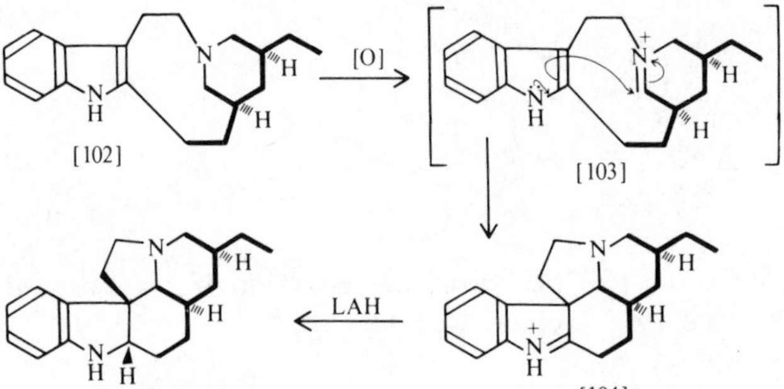

SCHEME 4.23

stereochemistry of the two new asymmetric centres being generated during such cyclizations is determined completely by the configuration at C-20. A similar cyclization of (−)-quebrachamine [91] with mercuric acetate and reduction of the crude indolenine intermediate afforded (±)-aspidospermidine [106] (Scheme 4.24).[81] On oxidation of

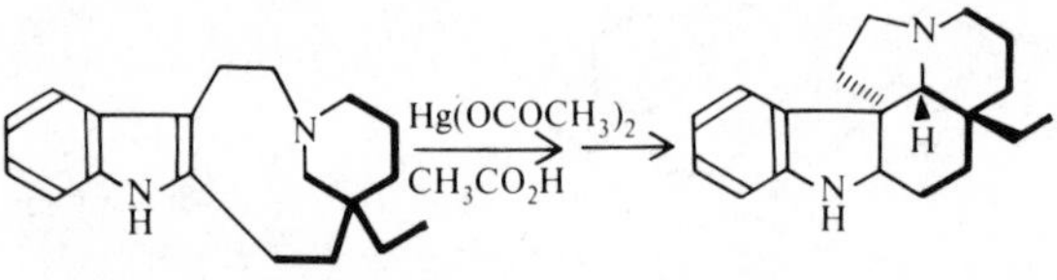

(−)-Quebrachamine [91]   (+)-Aspidospermidine [1]   SCHEME 4.24

(−)-quebrachamine [91] with potassium permanganate, Schmid and co-workers isolated the indolenine (−)-1,2-dehydroaspidospermidine [93] (Scheme 4.25).[82] A similar oxidative cyclization has been reported as the

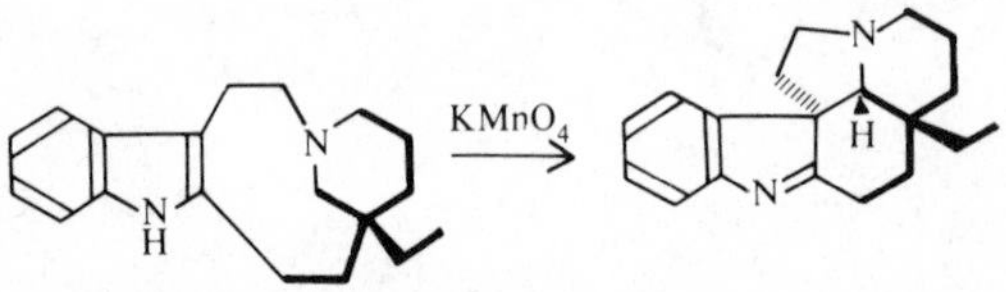

(−)-Quebrachamine [91]   (+)-1,2-Dehydroaspidospermidine [93]   SCHEME 4.25

final step in the syntheses of the *Iboga* alkaloids, coronaridine [56] and dihydrocatharanthine [107] (Scheme 4.26).[80]

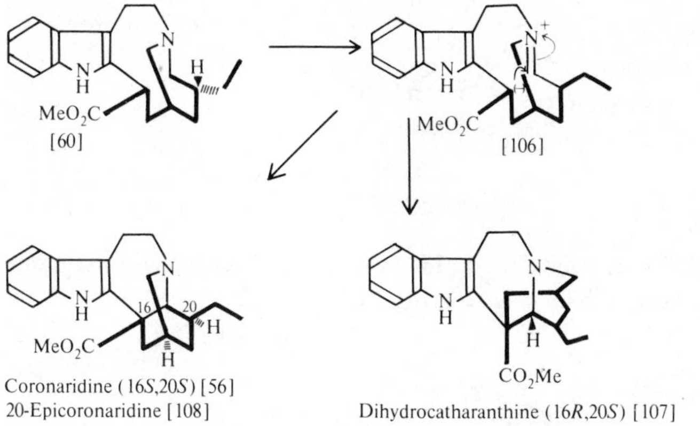
MeO2C
[60]
MeO2C
[106]
MeO2C
CO2Me
Coronaridine (16S,20S) [56]
20-Epicoronaridine [108]
Dihydrocatharanthine (16R,20S) [107]

SCHEME 4.26

An alternative possibility for the generation of the aspidospermine-type alkaloids exists from the corresponding secodine derivative [43] of the correct oxidation level. The β-position (C-7) of the indole nucleus could, by an internal Mannich reaction, afford the indolenine [89]. The C_{15}–C_{20} double bond of this intermediate could isomerize to the enamine [109] which may then attack the acrylate system to afford vincadifformine [48], a member of the aspidospermine group of indole alkaloids (Scheme 4.27).

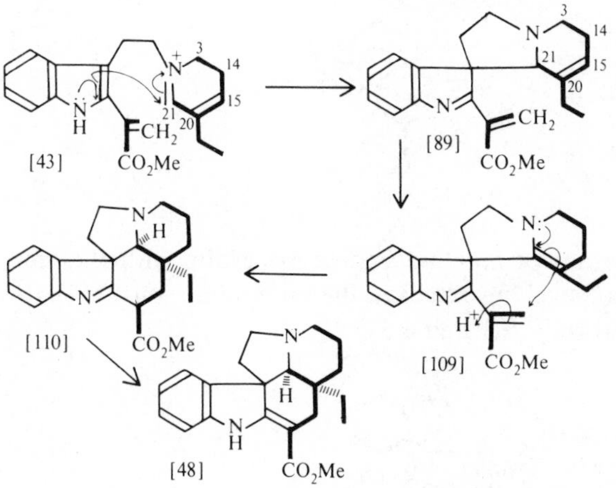
[43]
CO2Me
[89]
CO2Me
[110]
CO2Me
[109]
CO2Me
[48]
CO2Me

SCHEME 4.27

Indole alkaloids of the cathovaline group can arise by an intramolecular attack of the tertiary hydroxyl group on the olefin in vindorosine [111] with the formation of the ether bridge found in cathovaline [112] (Scheme 4.28). An *in vitro* conversion of desacetycathovaline from vindorosine N-oxide has been achieved by Potier and co-workers.[83] In vincoline [114],

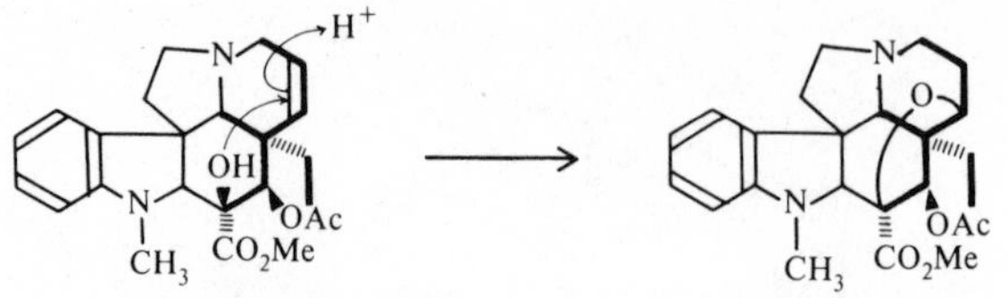

Vindorosine [111]   Cathovaline [112]   SCHEME 4.28

the hydroxyl function in ring C in [113] has cyclized with the ethylidene olefin to afford a new five-membered ring (Scheme 4.29).

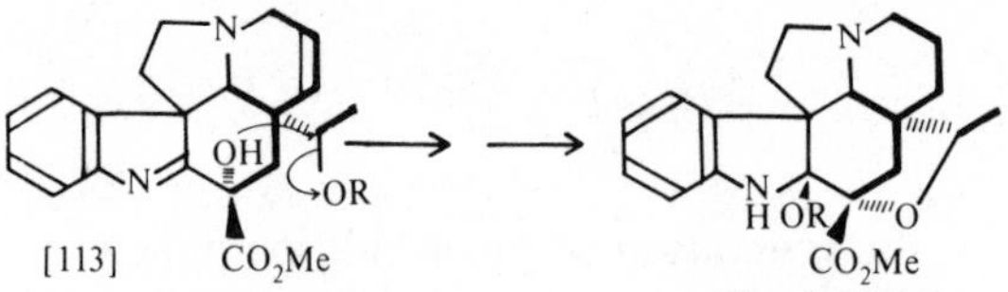

[113]   CO$_2$Me   Vincoline [114]   SCHEME 4.29

The apodine group of alkaloids may be considered to be generated from tabersonine derivative [115] in which the ethyl side-chain has been oxidized to the carboxylic-acid level and an intramolecular cyclization with the olefin has taken place (Scheme 4.30).

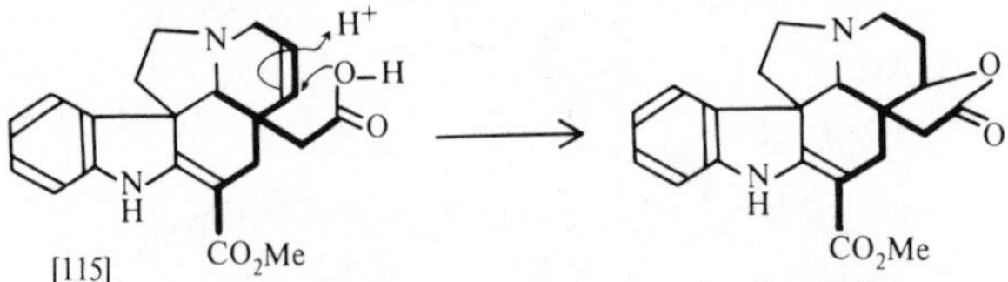

[115]   CO$_2$Me   Apodine [116]   SCHEME 4.30

In the aspidoalbidine group of alkaloids, the ethyl group with the terminal carbon at the alcohol level has attacked the immonium ion [117] to form a new five-membered ring (Scheme 4.31).

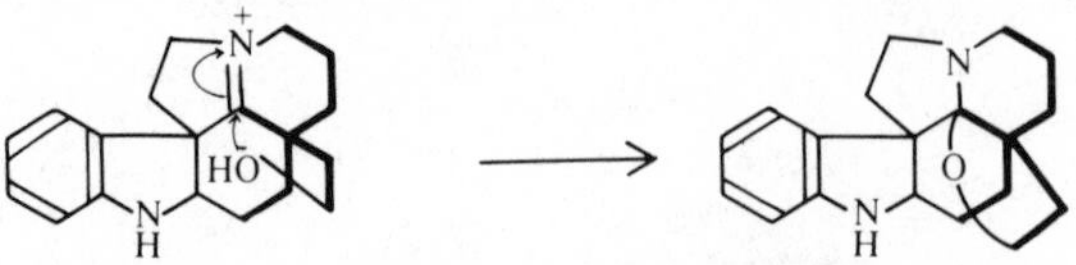

[117]   Aspidoalbidine [118]   SCHEME 4.31

The vincatine group of alkaloids can be considered to be formed from a vincadifformine-like precursor [119] by hydration (or epoxidation) of the olefin of the acrylate system followed by cleavage of ring C to afford vincatine [121] as illustrated in Scheme 4.32. The co-occurrence of vincatine

[121] with vincadifformine [48] in *Vinca minor* supports such a biogenetic scheme.

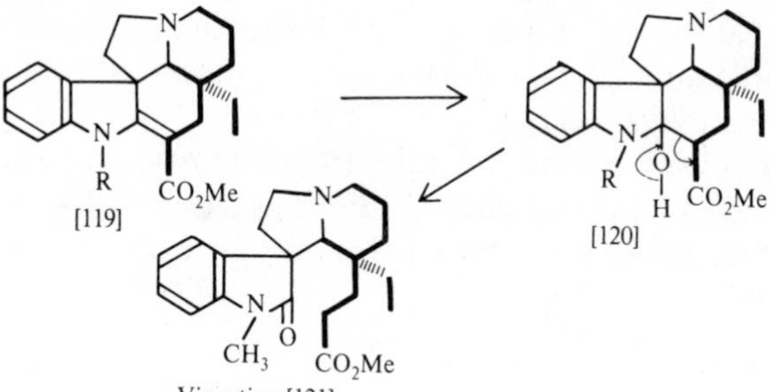

[119] [120] Vincatine [121]

SCHEME 4.32

The obscurinervine group of alkaloids can arise from a tabersonine-type precursor [122] by the oxidation of the ethyl groups to the carboxylic-acid level. This could be followed by attack at the adjacent carbon atom on ring C to afford obscurinervine [124] (Scheme 4.33).

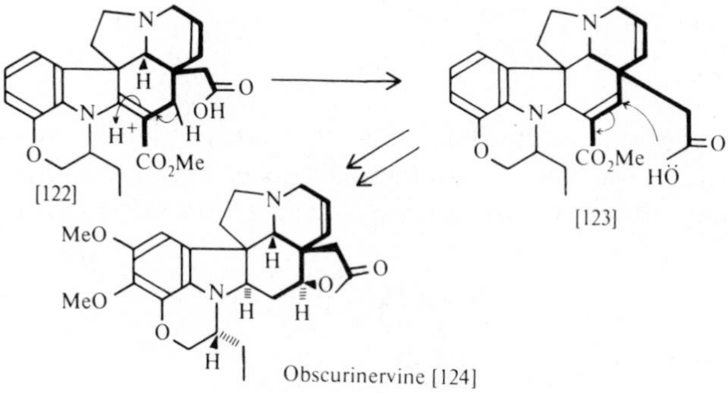

[122] [123] Obscurinervine [124]

SCHEME 4.33

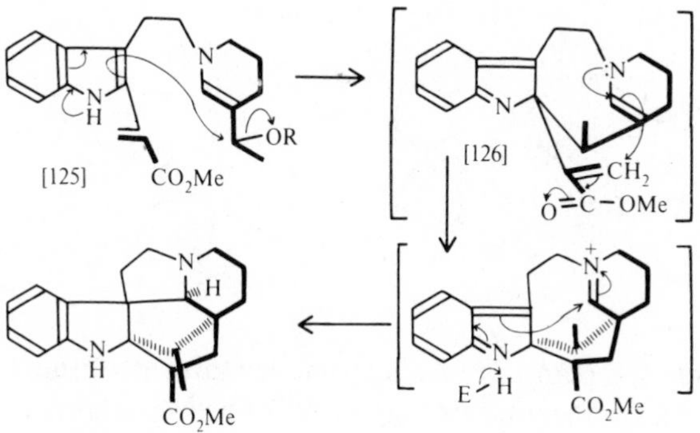

[125] [126] [127] Vindolinine [128]

SCHEME 4.34

The alkaloids of the vindolinine groups can arise from the secodine derivative [125] as shown in Scheme 4.34. An intramolecular attack of the indole 2 position on a suitable alkoxy substituent located at the ethyl group could afford the intermediate [126]. An enamine alkylation would lead to [127], which could undergo attack by a reducing enzyme system at the indole nitrogen to afford vindolinine [128].

The alkaloids of the pleiocarpine group can arise from minovincine [92] or a close analogue by attack of the side chain on the α-position of the indole nucleus. The formation of aspidofractine [131], a typical member of this group, is illustrated in Scheme 4.35.

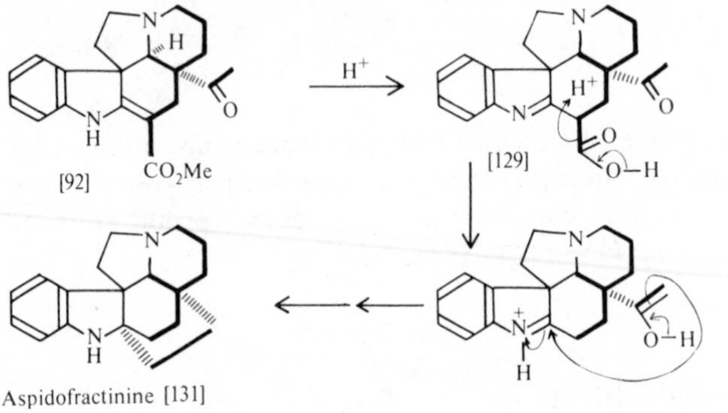

[92]   CO$_2$Me   [129]

Aspidofractinine [131]   SCHEME 4.35

The alkaloids of the kopsine group may arise from a pleiocarpine-type precursor [132] (Scheme 4.36). The *in vitro* thermal conversion of kopsinic acid [132] to kopsinone [134] in 50 per cent yields suggests the existence of

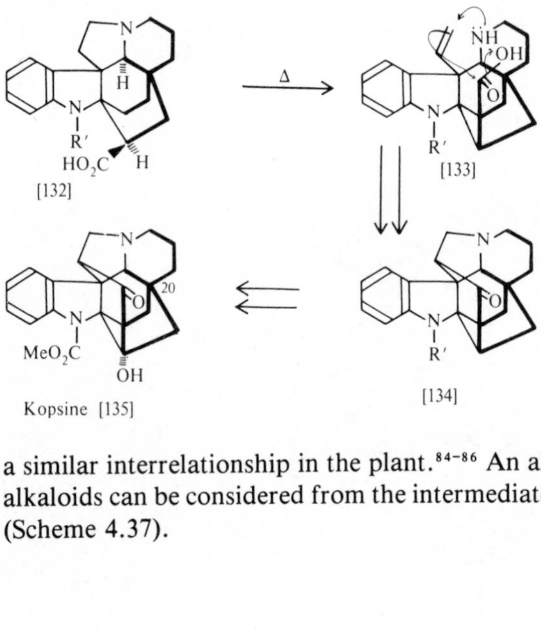

HO$_2$C   H
[132]   [133]

MeO$_2$C
OH
Kopsine [135]   [134]   SCHEME 4.36

a similar interrelationship in the plant.[84-86] An alternative pathway to these alkaloids can be considered from the intermediate [129] or a close analogue (Scheme 4.37).

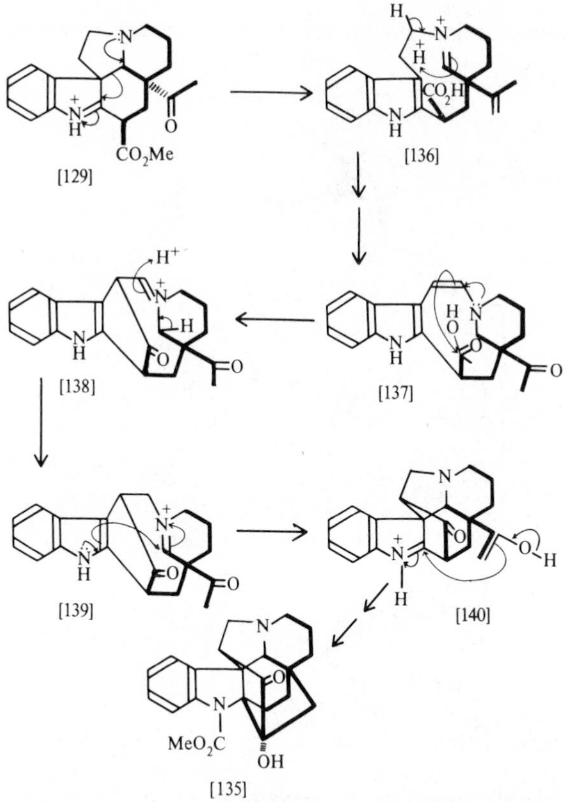

[129]

[136]

[138]

[137]

[139]

[140]

MeO₂C ÔH
[135]

SCHEME 4.37

The fruticosine group of alkaloids can be considered to be derived from those of the kopsine group. The co-occurrence of fruticosine [144] and fruticosamine [145] with kopsine [135], as well as the identity of the stereochemistry of these substances, supports such a conclusion. Kopsine [135] has been converted *in vitro* to fruticosine [144] and fruticosamine [145] as illustrated in Scheme 4.38, and a similar situation may prevail in nature. An acyloin reaction results in the formation of isokopsine [141].[87] Reduction to [142], oxidative cleavage to the keto aldehyde [143], and cyclization on heating affords fruticosine [144] and fruticosamine [145].[88] Fruticosine and fruticosamine are interconvertible on heating or treatment with acid or base.[88] Evidence against the involvement of such a pathway is provided, however, by the non-incorporation of labelled dihydroisokopsine [142] into fruticosine or fruticosamine in the plants.[89] It is therefore possible that the alkaloids of the kopsine group as well as those of the fruticosine group lie on different biogenetic branches arising from a common intermediate, e.g. pleiocarpine or a close derivative [132]. Another more plausible alternative is the direct conversion of isokopsine [141] to the ketoaldehyde [143] without the intermediacy of dihydroisokopsine as indicated in Scheme 4.38.

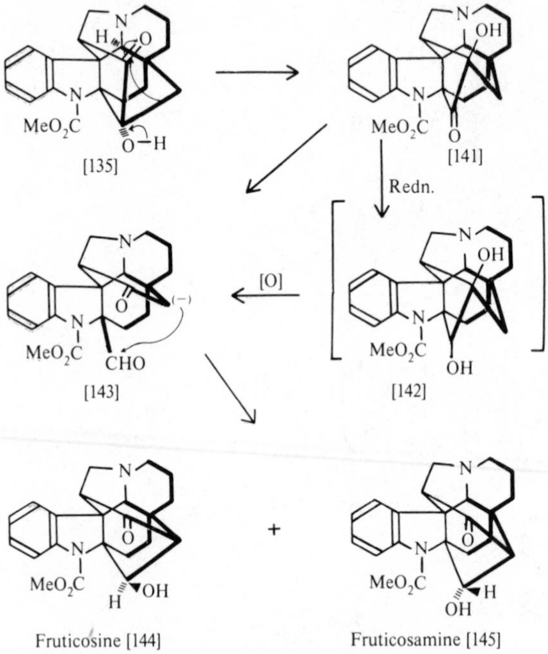
MeO$_2$C
[135]
[141]
Redn.
[O]
[143]
[142]
Fruticosine [144]
Fruticosamine [145]

SCHEME 4.38

The alkaloids of the schizophylline and schizozygine group probably arise from the aspidospermine group of alkaloids as illustrated in Scheme 4.39. A

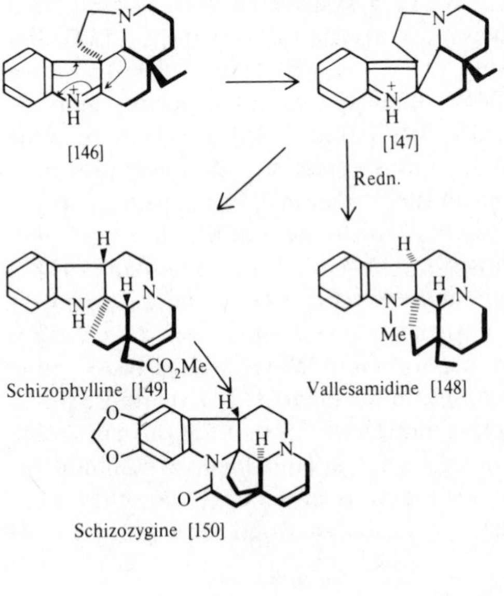
[146]
[147]
Redn.
Schizophylline [149]
Vallesamidine [148]
Schizozygine [150]

SCHEME 4.39

1,2-shift of C-21 in the indolenenium ion [146] can give rise to the intermediate [147], reduction of which can afford vallesamidine [148][90] a representative of the andrangine group. Evidence towards such a mechanism is provided by *in vitro* experiments in which tabersonine indolenine [151], on zinc–acetic acid reduction of the double bond and N-methylation, was found to afford the enantiomer [155] of vallesamidine [148] (Scheme 4.40).[91] Schizophylline [149] could arise by a similar

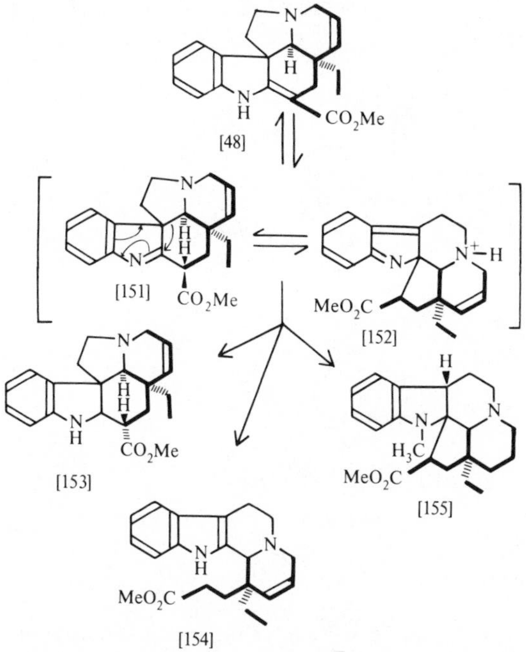

[48]

[151]

[152]

[153]

[155]

[154]

SCHEME 4.40

sequence of reactions starting from an aspidospermine-type alkaloid with the C-18 carbon of the ethyl group being present as an acid or ester (e.g. cylindrocarpidine). In alkaloids of the schizozygine [150] group, the ester group in schizophylline-type alkaloid is seen to have undergone cyclization with the indole nitrogen atom.

Vincamine [39] is a typical member of the *Hunteria* series of indole alkaloids. Incorporation studies with labelled precursors showed positive incorporations of geissoschizine [12],[48] 16,17-dihydrosecodine-17-ol [35],[48, 50] secodine, [34][48, 50] and tabersonine [17][48] into vincamine [39] demonstrating that vincamine was formed by the geissoschizine–stemmedenine–secodine–tabersonine route. Several modes of conversion of the aspidospermine-type alkaloids, e.g. tabersonine [17] or vincadifformine [48] to those of the vincamine group may be considered. It has been shown by LeMen and co-workers[92] that (−)-dehydro-1,2-aspidospermidine [156] on oxidation with peracids affords the corresponding N-oxide and the corresponding hydroxyindolenine-N-oxide [157]. Treatment of [157] with acetic acid in the presence of triphenylphosphine afforded (+)-vincanol [161] and (−)-eburnamenine [162] via the hydroxyindolenine [138] (Scheme 4.41).

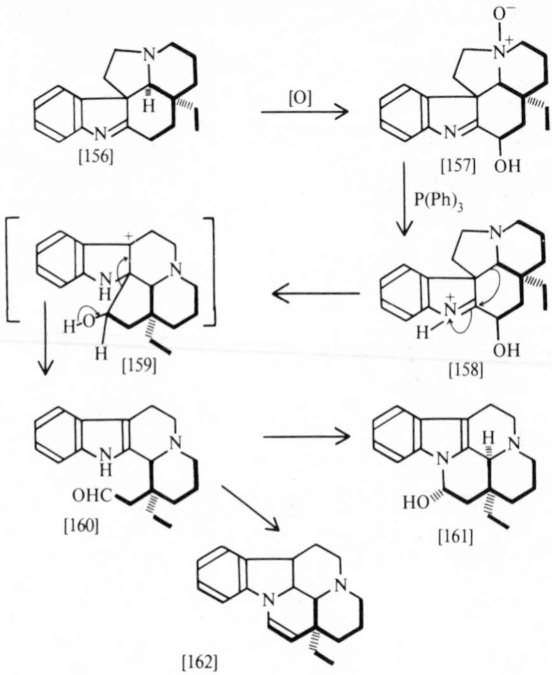

SCHEME 4.41

Similarly Poisson and co-workers have shown[93] that oxidation of vincadifformine with peracid followed by treatment with acid affords vincamine [39] (Scheme 4.42). An alternative route that can be considered

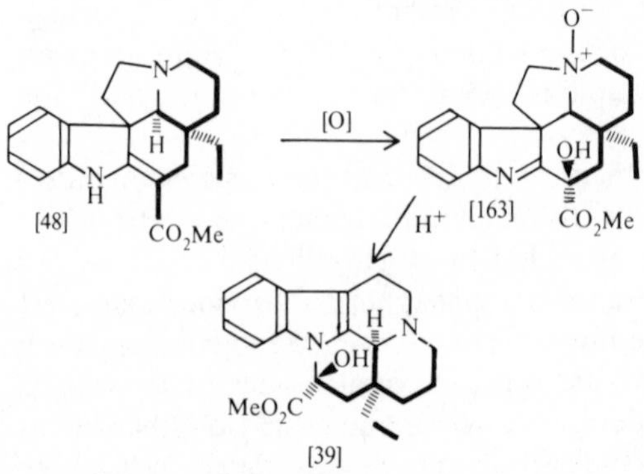

SCHEME 4.42

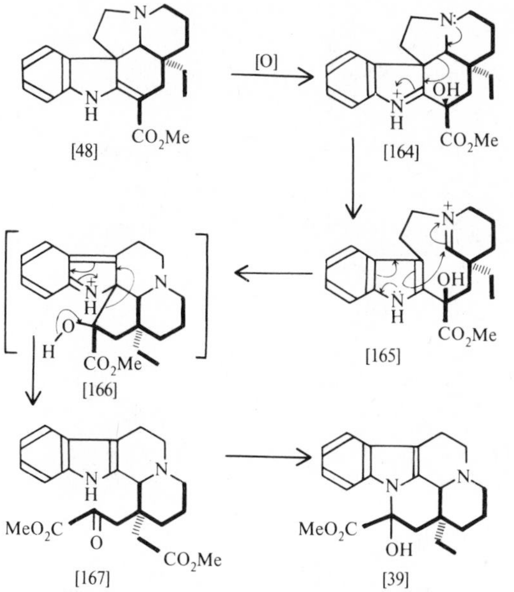

[48] [164] [165] [166] [167] [39]

SCHEME 4.43

starts from vincadifformine [48] as shown in Scheme 4.43. This can give rise to the keto ester [167] which on cyclization can lead to vincamine [39]. It is interesting that tabersonine [17] is converted in low yields on treatment with zinc/copper sulphate in acetic acid to [168], a close derivative of [167] (Scheme 4.44).[94] Recently it has been shown that vincadifformine can be converted directly to vincamine on ozonization.[95]

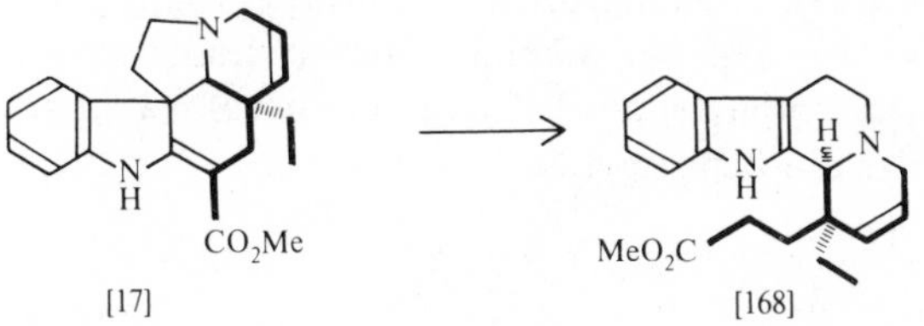

[17] [168]

SCHEME 4.44

An interesting reversal of these reactions has been demonstrated by Harley-Mason and co-workers, who synthesized aspidospermidine [106] by a novel rearrangement reaction (Scheme 4.45).[96, 97]

The alkaloids of the cuanzine group are closely related to those of the vincamine group, the ethyl group at the alcohol oxidation level having cyclized to afford a five-membered ring in cuanzine [172] (Scheme 4.46).

In the criocerine group, it is the tertiary hydroxyl group α-to the indole nitrogen atom that is involved in the cyclization step to form an ether bridge

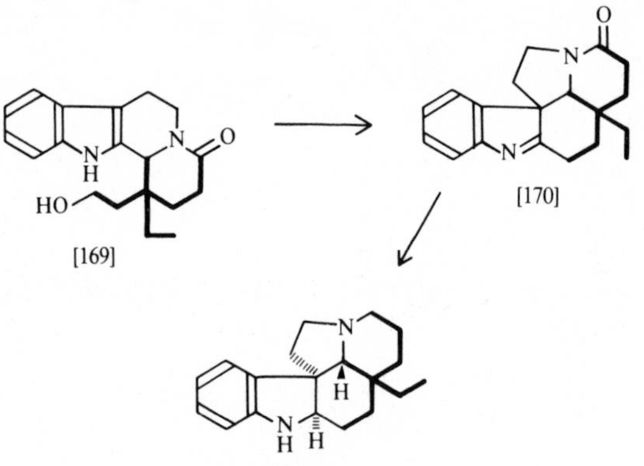

[169]   [170]   [106]

SCHEME 4.45

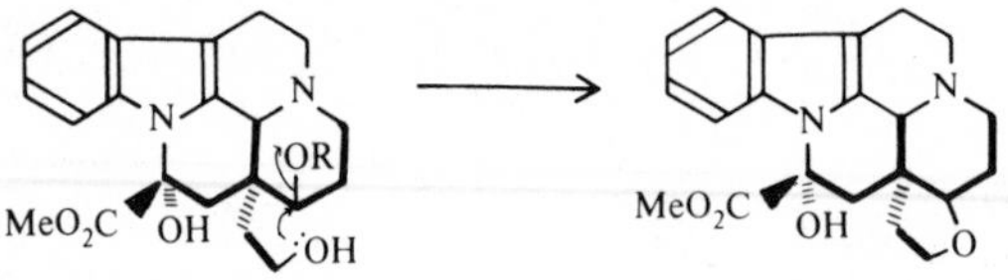

[171]   [172]

SCHEME 4.46

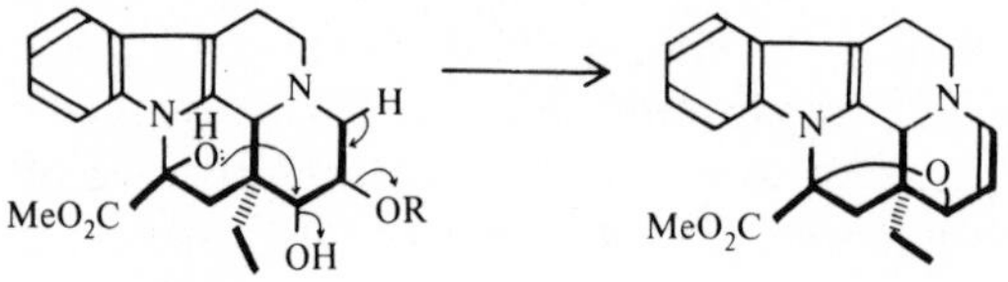

[173]   [174]

SCHEME 4.47

in criocerine [174] (Scheme 4.47). The biosynthetic pathways proposed for the *Aspidosperma* and *Hunteria* type of alkaloids are substantiated by labelling experiments with appropriately labelled geraniol, mevalonate, and loganin. These are summarized in Scheme 4.48.

Summary

Feeding experiments with suitably labelled precursors have established that the *Aspidosperma* and *Hunteria* alkaloids arise by the mevalonate → geraniol → deoxyloganin → loganin → secologanin pathway. There is strong evidence that one or more of the *Strychnos* alkaloids, e.g stemmadenine act as connecting links between the *Corynanthe* alkaloids and the *Aspidosperma–Hunteria* alkaloids, via secodine-type intermediates. The isolation of several 'seco' alkaloids such as secamines and secodine, supports such a biogenetic route. *Hunteria* alkaloids such as vincamine [39] arise from vincadifformine-like precursors (Schemes 4.41–4.44).

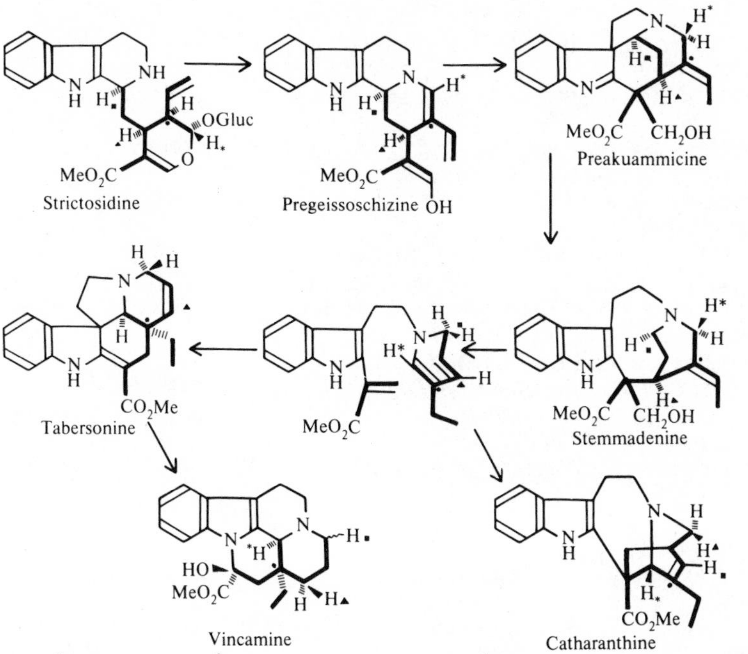

Strictosidine → Pregeissoschizine → Preakuammicine

Tabersonine ← ← Stemmadenine

Vincamine   Catharanthine

$* = [1-^3H_2]$-geraniol and $[S-^3H_R]$-mevalonate $\blacktriangle = [4-^3H_R]$-mevalonic acid

$\bullet = [2-^{14}C]$-geraniol $\blacksquare = [5-^3H]$-Loganin   SCHEME 4.48

References

1. THOMAS, R. *Tetrahedron Lett.* 544 (1961).
2. WENKERT, E. *J. Am. Chem. Soc.* **84**, 98 (1962).
3. McCAPRA, F., MONEY, T., SCOTT, A. I., and WRIGHT, I. G. *Chem. Commun.* **21**, 537 (1965).
4. GOEGEL, H. and ARIGONI, D. *Chem. Commun.* 538 (1965).
5. BATTERSBY, A. R., BROWN, R. T., KAPIL, R. S., PLUNKETT, A. O., and TAYLOR, J. B. *Chem. Commun.* 46 (1966).
6. BATTERSBY, A. R., BROWN, R. T., KAPIL, R. S., KNIGHT, J. A., MARTIN, J. A., and PLUNKETT, A. O. *Chem. Commun.* 819, 888 (1966).
7. MONEY, T., WRIGHT, I. G., McCAPRA, F., HALL, E. S., and SCOTT, A. I. *J. Am. Chem. Soc.* **90**, 4144 (1968).
8. BATTERSBY, A. R., BROWN, R. T., KNIGHT, J. A., MARTIN, J. A., and PLUNKETT, A. O. *Chem. Commun.* 346 (1966).
9. LOEW, P., GOEGGEL, H., and ARIGONI, D. *Chem. Commun.* 347 (1966).
10. HALL, E. S., McCAPRA, F., MONEY, T., FUKUMOTO, K., HANSON, T. R., MOOTOO, B. S., PHILLIPS, G. T., and SCOTT, A. I. *Chem. Commun.* 348 (1966).
11. LEETE, E. and UEDA, S. *Tetrahedron Lett.* 4915 (1966).
12. BATTERSBY, A. R., BYRNE, T. C., KAPIL, R. S., MARTIN, J. A., PAYNE, T. G., ARIGONI, D., and LOEW, P. *Chem. Commun.* 951 (1968).

13. ESCHER, S., LOEW, P., and ARIGONI, D. *Chem. Commun.* 823 (1970).
14. BATTERSBY, A. R., BROWN, S. H., and PAYNE, T. G. *Chem. Commun.* 827 (1970).
15. BATTERSBY, A. R., BURNETT, A. R., and PARSONS, P. G. *Chem. Commun.* 826 (1970).
16. BATTERSBY, A. R., BROWN, R. T., KAPIL, R. S., MARTIN, J. A., and PLUNKETT, A. O. *Chem. Commun.* 812, 890 (1966).
17. BATTERSBY, A. R., KAPIL, R. S., MARTIN, J. A., and MO, L. *Chem. Commun.* 133 (1968).
18. LOEW, P. and ARIGONI, D. *Chem. Commun.* 137 (1968).
19. BATTERSBY, A. R. and GIBSON, K. H. *Chem. Commun.* 902 (1971).
20. BATTERSBY, A. R., BURNETT, A. R., HALL, E. S., and PARSONS, P. G. *Chem. Commun.* 1582 (1968).
21. BATTERSBY, A. R., HUTCHINSON, C. R., and LARSON, R. A. *A.C.S. National Meeting, Boston, Mass, Abstracts, Organic Section No. II* (1972).
22. BATTERSBY, A. R., BURNETT, A. R., and PARSONS, P. G. *Chem. Commun.* 1280 (1968).
23. BATTERSBY, A. R., BURNETT, A. R., and PARSONS, P. G. *J. Chem. Soc.* 1187 (1969).
24. STUART, K. L., KUTNEY, J. P., HONDA, T., LEWIS, N. G., and WORTH, B. R. *Heterocycles* 9, 647 (1978).
25. BATTERSBY, A. R., BURNETT, A. R., and PARSONS, P. G. *J. Chem. Soc.* 1282 (1968).
26. BATTERSBY, A. R., BURNETT, A. R., and PARSONS, P. G. *J. Chem. Soc.* 1193 (1969).
27. BATTERSBY, A. R. and HALL, E. S. *Chem. Commun.* 793 (1969).
28. STOCKIGT, J. and ZENK, M. H. *J. Chem. Soc. Chem. Commun.* 646 (1977).
29. BROWN, R. T., LEONARD, J., and SLEIGH, S. K. *Phytochemistry* 17, 899 (1978).
30. NAGAKURA, N., RUFFER, M., and ZENK, M. H. *J. Chem. Soc. Perkin Trans. I* 9, 2309 (1979).
31. SCOTT, A. I. *Acc. Chem. Res.* 3, 151 (1970).
32. QURESHI, A. A. and SCOTT, A. I. *Chem. Commun.* 948 (1968).
33. SCOTT, A. I., CHERRY, P. C., and QURESHI, A. A. *J. Am. Chem. Soc.* 91, 4932 (1969).
34. WALSER, A. and DJERASSI, C. *Helv. Chem. Acta* 48, 391 (1965).
35. BATTERSBY, A. R. *Pure Appl. Chem.* 14, 117 (1967).
36. BATTERSBY, A. R. *Chimia* 22, 310 (1968).
37. EVANS, D. A., SMITH, G. F., SMITH, G. N., and STAPLEFORD, K. S. J. *Chem. Commun.* 859 (1968).
38. EVANS, D. A., JOULE, J. A., and SMITH, G. F. *Phytochemistry* 7, 1429 (1968).
39. CORDELL, G. A. *Lloydia* 37, 219 (1974).
40. SAKAI, S. N., AIMI, K., KOTO, K., IDO, H., and HAGINIWA, J. *Chem. Pharm. Bull. (Tokyo)* 19, 1503 (1971).
41. SAKAI, S., AIMI, N., KATO, K., IDO, H., MASUDA, K., WATANABE, Y. and HAGINIWA, J. *Yakugaku Zashi* 95, 1152 (1975).
42. CORDELL, G. A., SMITH, G. F., and SMITH, G. N. *Chem. Commun.* 189 (1970).
43. CROOKS, P. A., ROBINSON, B. J., and SMITH, G. F. *Chem. Commun.* 1210 (1968).
44. KOMPIS, I. *Symposium on the chemistry of the alkaloids*, Manchester, 1969.

45. MOKRY, J., KOMPIS, I., and SPITELLER, G. *Coll. Czech. Chem. Commun.* **32**, 2523 (1967).
46. BATTERSBY, A. R. and BHATNAGAR, A. K. *Chem. Commun.* 193 (1970).
47. CORDELL, G. A., SMITH, G. F., and SMITH, G. N. *Chem. Commun.* 191 (1970).
48. KUTNEY, J. P., BECK, J. F., NELSON, V. R., and SOOD, R. S. *J. Am. Chem. Soc.* **93**, 255 (1971).
49. KUTNEY, J. P., BECK, J. F., EGGERS, N. J., HANSEN, H. W., SOOD, R. S., and WESTCOTT, N. D. *J. Am. Chem. Soc.* **93**, 7322 (1971).
50. KUTNEY, J. P., BECK, J. F., EHRET, C., POULTON, G., SOOD, R. S., and WESTCOTT, N. D. *Bio-org. Chem.* **1**, 194 (1971).
51. QURESHI, A. A. and SCOTT, A. I. *Chem. Commun.* 945 (1968).
52. QURESHI, A. A. and SCOTT, A. I. *Chem. Commun.* 947 (1968).
53. BROWN, R. T., HILL, J. S., SMITH, G. F., STAPLEFORD, K. S. J., POISSON, J., MUQUET, M., and KUNESCH, N. *Chem. Commun.* 1475 (1969).
54. BROWN, R. T., HILL, J. S., SMITH, G. F., and STAPLEFORD, K. S. J. *Tetrahedron Lett.* 5217 (1971).
55. MUQUET, M., KUNESCH, N., and POISSON, J. *Tetrahedron* **28**, 1363 (1972).
56. SCOTT, A. I. and WEI, C. C. *J. Am. Chem. Soc.* **94**, 8266 (1972).
57. GORMAN, M., NEUSS, N., and CONE, N. J. *J. Am. Chem. Soc.* **87**, 93 (1965)..
58. RAHMAN, ATTA-UR-, FIRDOUS, S., and BASHA, A. *Z. Naturforsch.* **33b**, 469 (1978).
59. SCOTT, A. I. and WEI, C. C. *J. Am. Chem. Soc.* **94**, 8263 (1972).
60. SCOTT, A. I. and WEI, C. C. *J. Am. Chem. Soc.* **94**, 8264 (1972).
61. SCOTT, A. I. *J. Am. Chem. Soc.* **94**, 8262 (1972).
62. SCOTT, A. I. *M.P.T. Int. Rev. Sci.* **9**, 105 (1973).
63. SCOTT, A. I. and CHERRY, P. C. *J. Am. Chem. Soc.* **91**, 5872 (1969).
64. RETEY, J., STETTEN, E. VON, COY, U., and LYNEN, F. *Eur. J. Biochem.* **15**, 72 (1970).
65. SEILER, M., ACKLIN, W., and ARIGONI, D. *Chem. Commun.* 1394 (1970).
66. CORNFORTH, J. W. and ROSS, F. P. *Chem. Commun.* 1395 (1970).
67. SCOTT, A. I., PHILLIPS, G. T., REICHARDT, P. B., and SWEENEY, J. G. *Chem. Commun.* 1396 (1970).
68. ZSADON, B., EGRY, E., and SARKOZI, M. *Acta Chim. Acad. Sci. Hung.* **67**, 77 (1971); *Chem. Abs.* **74**, 121331.
69. PANAS, J. M., MORFAUX, A. M., OLIVIER, L., and LE MEN, J. *Ann. Pharm. Fr.* **30**, 273 (1972).
70. ZSADON, B. and KAPOSI, P. *Tetrahedron Lett.* 4615 (1970).
71. ZSADON, B. and KAPOSI, P. *Acta Chim. Acad. Sci. Hung.* **71**, 115 (1972); *Chem. Abs.* **76**, 138213.
72. ZSADON, B., HUBAY, R., EGRY, E., RAKLI, M., and SARKOSI, M. *Magy. Kem. Foly.* **76**, 466 (1970); *Chem. Abs.* **74**, 13312.
73. JANOT, M. M., POURRAT, H., and LEMEN, J. *Bull. Soc. Chim. Fr.* 707 (1954).
74. INOUYE, H., YOSHIDA, T., NAKAMURA, T., and TOBITA, S. *Tetrahedron Lett.* 4429 (1968).
75. KLYNE, W., SWAN, R. J., BYCROFT, B. W., SCHUMANN, D., and SCHMID, H. *Helv. Chim. Acta* **48**, 443 (1965).
76. WENKERT, E. and BRINGI, N. V. *J. Am. Chem. Soc.* **81**, 1474 (1959).
77. SMITH, G. F. and WROBEL, J. T. *J. Chem. Soc.* 792 (1970).
78. BIEMANN, K. and SPITELLER, G. *Tetrahedron Lett.* 299 (1961).
79. HARLEY-MASON, J. and RAHMAN, ATTA-UR-, *Chem. Commun.* 743 (1966).

80. KUTNEY, J. P. and PIERS, E. *J. Am. Chem. Soc.* **86**, 953 (1964).
81. CAMERMAN, A., CAMERMAN, N., KUTNEY, J. P., PIERS, E., and TROFFERS, T. *Tetrahedron Lett.* p. 637 (1965).
82. BYCROFT, B. W., SCHUMANN, D., PATEL, M. B., and SCHMID, H. *Helv. Chim. Acta* **47**, 1147 (1964).
83. DIATTA, L., LANGLOIS, Y., LANGLOIS, N., and POTIER, P. *Bull. Soc. Chim. Fr.* p. 671 (1975).
84. GORMAN, A. A., DASTOOR, N. J., HESSE, M., VON PHILIPSBORN, W., RENNER, U., and SCHMID, H. *Helv. Chim. Acta* **52**, 33 (1969).
85. GUGGISBERG, A., GORMAN, A. A., BYCROFT, B. W., and SCHMID, H. *Helv. Chim. Acta* **52**, 76 (1969).
86. SCHNOES, H. K. and BIEMANN, K. *J. Am. Chem. Soc.* **86**, 5693 (1964).
87. GOVINDACHARI, T. R., NAGARAJAN, K., and SCHMID, H. *Helv. Chim. Acta* **46**, 433 (1963).
88. GUGGISBERG, A., HESSE, M., VON PHILIPSBORN, W., NAGARAJAN, K., and SCHMID, H. *Helv. Chim. Acta* **49**, 2322 (1966).
89. KOMPIS, K., HESSE, M., and SCHMID, H. *Lloydia* **34**, 269 (1971).
90. BROWN, S. H., DJERASSI, C., and SIMPSON, P. G. *J. Am. Chem. Soc.* **90**, 2445 (1968).
91. LEVY, J., MAUPERIN, P., DOE DE MAINDREVILLE, M., and LE MEN, J. *Tetrahedron Lett.* 1003 (1971).
92. HUGEL, G., LEVY, J., and LE MEN, J. *Tetrahedron Lett.* 3109 (1974).
93. CROQUELOIS, G., KUNESCH, N., and POISSON, J. *Tetrahedron Lett.* 4427 (1974).
94. MAUPERIN, P., LEVY, J., and LEMEN, J. *Tetrahedron Lett.* 999 (1971).
95. DANIELI, B., LESMA, G., and PALMISANO, G. *Chem. Commun.* 908 (1981).
96. BARTON, J. E. D. and HARLEY-MASON, J. *Chem. Commun.* 298 (1965).
97. HARLEY-MASON, J. and KAPLAN, M. *Chem. Commun.* 915 (1967).

BIOSYNTHESIS OF CLASS III ALKALOIDS

5.1 *Iboga* alkaloids

Over a dozen known alkaloids of the *Iboga* sub-class bear a rearranged secologanin unit. In these alkaloids, the non-tryptophan C-10 unit [1] found in the *Yohimbe–Strychnos* alkaloids (Class I) has undergone cleavage between C-4 and C-3, and the three-carbon unit has joined up with C-5 giving rise to the *Iboga*-type non-tryptophan unit [2] (Scheme 5.1).

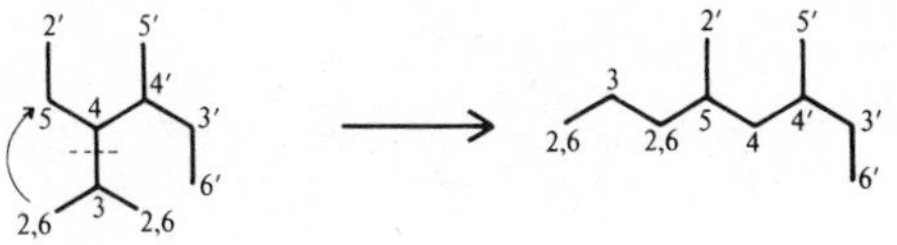

[1] [2] SCHEME 5.1

This unit can be recognized in catharanthine [3], a typical member of this class. The *Iboga* alkaloids have been isolated only from the family Apocynaceae particularly from the tribe Tabernaemontana.

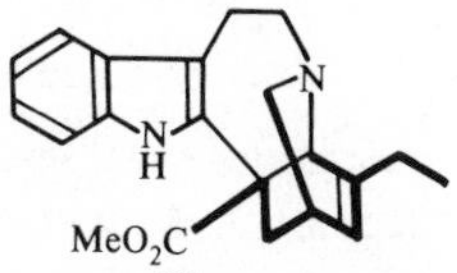

MeO$_2$C
[3]

The positive incorporations of labelled mevalonate [4],[1-4] geraniol [5],[2, 5, 6] 10-hydroxygeraniol [6],[4] geraniol pyrophosphate [7],[6] nerol [9],[3] 10-hydroxynerol [10],[4] deoxyloganin [13],[7] loganin [14],[8-13] and secologanin [16][14, 15] into catharanthine [3] conclusively showed that the *Iboga* alkaloids follow the same pathway as those of the *Yohimbe–Strychnos* series or the *Aspidosperma–Hunteria* series in the earlier stages (Scheme 5.2).

This was strong evidence in favour of the generation of the rearranged *Iboga* system at some stage after condensation of secologanin [16] with tryptophan. Positive incorporations were also observed with labelled specimens of strictosidine [17a],[12, 16-22] geissoschizine [18],[17, 23] stemmadenine,[24] 16, 17-dihydrosecodine-17-ol [26],[25] secodine,[25] and tabersonine [24].[24, 26] These studies greatly clarified the biosynthetic pathway to the *Iboga* alkaloids as they served to establish that strictosidine [17a], geissoschizine [18], and stemmadenine [20] were precursors to these

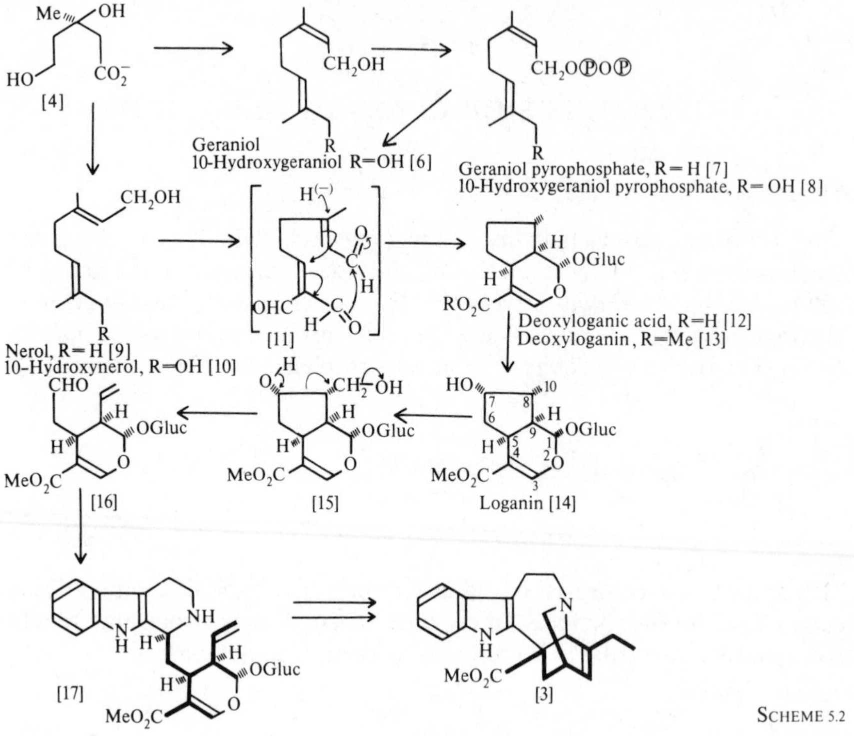

SCHEME 5.2

systems. The positive incorporations with secodine derivatives provided the necessary insight into the mechanism by which the rearrangement occurred. The mechanism proposed[27, 28] for the generation of the secodine intermediate [22] involves an initial isomerization of the olefin [20] to the enamine [21]. An alternative mechanism may involve the direct delivery of hydride from an appropriate enzyme to the olefinic carbon (C-19) in [20]. This could be accompanied by the generation of either secodine [25] or 16,17-dihydrosecodine-17-ol [26] (Scheme 5.3). An attractive test for the correct mechanism would be to feed stemmadenine tritiated at C-21 and to check whether 50 per cent tritium is lost in [26] as predicted by Scott's mechanism. It was clear from these experiments that the *Iboga* alkaloids lie at the end of the biosynthetic pathway, the *Strychnos* alkaloids giving rise to the *Aspidosperma* alkaloids on the one hand, and the *Iboga* alkaloids on the other, through the intermediacy of a common secodine derivative [22].

In an attempt to substantiate their earlier results[27, 28] (which were found to be non-reproducible by British and French workers)[29, 30, 33] Scott and Qureshi have published new results, but under entirely different experimental conditions on the reactions exhibited by dihydrostemmadenine acetate [29] and dihydropreakuammicine acetate [33] on heated silica.[31] Scott

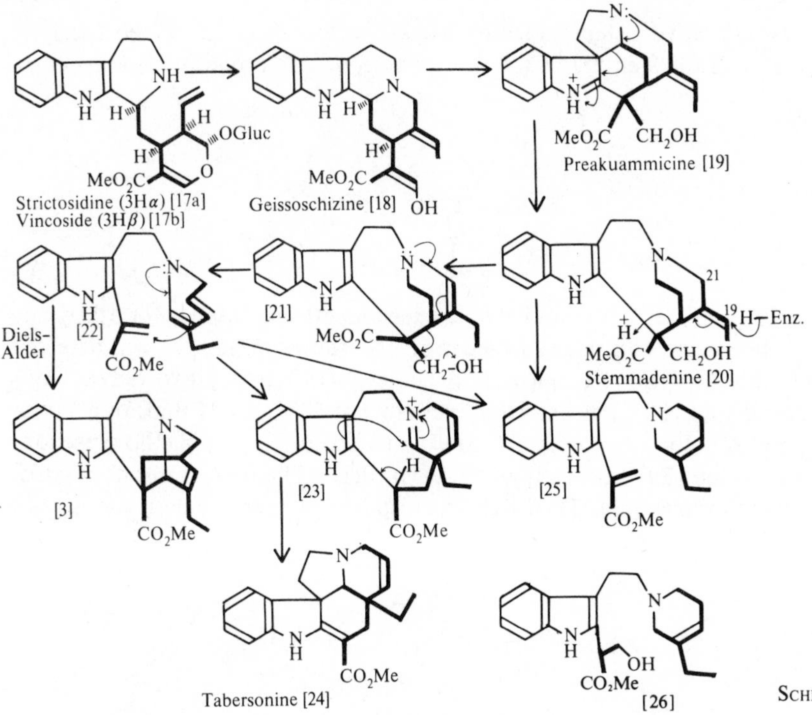

Strictosidine (3Hα) [17a]
Vincoside (3Hβ) [17b]

Geissoschizine [18]

MeO$_2$C   CH$_2$OH
Preakuammicine [19]

Diels-Alder [22]

CO$_2$Me

[21]

MeO$_2$C   CH$_2$OH
Stemmadenine [20]

[3]

CO$_2$Me

[23]

CO$_2$Me

[25]

CO$_2$Me

Tabersonine [24]

CO$_2$Me   OH

[26]

SCHEME 5.3

MeO$_2$C   CH$_2$OR
Stemmadenine, R=H [20]
Stemmadenine acetate, R=Ac [27]

MeO$_2$C   CH$_2$OR
Dihydrostemmadenine, R=H [28]
Dihydrostemmadenine acetate, R=Ac [29]

CO$_2$Me
Pseudocatharanthine [30]
15,20- Dihydropseudocatharanthine [31]

CO$_2$Me   CH$_2$
Dehydrosecodine [32]

CO$_2$Me
(+)-15-Methoxypseudocatharanthine [34]

MeO$_2$C   CH$_2$OAc
Dihydropreakuammicine
acetate [33]

OMe
CO$_2$Me
(−)-15-Methoxypseudocatharanthine [35]

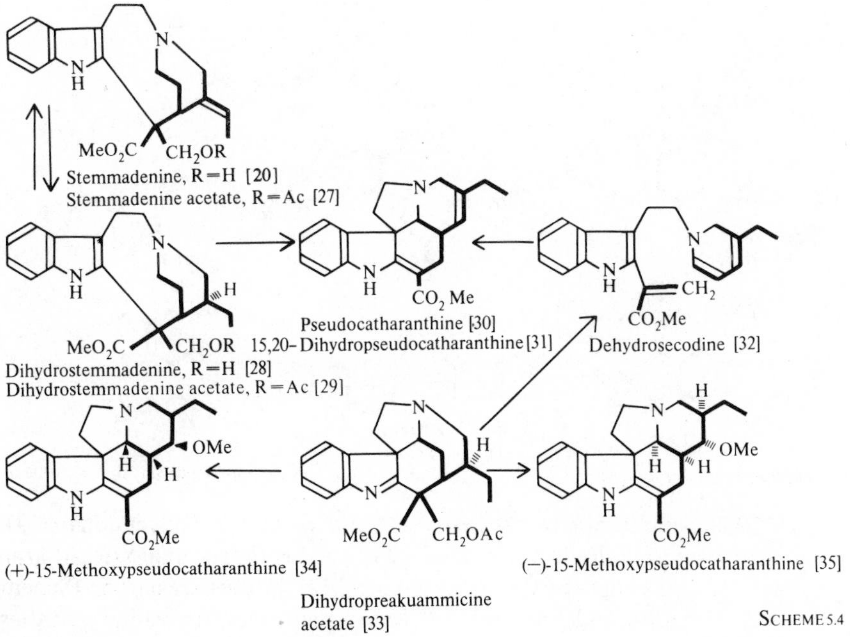

SCHEME 5.4

found that dihydrostemmadenine acetate [29] prepared by the catalytic reduction of the stemmadenine derivative [27], afforded pseudocatharan-

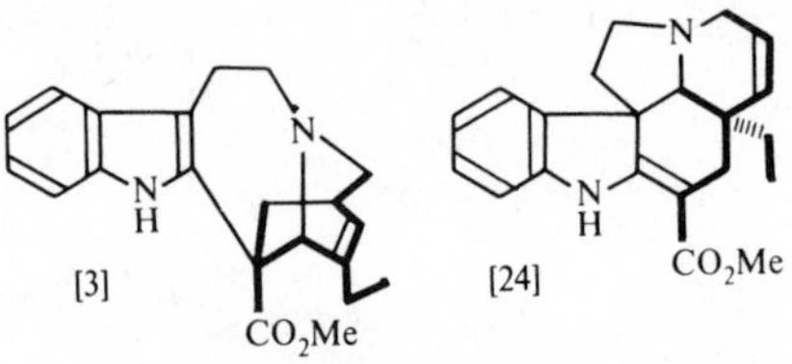

thine [30] (1 per cent) and dihydropseudocatharanthine [31] (0.05 per cent) when heated on silica surface in air. No tabersonine [24] was detected under the conditions employed (Scheme 5.4). When dihydropreakuam-micine acetate [33] was heated in methanol at 80°C for 15 min, or at room temperature for 4 h, it afforded (+)- and (−)-15-methoxypseudo-catharanthine [34] and [35] in a ratio of 9:1.[31] The formation of dihydro-pseudocatharanthine [31] and the corresponding methoxy derivatives [34] and [35] is rationalized in Scheme 5.5.

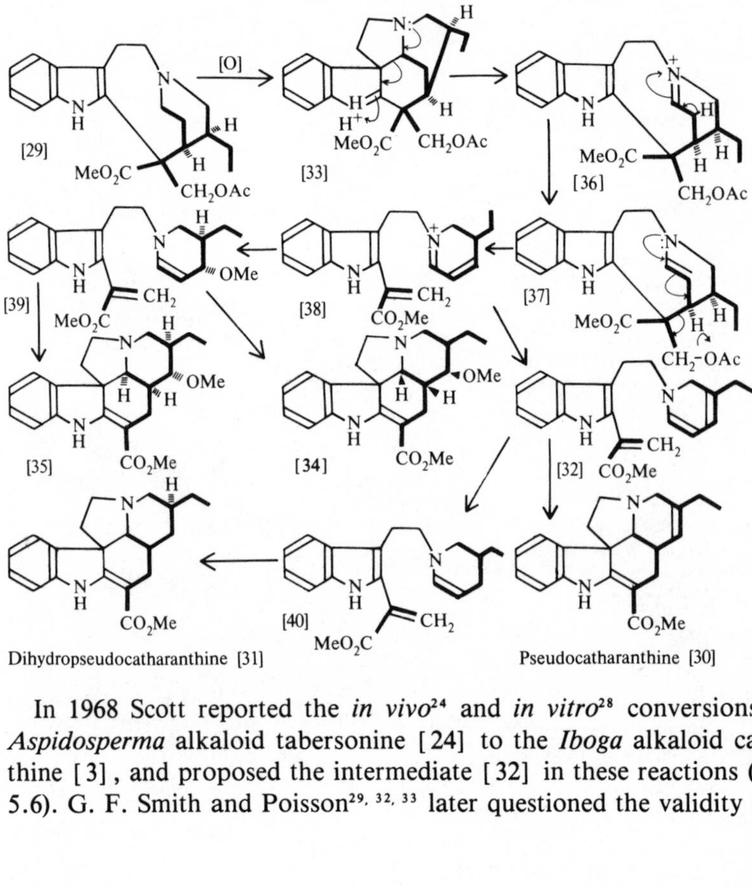

Dihydropseudocatharanthine [31]   Pseudocatharanthine [30]   SCHEME 5.5

In 1968 Scott reported the *in vivo*[24] and *in vitro*[28] conversions of the *Aspidosperma* alkaloid tabersonine [24] to the *Iboga* alkaloid catharan-thine [3], and proposed the intermediate [32] in these reactions (Scheme 5.6). G. F. Smith and Poisson[29, 32, 33] later questioned the validity of these

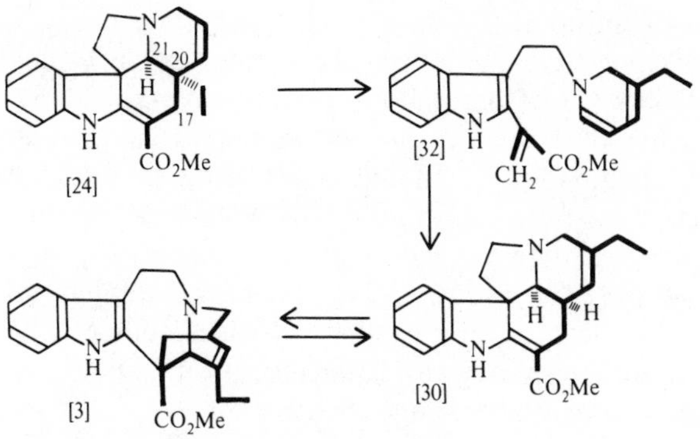

SCHEME 5.6

results and reported that tabersonine [24] did not undergo the transformations demonstrated by Scott *et al.*, but instead affords only dihydrotaber-
sonine (vincadifformine) [42] (14 per cent), allocatharanthine [43] (17 per
cent), acetoallocatharanthine [44] (6 per cent), and dihydroallocatharanthine [45] (15 per cent), G. F. Smith and co-workers suggested that the
reaction proceeds not by the rupture of both C_7-C_{21} and $C_{17}-C_{20}$ bonds as

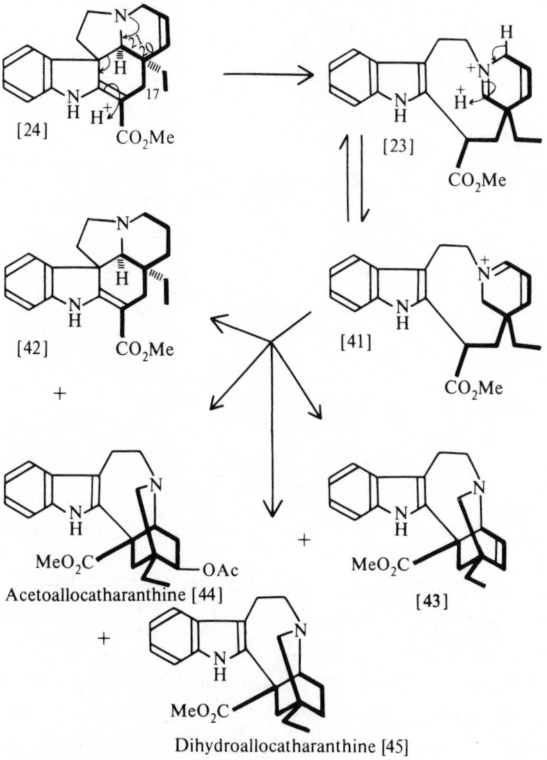

Acetoallocatharanthine [44]

[43]

Dihydroallocatharanthine [45]

SCHEME 5.7

suggested by Scott, but by rupture of the C_7-C_{21} bond only (Scheme 5.7).

Scott has subsequently reported [34] that when tabersonine [24] is heated in acetic acid for 15 h (140°C) it affords allocatharanthine [43] (11 per cent). Allocatharanthine thus obtained was heated on a silica gel plate at 150°C for 30 min, and examination of the reaction products afforded (+)-pseudocatharanthine [30] (4 per cent) and *optically pure* tabersonine [24] (4 per cent). The absence of racemization of tabersonine [24] suggested that the pathway [43] → [41] → [23] → [24] was operative. Since no racemic tabersonine was produced, it appears that dehydrosecodine [32] goes directly to [30] without isomerizing to [22], since the cyclization of [22] would have afforded racemic tabersonine (Scheme 5.8)

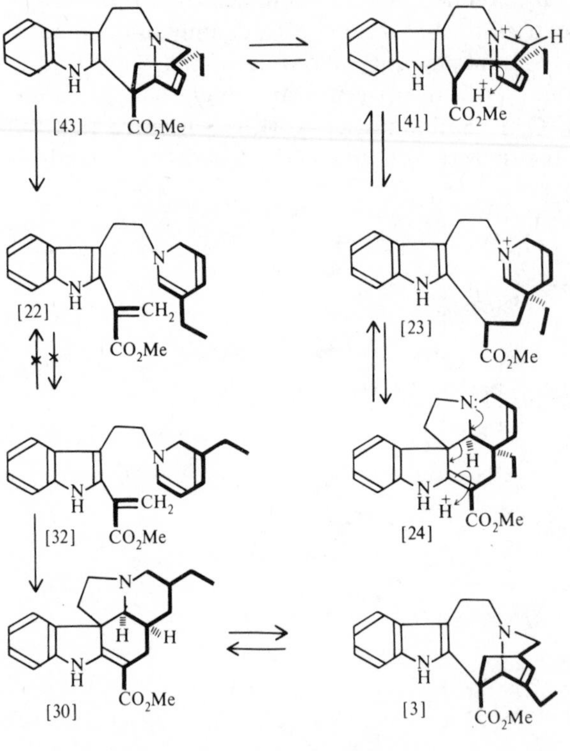

SCHEME 5.8

While it has been shown[24] that tabersonine [24] is converted to catharanthine [3] in the plant, Scott found that tabersonine does not serve as a precursor for coronaridine (15,20-dihydrocatharanthine) [46].[34] It was suggested that the dehydrosecodine [32] as the immonium salt [38] can undergo 1,4 reduction to the secodine isomer [40] which could then cyclize to coronaridine [46] without involving catharanthine [3] (Scheme 5.9).

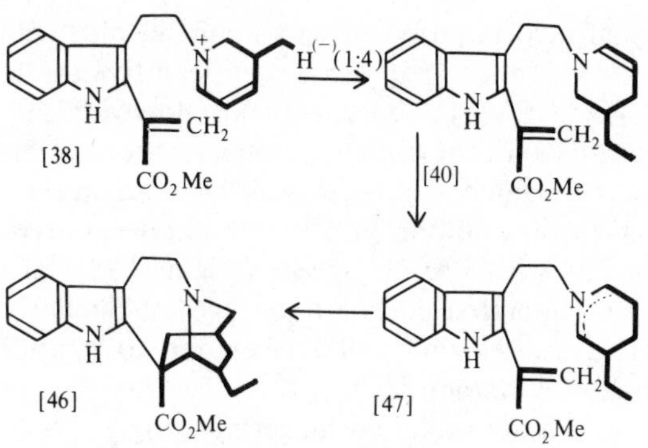

SCHEME 5.9

The 1,3 shift of the olefin in [47] appears, therefore, to be of considerable importance, in view of the revision of *Iboga* alkaloid stereochemistry.[35]

An *in vitro* relationship has also been demonstrated[28] between the *Corynanthe* and *Iboga* alkaloids. When corynantheine aldehyde [48] or geissoschizine [18] were heated in glacial acetic acid under nitrogen for 72 h, they were found to afford pseudocatharanthine [30] (15 per cent) and catharanthine [3] (5 per cent). The reaction was suggested to proceed via the intermediacy of the *Strychnos* derivative [51] (Scheme 5.10). It has been suggested that the essential requirements of these conversions was the presence of both the β-dicarbonyl system as well as the correct oxidation level of the two-carbon side chain since corynantheine [49] as well as the dihydroaldehyde [54] failed to rearrange under the conditions employed.

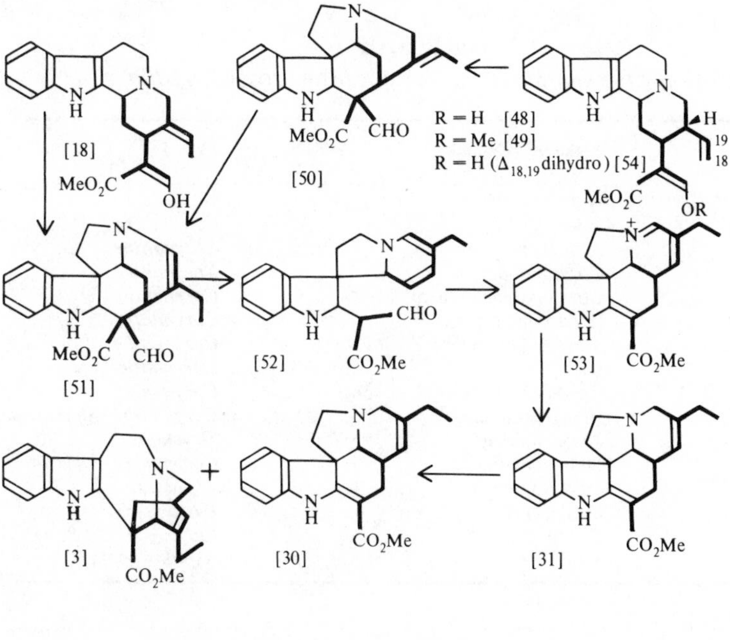

SCHEME 5.10

Interestingly, the efficiency of incorporation of corynantheine aldehyde [48] into catharanthine [3] and vindoline [55] varies significantly with the age of *V. rosea* plants. When [O-methyl-³H]-corynantheine aldehyde [49] was fed into *V. rosea* seedlings, 0.3 per cent incorporation was observed into catharanthine [3].[24] When the same substance was fed into mature plants, however, much lower incorporations into indole alkaloids were observed.[2, 24] Feeding experiments with [Ar-²H]-geissoschizine [18] into *V. rosea* seedlings have led to deuterated coronaridine [46].[36] Similarly [Ar-³H]- and [O-methyl-³H]-geissoschizine [18], when fed to mature *V. rosea* plants, afforded labelled catharanthine [3].[17]

Sequential formation of various indole alkaloids in germinating *V. rosea* seeds has afforded important information regarding the biosynthetic pathway to the *Iboga* alkaloids.[23, 36] Initially, alkaloids are absent from the seeds, but as the seeds germinate a changing spectrum of alkaloid-type and content is observed until the seedlings reach an age of 10–12 days, when the alkaloid composition closely resembles that found in the mature plants. With 24–26 h of germination, alkaloids of the *Corynanthe* group such as vincoside [17], and corynantheine [49] are detectable by thin-layer chromatography. Between 28 and 40 h after germination, corynantheine aldehyde [48], and geissoschizine [18] make their appearance. After 50 h, the *Strychnos* alkaloid stemmadenine [20] and the *Aspidosperma* alkaloid tabersonine [24] are detectable. It is not until 100–160 h have elapsed that the *Iboga* alkaloids catharanthine [3] and coronaridine [46] are found. These data are summarized in Table 5.1.

Table 5.1

Isolation of alkaloids from Catharanthus roseus *seedlings*

Time after germination (h)	Alkaloid isolated	Type
0	None	—
26	Vincoside	*Corynanthe*
	Ajmalicine	*Corynanthe*
	Corynantheine	*Corynanthe*
28–40	Corynantheine aldehyde	*Corynanthe*
	Geissoschizine	*Corynanthe*
	β-Hydroxyindolenine [56]	*Corynanthe*
	Diol [57]	*Corynanthe*
	Geissoschizine oxindole [58]	*Corynanthe*
40–50	Preakuammicine	*Corynanthe–Strychnos*
	Akuammicine	*Strychnos*
	Stemmadenine	*Corynanthe–Strychnos*
	Tabersonine	*Aspidosperma*
100–160	Catharanthine	*Iboga*
	Coronaridine	*Iboga*
200	Vindoline	*Aspidosperma*

The β-hydroxyindolenine [56], the diol [57], and geissoschizine oxin-dole [58] have also been detected 28–40 h after germination and provide strong evidence towards the mechanism of formation of the *Iboga* alkaloids shown in Scheme 5.11.

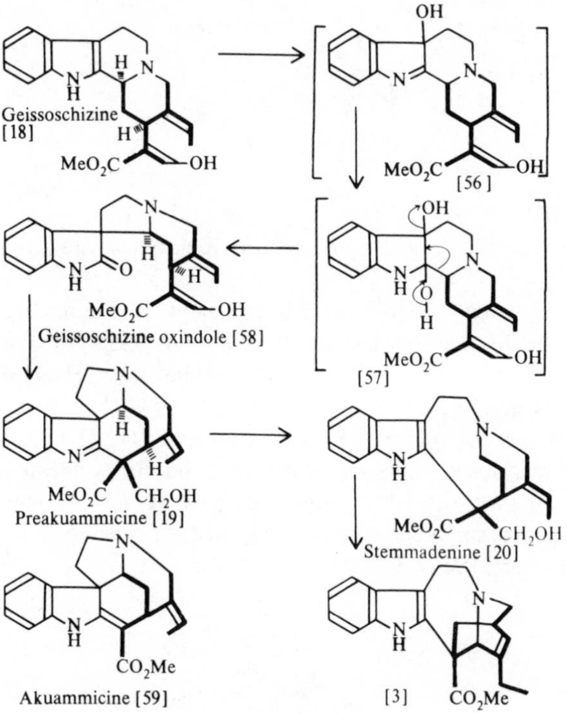

SCHEME 5.11

It is interesting that in the sequential studies, coronaridine [46] appears after 45 h, much earlier than catharanthine [3]. Even more interesting is the non-incorporation of [O-methyl-³H] -and [ar-³H] -coronaridine [46] into catharanthine [3].[37] Similarly [ar-³H] -catharanthine [3] was not con-verted to [ar-³H] -coronaridine [46] in *V. rosea*. This suggests that both coronaridine and catharanthine may arise from a common dehydrosecodine intermediate [32]. Subsequently it has been shown[37] that (+)-catharanthine is converted *in vitro* to (+)-coronaridine. X-ray crystallography of coronaridine hydrobromide has confirmed that the coronaridine with a negative rotation obtained from *C. roseus* belongs to the opposite antipodal series to catharanthine.[38] This seems logical if both are formed from a com-mon achiral intermediate [32] (Scheme 5.12).

Within the *Iboga* alkaloid family the two stereochemical series are seen in (+)-catharanthine, (+)-ibogamine, and (+)-coronaridine which belong to

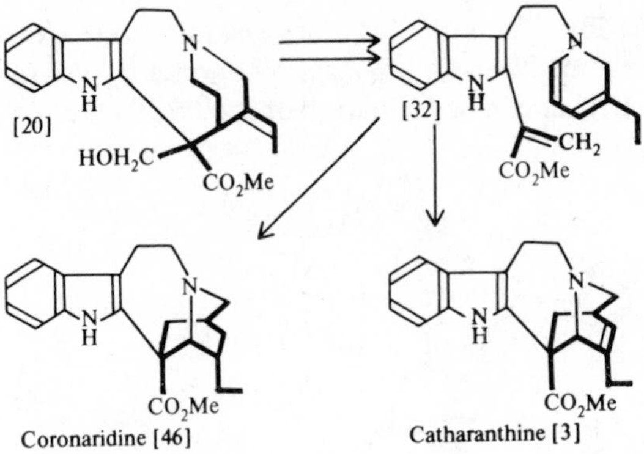

Coronaridine [46]   Catharanthine [3]   SCHEME 5.12

one series, while (−)-coronaridine and (−)-ibogamine belong to the anti-podal series.[38, 41]

It has been observed that when 5R, S- and [5S-³H]-mevalonic acids are fed to *V. rosea* plants, a certain degree of stereospecificity prevails for one of the protons in the conversion of geissoschizine [18] to vindoline (H_R,H_R) [55] and catharanthine (H_R,H_S) [3].[39] This implies that (−)-tabersonine (H_R,H_S) [24] cannot act as a true precursor for catharanthine (H_R,H_S) [3]. To explain this it has been proposed that (−)-tabersonine may be converted to catharanthine by a minor pathway[37, 40] or it may be only the (+)-isomer (H_R,H_S)[W] which acts as a precursor for catharanthine, the (−)-isomer being converted to vindoline, (H_R,H_S) (Scheme 5.13).

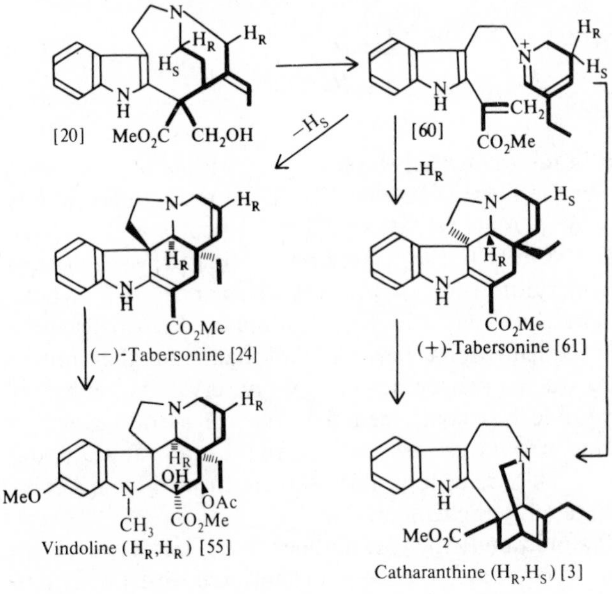

Vindoline (H_R,H_R) [55]   Catharanthine (H_R,H_S) [3]   SCHEME 5.13

In the rupicoline group of alkaloids, a ring contraction has occurred to afford the 3-keto compounds. This reaction probably proceeds via the 3-hydroxyindolenine as shown in Scheme 5.14. The intermediate [62] could undergo further hydration to afford the diol [64], which on rearrangement would lead to the 2-keto series exemplified by the crassanine [65] group (Scheme 5.14).

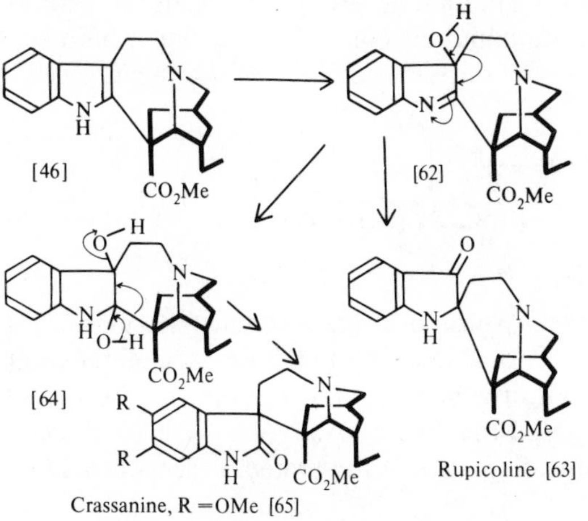

[46]

[62]

[64]

Crassanine, R = OMe [65]

Rupicoline [63]

SCHEME 5.14

In the elegandine group of alkaloids, an ether bridge is seen to be formed to afford a new five-membered heterocyclic ring. Several 19-oxo-coronaridine derivatives are found in nature, and elegandine [66] can be considered to arise from one of these.

The interesting capuronine [67] group bears a nitrogen-containing nine-membered ring identical to that found in the indolic 'cleavamine' moiety of the binary anti-leukaemic alkaloids, vinblastine and vincristine.[42]

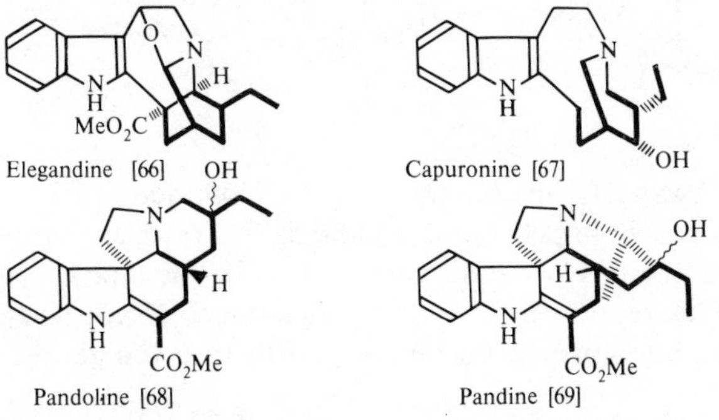

Elegandine [66]

Capuronine [67]

Pandoline [68]

Pandine [69]

The pandoline group [68] carries the pseudocatharanthine type skeleton, which has been shown[43] to be generated *in vitro* on heating catharanthine in glacial acetic acid.

In the pandine [69] group an additional C-17 to C-21 bond formation has taken place.

5.2 Other diverse skeletons derived from tryptophan and secologanin

Some structurally novel indole alkaloids have been isolated from *Ervatamia orientalis*.[44] The alkaloids of the ervatamine group contain three carbon atoms between the indole β-position and the basic nitrogen, and it has been suggested that these are products of fairly extensive rearrangement.

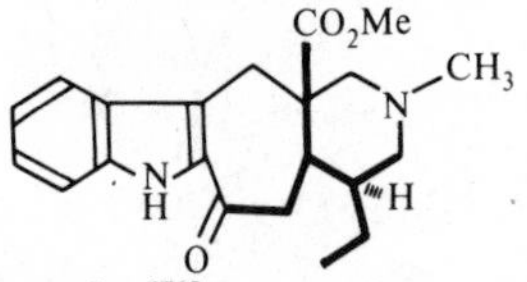

Ervatamine [70]

An examination of the carbon skeleton of the non-indole C-10 unit of ervatamine [70] shows that it is identical to that found in secologanin, except for one additional carbon atom attached at the β-position of the indole nucleus. It is, therefore, proposed that these alkaloids may arise directly from secologanin and a gramine derivative as illustrated in Scheme 5.15.

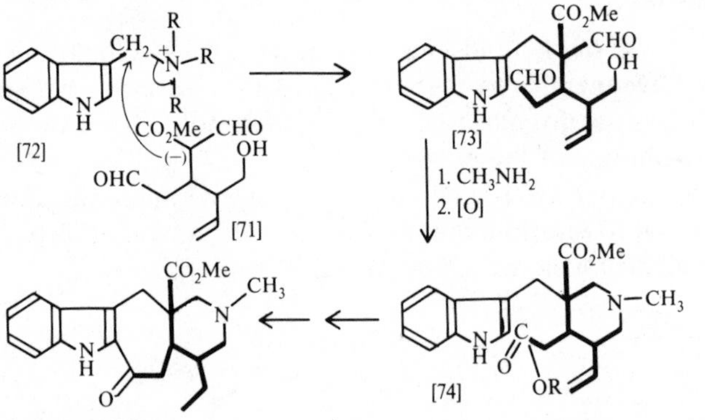

Ervatamine [70]   SCHEME 5.15

Other pathways to ervatamine can be considered from a vobasine derivative as shown in Schemes 5.16 and 5.17. Potier and co-workers have demonstrated the *in vitro* conversion of a vobasine-type compound to ervatamine.[45] The recent conversion of dregamine to 20-epi-ervatamine by an enzyme preparation provides strong support for such a pathway.[67]

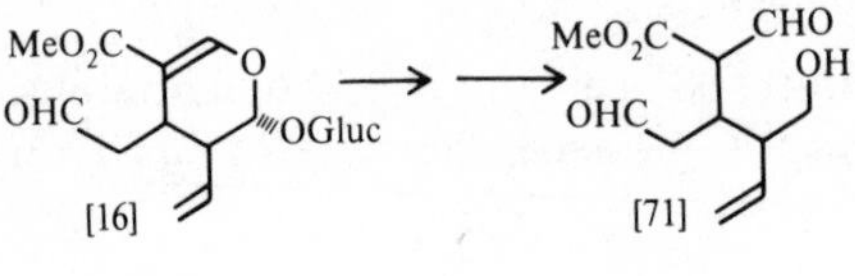

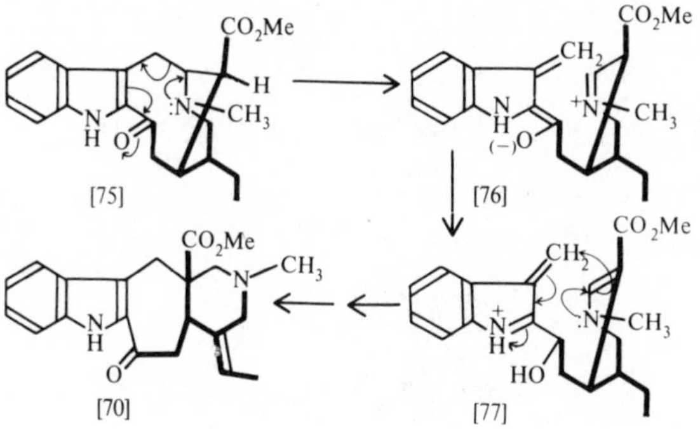

SCHEME 5.16

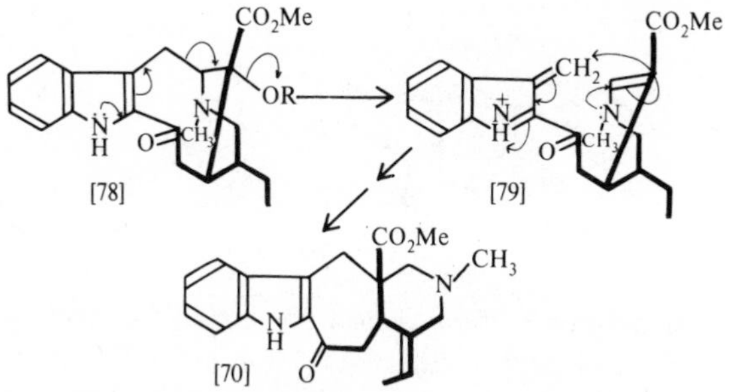

SCHEME 5.17

Two more complex routes to ervatamine have been proposed by Cordell and are illustrated in Schemes 5.18 and 5.19.[46]

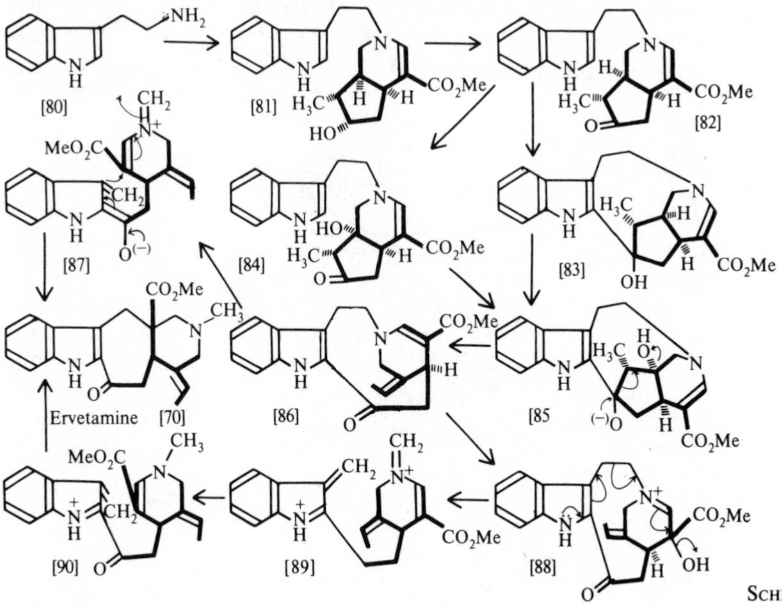

SCHEME 5.18

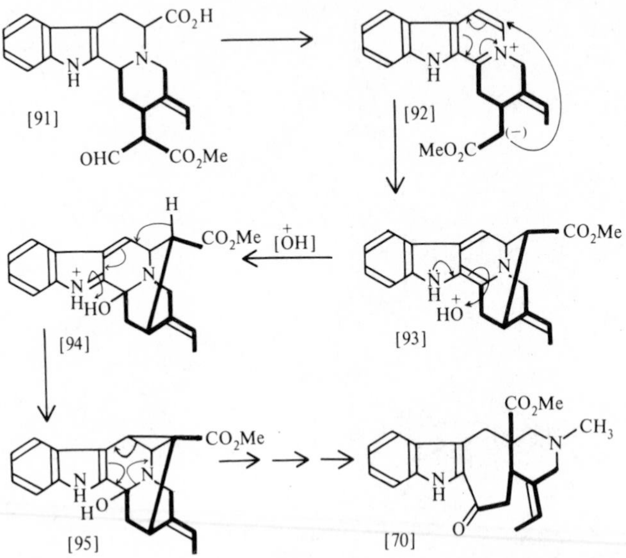

SCHEME 5.19

The andranginine group of alkaloids show a non-tryptophan C-10 unit which is different from any of the alkaloids described previously. However, andranginine [100] can be considered to arise from a secodine derivative [98] as shown in Scheme 5.20 and they may therefore be considered along with the secodines. Andranginine is obtained in 28 per cent yield by the thermolysis of precondylocarpine acetate [97] at 100°C in ethyl acetate. It is probable that the same achiral secodine [98] is generated in this conversion (Scheme 5.20) which also accounts for the lack of optical activity in the product.

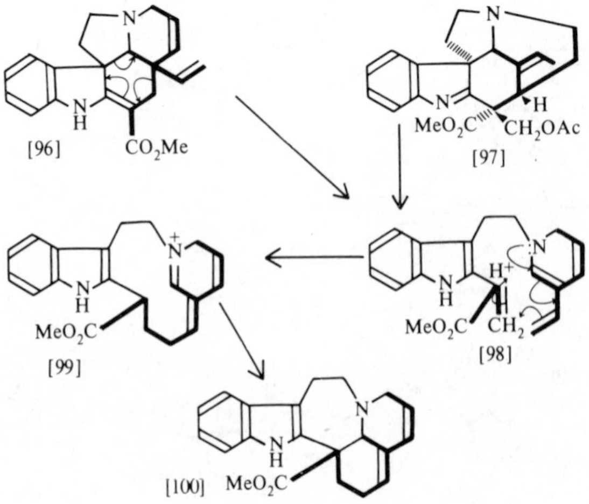

SCHEME 5.20

Potier and co-workers have demonstrated that tabersonine in methanol solution affords (±)-andranginine [100], epiandranginine, and 15-methoxy-14, 15-dihydro- Δ^{18}-allocatharanthine [101] (Scheme 5.21).[47]

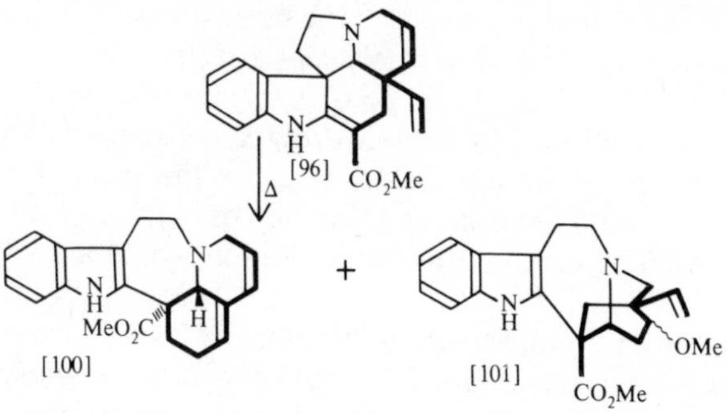

SCHEME 5.21

Brief mention may be made of two other types of alkaloids which, though they do not bear the indole nucleus, may be considered along with the indole alkaloids as they arise from the same tryptophan and loganin precursors. These are the *Cinchona* alkaloids, e.g. quinine [110] and the *Camptotheca* alkaloid camptothecin.

Feeding experiments with tryptophan,[48] geraniol,[3, 49] and loganin[50] established that they were the biosynthetic precursors to quinine. Feeding of a mixture of labelled strictosidine and vincoside afforded radioactive cinchonidine [111], quinine [110], and cinchonine [108].[51] Tritiated cory-

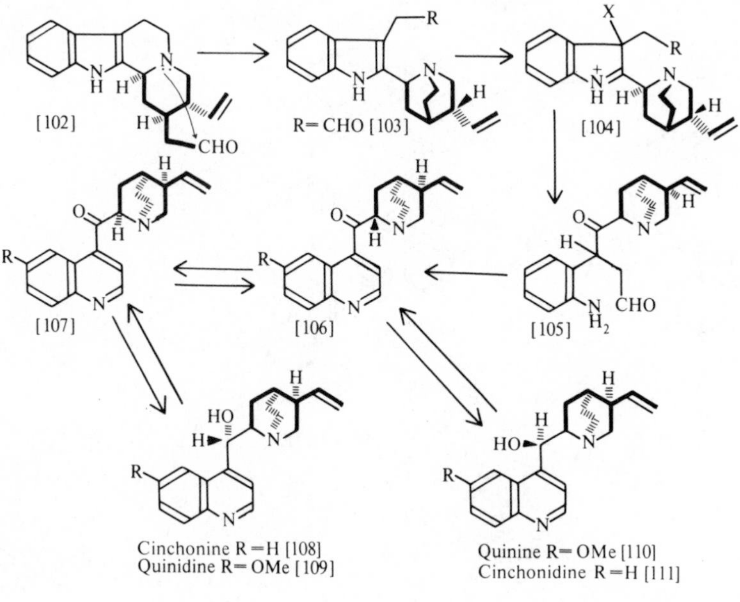

Cinchonine R = H [108]
Quinidine R = OMe [109]

Quinine R = OMe [110]
Cinchonidine R = H [111]

SCHEME 5.22

nantheal [102] was also found to be incorporated. It was shown by feeding [1-^{15}N, indole α-^{14}C] - and [1-^{15}N, 2-^{14}C] -tryptophans into quinine in *Cinchona succirubra* that the ^{15}N was specifically incorporated in the quinoline nitrogen and the label at C-2 was incorporated into the carbon atom between the quinoline and quinuclidine systems.[52, 53] The biosynthetic route to quinine is presented in Scheme 5.22. An intramolecular attack of the basic nitrogen on to the aldehyde function in [102], accompanied by cleavage of C—N bond can afford the aldehyde [103]. A hydrolytic cleavage of the imine [104] followed by re-cyclization could lead to the quinoline derivative [106]. Reduction would result in the formation of quinine [110] (or cinchonidine) [111] while reduction after isomerisation would afford cinchonine (or quinidine).

Camptothecin [116] is a pyrrolo-(3,4-b)-quinoline alkaloid isolated from the Chinese tree *Camptotheca acuminata*[54, 55] and later from *Mappia foetida*,[56, 57] and *Ophiorrhiza mungos*.[58] A number of feeding experiments have established that camptothecin is derived from tryptamine and secologanin.[59, 60] Wenkert had earlier proposed that camptothecin may be a monoterpene indole alkaloid[61] and an efficient biomimetic synthesis of [116] has been published.[62-64] The biosynthetic route to [116] is presented

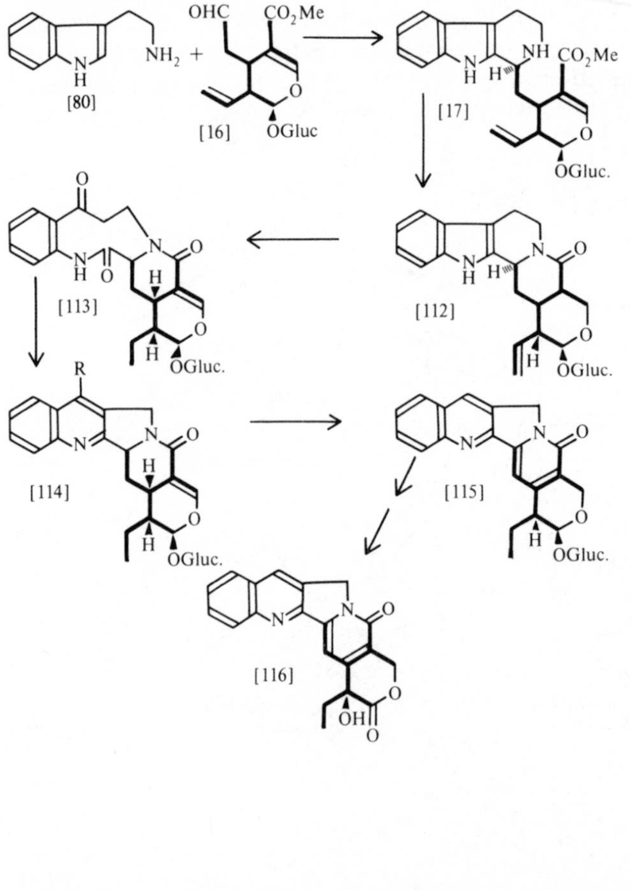

SCHEME 5.23

in Scheme 5.23. The condensation of tryptamine with secologanin would afford strictosidine [17] which could undergo an intramolecular cyclization to afford strictosamide [112] (recently isolated from *Nauclea latifolia*[65]). Feeding of labelled strictosamide has afforded positive incorporations into camptothecin.[66] Oxidation of ring B followed by recyclization would afford [114] which on successive oxidation would afford camptothecin [116]. A detailed discussion on the probable mechanistic steps involved in these oxidations have been recently published.[66]

5.3 Summary

The *Iboga* alkaloids arise via *Corynanthe* and *Strychnos* alkaloids. A common secodine-type intermediate could give rise to both the *Aspidosperma* and the *Iboga* alkaloids by appropriate intramolecular cyclizations (Scheme 5.4).

The biosynthesis of other diverse skeleta derived from tryptophan and secologanin is also discussed.

References

1. BATTERSBY, A. R., BROWN, R. T., KAPIL, R. S., PLUNKETT, A. O., and TAYLOR, J. B. *Chem. Commun.* 46 (1966).
2. BATTERSBY, A. R., BYRNE, T. C., KAPIL, R. S., MARTIN, J. A., PAYNE, T. G., ARIGONI, D., and LOEW, P. *Chem. Commun.* 951 (1968).
3. BATTERSBY, A. R., BROWN, R. T., KAPIL, R. S., KNIGHT, J. A., MARTIN, J. A., and PLUNKETT, A. O. *Chem. Commun.* 810, 888 (1966).
4. ESCHER, S., LOEW, P., and ARIGONI, D. *Chem. Commun.* 823 (1970).
5. LEETE, E. and UEDA, S. *Chem. Commun.* 348 (1966).
6. BATTERSBY, A. R., BROWN, R. T., KNIGHT, J. A., MARTIN, J. A., and PLUNKETT, A. O. *Chem. Commun.* 346 (1966).
7. BATTERSBY, A. R., BURNETT, A. R., and PARSONS, P. G. *Chem. Commun.* 826 (1970).
8. BATTERSBY, A. R., BROWN, R. T., KAPIL, R. S., MARTIN, J. A., and PLUNKETT, A. O. *Chem. Commun.* 812, 890 (1966).
9. BATTERSBY, A. R. and GIBSON, K. H. *Chem. Commun.* 902 (1971).
10. BATTERSBY, A. R., KAPIL, R. S., MARTIN, J. A., and MO, L. *Chem. Commun.* 133 (1968).
11. LOEW, P. and ARIGONI, D. *Chem. Commun.* 137 (1968).
12. BATTERSBY, A. R., BURNETT, A. R., HALL, E. S., and PARSONS, P. G. *Chem. Commun.* 1582 (1968).
13. BATTERSBY, A. R., HUTCHINSON, C. R., and LARSON, R. A. *A.C.S.National Meeting, Boston, Mass., Abstracts, Organic Section No. 11*, (1972).
14. BATTERSBY, A. R., BURNETT, A. R., and PARSONS, P. G. *Chem. Commun.* 1280 (1968).
15. BATTERSBY, A. R., BURNETT, A. R., and PARSONS, P. G. *J. Chem. Soc.* 1187 (1969).

16. BATTERSBY, A. R., BURNETT, A. R., and PARSONS, P. G. *J. Chem. Soc.* 1193 (1969).
17. BATTERSBY, A. R. and HALL, E. S. *Chem. Commun.* 793 (1969).
18. BATTERSBY, A. R., BURNETT, A. R., and PARSONS, P. G. *Chem. Commun.* 1282 (1968).
19. STOCKIGT, J. and ZENK, M. H. *J. Chem. Soc. Chem. Commun.* 546 (1977).
20. BATTERSBY, A. R., LEWIS, N. G., and TIPPETT, J. M. *Tetrahedron Lett.*, 4849 (1978).
21. BROWN, R. T., LEONARD, J., and SLEIGH, S. K. *Phytochemistry* **17**, 899 (1978).
22. NAGAKURA, N., RUFFER, M., and ZENK, M. H. *J. Chem. Soc. Perkin Trans. I*, **9**, 2309 (1979).
23. SCOTT, A. I. *Acc. Chem. Res.* **3**, 151 (1970).
24. QURESHI, A. A. and SCOTT, A. I. *Chem. Commun.* 948 (1968).
25. KUTNEY, J. P., BECK, J. F., EHRET, C., POULTON, G., SOOD, R. S., and WESTCOTT, N. D. *Bio-org. Chem.* **1**, 194 (1971).
26. KUTNEY, J. P., CRETNEY, W. J., HADFIELD, J. R., HALL, E. S., NELSON, V. R., and WIGFIELD, D. C. *J. Am. Chem. Soc.* **90**, 3566 (1968).
27. QURESHI, A. A. and SCOTT, A. I. *Chem. Commun.* 945 (1968).
28. QURESHI, A. A. and SCOTT, A. I. *Chem. Commun.* 947 (1968).
29. BROWN, R. T., HILL, J. S., SMITH, G. F., STAPLEFORD, K. S. J., POISSON, J., MUQUET, M., and KUNESCH, N. *Chem. Commun.* 1475 (1969).
30. BROWN, R. T., HILL, J. S., SMITH, G. F., and STAPLEFORD, K. S. J. *Tetrahedron Lett.*, p. 5217 (1971).
31. SCOTT, A. I. and WEI, C. C. *J. Am. Chem. Soc.* **94**, 8262 (1972).
32. CORDELL, G. A., SMITH, G. F., and SMITH, G. N. *Chem. Commun.* 189 (1970).
33. MUQUET, M., KUNESCH, N., and POISSON, J. *Tetrahedron* **28**, 1363 (1972).
34. SCOTT, A. I. and WEI, C. C. *J. Am. Chem. Soc.* **94**, 8266 (1972).
35. BLAHA, K., KOBLICOVA, Z., and TROJANEK, J. *Tetrahedron Lett.* 2763 (1972).
36. SCOTT, A. I., CHERRY, P. C., and QURESHI, A. A. *J. Am. Chem. Soc.* **91**, 4932 (1969).
37. SCOTT, A. I. *International Reviews of Science (alkaloids)* **9**, 105. M.T.P. Press, Lancaster (1973).
38. KUTNEY, J. P., FUJI, K., TREASURYWALA, A. M., FAYOS, J., CLARDY, J., SCOTT, A. I., and WEI, C. C. *J. Am. Chem. Soc.* **95**, 5408 (1973).
39. BATTERSBY, A. R. Specialist Periodical Report, *The Alkaloids* **1**, 39 (1971).
40. SCOTT, A. I., SLAYTOR, M. B., REICHARDT, P. B., and SWEENY, J. G. *Bio-org. Chem.* **1**, 157 (1971).
41. BLAHA, K., KOBLICOVA, Z., and TROJANEK, J. *Tetrahedron Lett.* 2763 (1972).
42. LORIAUX-CHARDON, I. and HUSSON, H. P. *Tetrahedron Lett.* 1845 (1975).
43. GORMAN, M., NEUSS, N., and CONE, N. J. *J. Am. Chem. Soc.* **87**, 93 (1965).
44. KNOX, J. R. and SLOBBE, J. *Tetrahedron Lett.*, 2149 (1971).
45. HUSSON, A., LANGLOIS, Y., RICHE, C., HUSSON, H. B., and POTIER, P. *Tetrahedron* **29**, 3095 (1973).
46. CORDELL, G. A. *Lloydia* **37**, 219 (1974).
47. ANDRIAMIALISOA, R. Z., DIATTA, L., RASOANIVO, P., LANGLOIS, N., and POTIER, P., *Tetrahedron* **31**, 2347 (1975).
48. KOWANKO, N. and LEETE, E. *J. Am. Chem. Soc.* **84**, 4919 (1962).
49. LEETE, E. and WEMPLE, J. N. *J. Am. Chem. Soc.* **91**, 2698 (1969).
50. BATTERSBY, A. R. and HALL, E. S. *Chem. Commun.* 194 (1970).

51. BATTERSBY, A. R. and PARRY, R. J. *Chem. Commun.* 30 (1971).
52. LETTE, E. and WEMPLE, J. N. *J. Am. Chem. Soc.* **88**, 4743 (1966).
53. McGINN, D. G. *Diss. Abst.* **32**, 177 (1971).
54. McPHAIL, A. T. and SIM, G. A. *J. Chem. Soc.*, 923 (1968).
55. WALL, M. E., WANI, M. C., COOK, C. E., PALMER, K. H., McPHAIL, A. T., and SIM, G. A. *J. Am. Chem. Soc.* **88**, 3888 (1966).
56. GOVINDACHARI, T. R. and VISWANATHAN, N. *Indian. J. Chem.* **10**, 453 (1972).
57. GOVINDACHARI, T. R. and VISWANATHAN, N. *Phytochemistry* **11**, 3529 (1972).
58. TAFUR, S., NELSON, J. D., DELONG, D. C., and SVOBOOA, G. H. *Lloydia* **39**, 261 (1976).
59. HUTCHINSON, C. R., HECKENDORF, A. M., DADDONA, P. E., HAGAMAN, E., and WENKERT, E. *J. Am. Chem. Soc.* **96**, 5609 (1974) and references therein.
60. SHERIHA, G. M. and RAPOPORT, H. *Phytochemistry* **15**, 505 (1976).
61. WENKERT, E., DAVE, K. G., LEWIS, R. G., and SPRAGUE, P. W. *J. Am. Chem. Soc.* **89**, 6741 (1967).
62. BOCH, M., KORTH, T., NELKE, J. M., PIKE, D., RADUNZ, H., and WINTERFELDT, E. *Chem. Ber.* **105**, 2126 (1972).
63. WINTERFELDT, E. *Justus Liebigs Ann. Chem.* **745**, 23 (1971).
64. WARNKEKE, J. and WINTERFELDT, E. *Chem. Ber.* **105**, 2120 (1972).
65. BROWN, R. T., CHAPPLE, C. L., and LASHFORD, A. G. *Phytochemistry* **16**, 1619 (1977).
66. HUTCHINSON, C. R., HECKENDORF, A. H., STRAUGHN, J. C., DADDONA, P. E., and CANE, D. E. *J. Am. Chem. Soc.* **101**, 3358 (1979).
67. THAL, C., DUFOUR, M., POTIER, P., JAOUEN, M., and MANSUY, D. *J. Am. Chem. Soc.* **103**, 4956 (1981).

6

BIOSYNTHESIS OF CLASS IV ALKALOIDS

There are several alkaloids which are not derived by the combination of secologanin with tryptophan, as is the case with the first three classes. These alkaloids include some of the simplest indoles found in nature, such as the carbazole alkaloids in which a tryptophan moiety is not involved, the *Peganum harmala* alkaloids in which a monoterpene has not participated, and several alkaloids in which tryptophan has taken part in the biosynthesis although the non-tryptophan moiety is an isoprenoid unit. This class of alkaloids is therefore conveniently discussed in three subsections below.

6.1 Non-tryptophan indole alkaloids

While the major pathways to the more complicated indole alkaloids are now fairly well understood, little is known about the biosynthesis of the carbazole alkaloids. Several interesting features of these alkaloids warrant closer investigation, including the question of the mode of formation of the indole nucleus, the formation of ring C, and the origin of the methyl group at positions 3 or 6.

Several theories have been advanced for the origin of these alkaloids,[1-7] of which anthranilic acid [1] has been suggested to be a precursor.[1,2] Formation of the amino acid [3] from anthranilic acid [1] followed by decarboxylation and cyclization could give rise to the carbazole nucleus [4] (Scheme 6.1). Chakraborty has suggested[3] that the methyl group at C-3 or C-6 (often in an oxidized form) can be generated by C-methylation before or after the formation of the carbazole nucleus.

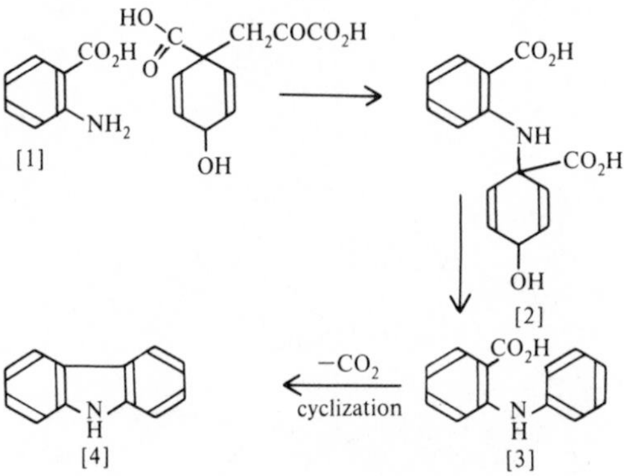

SCHEME 6.1

Kureel and co-workers have put forward an alternative proposal suggesting the mevalonoid origin of ring C.[4] Since carbazole [4] is a symmetrical molecule, the 3- and 6- positions are identical so that it is only upon further substitution that the differentiation arises. It was proposed by the same workers that the indole ring in these systems could originate from anthranilic acid [1] via 3-dimethylallylquinolines [5]. Ring contraction and further modification could give rise to the carbazole system [6] (Scheme 6.2).

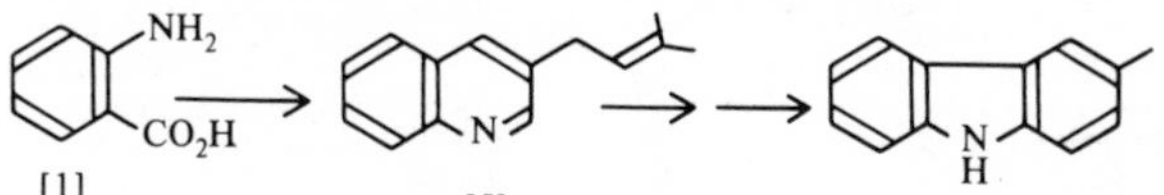

SCHEME 6.2

Erdtman has also proposed[5] that carbazole alkaloids are of mevalonoid origin. He suggested that the 3-prenylated quinoline [7] undergoes ring contraction to the 2-prenylated indoline [8], which could then give rise to murrayanine [9] (Scheme 6.3).

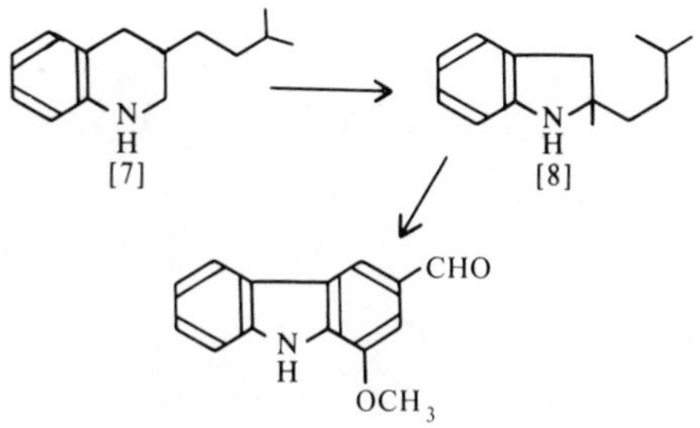

Murrayanine [9]

SCHEME 6.3

Feeding experiments carried out by Kapil and co-workers[6] using ^{14}C-methyl-L-methionine in *Murraya koenigii* leaves led to good incorporation into koenigicine [10]. The activity was found to be located only in the

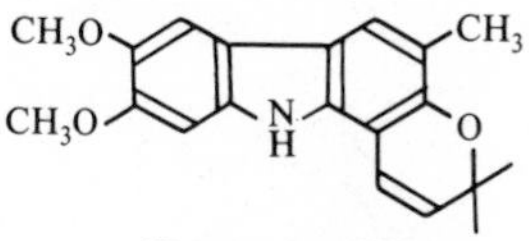

Koenigicine [10]

two methoxy groups and the extent of incorporation was in agreement with the predicted values. Since no label was incorporated into the C-methyl group, this experiment ruled out the involvement of a one-carbon unit as the origin of C-3 or C-6 methyl groups as proposed by Chakraborty and Das.[3]

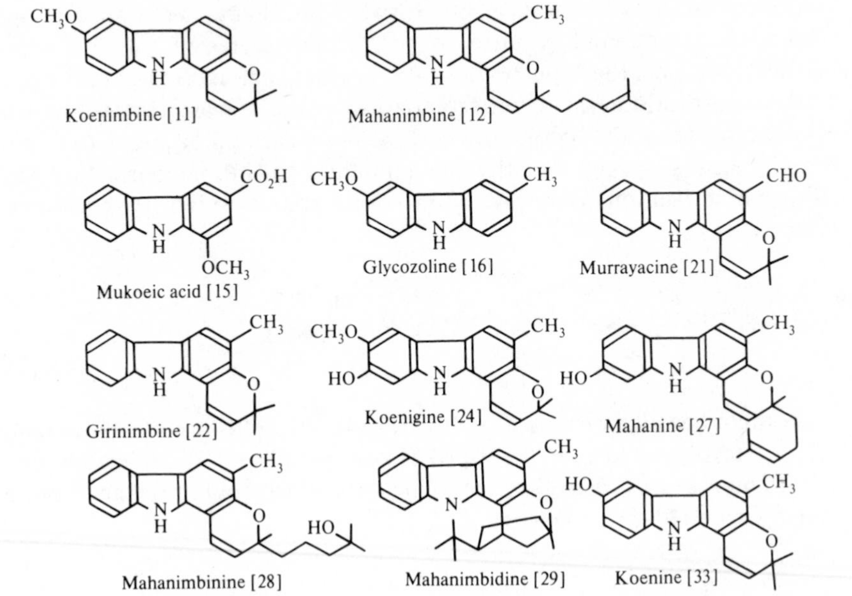

Koenimbine [11]   Mahanimbine [12]

Mukoeic acid [15]   Glycozoline [16]   Murrayacine [21]

Girinimbine [22]   Koenigine [24]   Mahanine [27]

Mahanimbinine [28]   Mahanimbidine [29]   Koenine [33]

Kapil and Popli have obtained[6] successful incorporations of $[2\text{-}^{14}C]$ - and $[2\text{-}^{3}H]$ -mevalonic acid lactone into koenimbine [11], koenigicine [10], and mahanimbine [12] in *M. koenigii* plants. The position of labelling re-

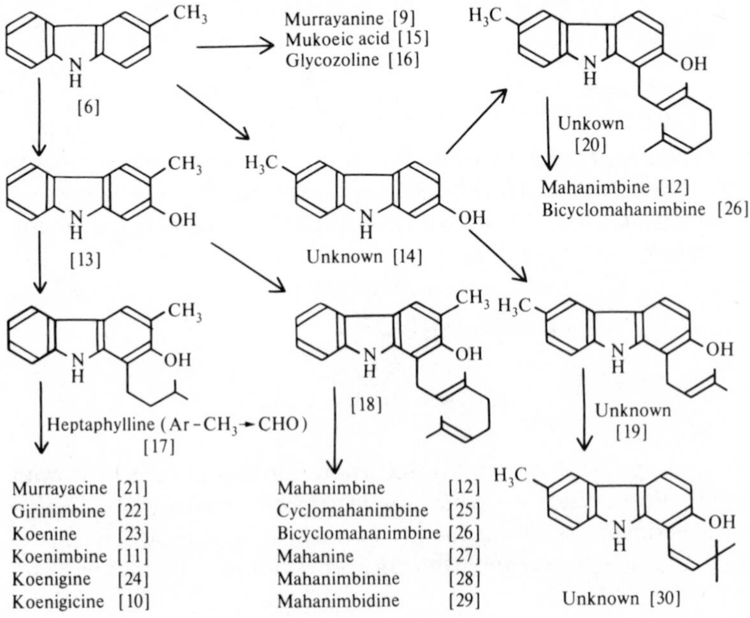

[6] → Murrayanine [9], Mukoeic acid [15], Glycozoline [16]

Unkown [20]

Mahanimbine [12], Bicyclomahanimbine [26]

[13]

Unknown [14]

Heptaphylline (Ar-CH₃→CHO) [17]

[18]

Unknown [19]

Murrayacine [21]	Mahanimbine [12]	
Girinimbine [22]	Cyclomahanimbine [25]	
Koenine [23]	Bicyclomahanimbine [26]	
Koenimbine [11]	Mahanine [27]	
Koenigine [24]	Mahanimbinine [28]	
Koenigicine [10]	Mahanimbidine [29]	Unknown [30]

SCHEME 6.4

mains to be ascertained. 3-Methyl carbazole is probably generated at some stage and gives rise to all the known carbazole alkaloids as outlined in Scheme 6.4.

In cyclomahanimbine [25] the isoprene unit attached to ring D in mahanimbine [12] has undergone cyclization to afford a fifth ring. This conversion has been demonstrated *in vitro*,[8, 9] mahanimbine [12] being converted to cyclomahanimbine [25] quantitatively on treatment with acid (Scheme 6.5). When mahanimbine [12] is shaken with silica gel in benzene it is found to be converted to cyclomahanimbine [25]. This reaction may be represented as shown in Scheme 6.5.[8, 9]

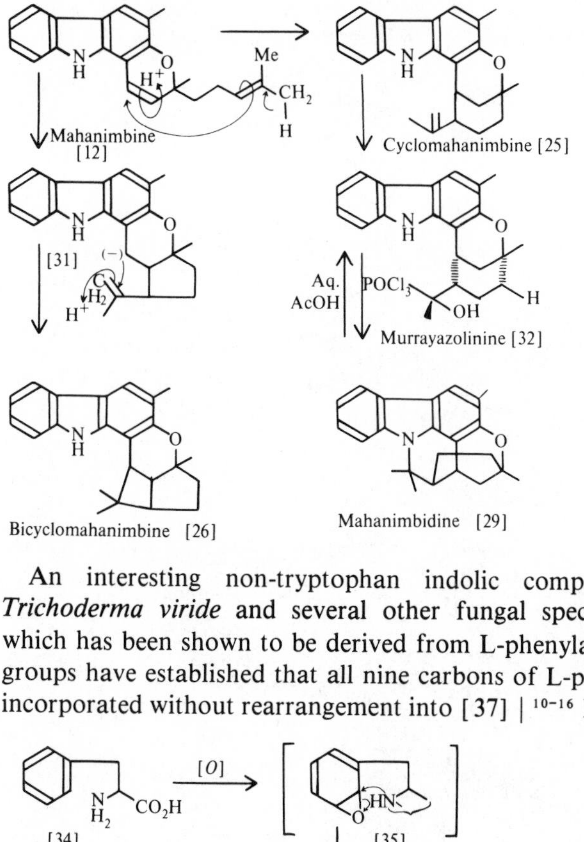

Mahanimbine [12]

Cyclomahanimbine [25]

[31]

Aq. AcOH | POCl$_3$

Murrayazolinine [32]

Bicyclomahanimbine [26]

Mahanimbidine [29]

SCHEME 6.5

An interesting non-tryptophan indolic compound isolated from *Trichoderma viride* and several other fungal species is gliotoxin [37], which has been shown to be derived from L-phenylalanine. Two different groups have established that all nine carbons of L-phenylalanine [34] are incorporated without rearrangement into [37][10-16] It has also been shown

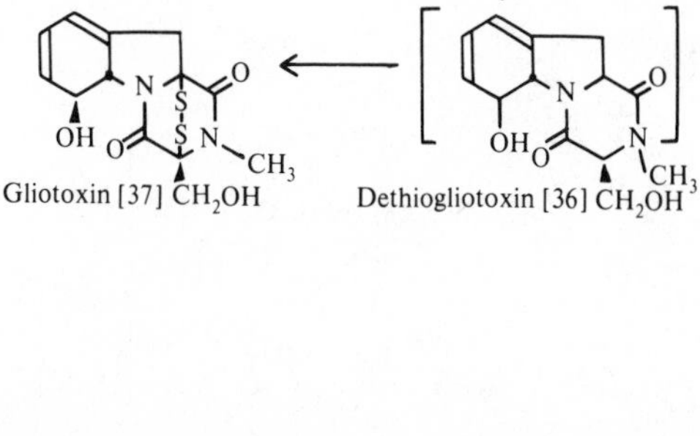

[34]

[0]

[35]

Gliotoxin [37] CH$_2$OH

Dethiogliotoxin [36] CH$_2$OH

SCHEME 6.6

that no loss of the five aromatic hydrogen atoms of phenylalanine accompanies its incorporation into gliotoxin. Taking these results into consideration, along with the *trans* disposition of the nitrogen and hydroxyl groups in rings A and B of gliotoxin, it was proposed that gliotoxin arises by an intramolecular cyclization of the arene epoxide [35] (Scheme 6.6).[14, 15] Subsequent work of Bu'Lock suggests that gliotoxin may be formed from the dipeptide cyclo-(phenylalanyl seryl) [38] in *T. viride*.[17] Moreover, it has also been shown the 3-deoxyanalogue of gliotoxin [42] is produced by the same organism when fed with ^{14}C-labelled cyclo-(L-alanyl-L-phenyl-alanyl) [40] (Scheme 6.7).[18]

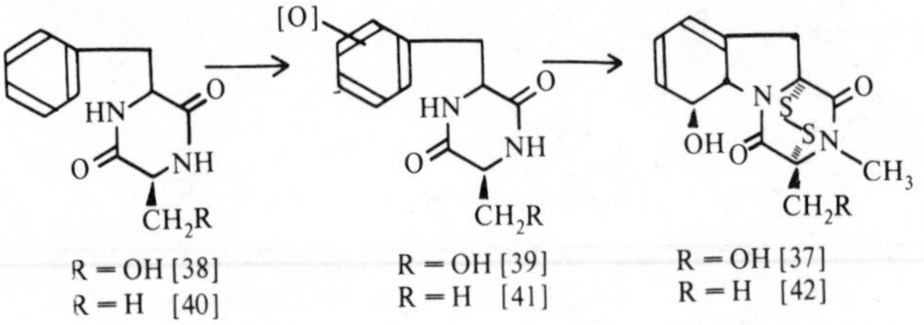

R = OH [38]   R = OH [39]   R = OH [37]
R = H [40]   R = H [41]   R = H [42]   SCHEME 6.7

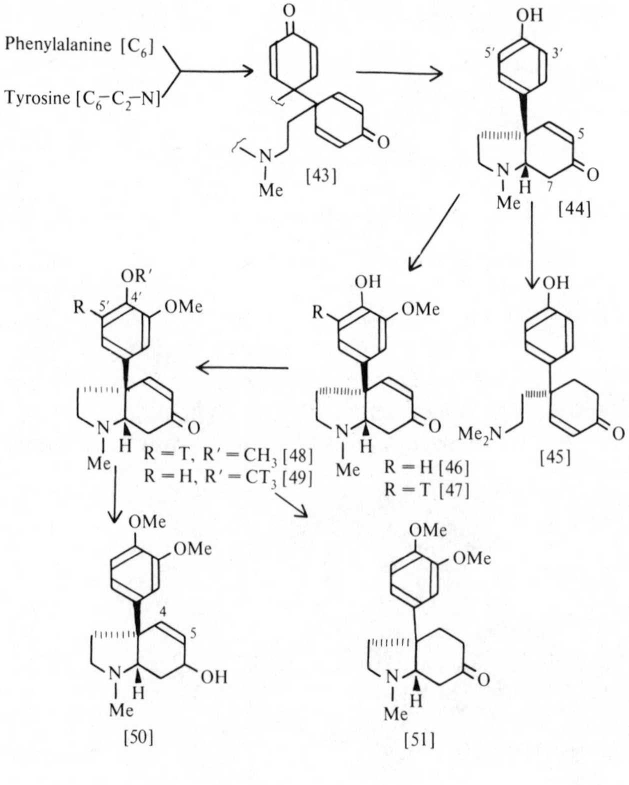

SCHEME 6.8

The mesembrine group of alkaloids arise from the amino acids tyrosine and phenylalanine[19], probably via the intermediacy of the dienone system [43] (Scheme 6.8) which can undergo an intermolecular Michael addition to afford the octahydroindole ring system. Radio-labelled tracer experiments with sceletenone [44], 4-0-demethylmesembrenone [47], and mesembrenones [48] and [49] in *Sceletium strictum* have shown that alkaloids such as mesembrine [51] and mesembrenol [50] arise by the pathway shown in Scheme 6.8.[19-23]

6.2 Non-isoprenoid tryptophan alkaloids

These alkaloids include simple derivatives of tryptophan, such as tryptamine and serotonin. The biosynthesis of tryptophan [66] has been intensively studied and it has been established that shikimic acid [57] plays a key role in the biosynthesis of the aromatic amino acids, phenylalanine,

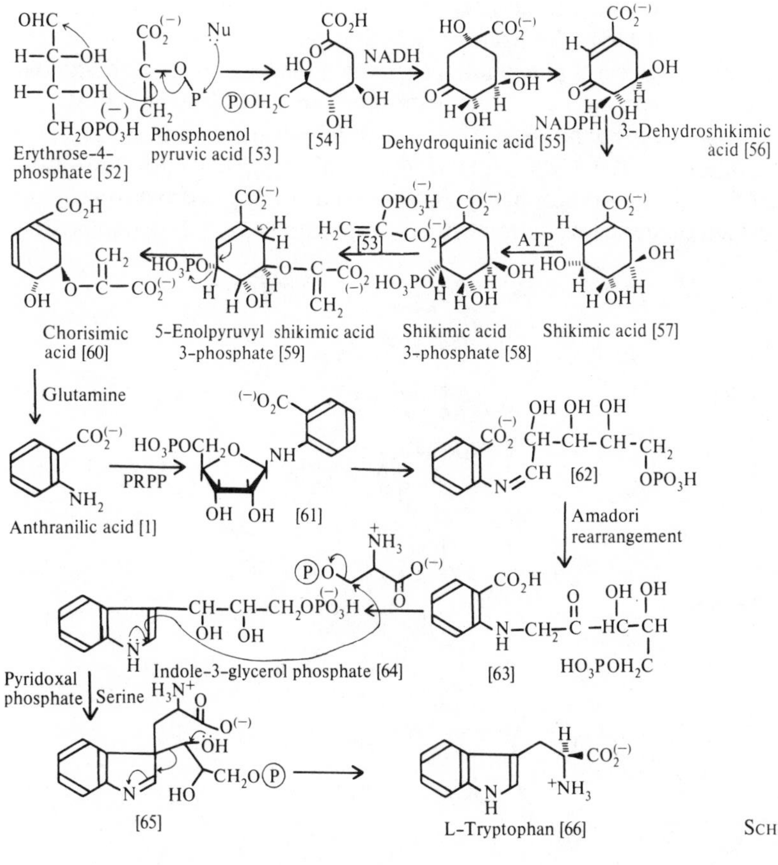

SCHEME 6.9

tyrosine, and tryptophan [66] .[24] Shikimic acid is formed by the initial combination of erythrose-4-phosphate [52] with phosphoenol pyruvic acid
[53] to afford 3-deoxy-D-arabinoheptulosonic acid-7-phosphate (DAHP)
[54], the seven-carbon precursor of shikimic acid [57]. This apparently
simple aldol-type condensation has been the subject of intensive
mechanistic investigations and several mechanisms have been proposed.[25-42]
Cyclization of this affords 3-dehydroquinic acid [55][43, 44] which on
dehydration,[45-48] oxidation, and phosphorylation affords shikimic acid
[57]. This is readily phosphorylated with ATP to afford shikimic acid-3-
phosphate [58]. The condensation of shikimic acid-3-phosphate [58] with
phosphoenolpyruvic acid [53] is catalysed by 5-enolpyruvylshikimate-3-
phosphate synthetase to afford 5-enolpyruvylshikimic acid-3-phosphate
[59][49-55] which loses a molecule of phosphoric acid to give chorismic acid
[60].[56-63] Reaction with glutamine and dehydration then leads to anthranilic acid [1].[64-68] Anthranilic acid reacts with ribose-5-phosphate to
afford N-(5'-phosphoribosyl)-anthranilic acid [61]. This ring-opens to the
Schiff base [62] which undergoes an Amadori-type rearrangement to afford the ketodiol [63]. Cyclization/dehydration and decarboxylation then
afford indole-3-glycerol phosphate [64] which reacts with serine (catalysed
by pyridoxal phosphate) to afford L-tryptophan [66] (Scheme 6.9).[69]

The simple tryptophan derivatives are widely distributed in nature. The
hydroxylation of the tryptamine ring usually occurs at the 5-position.
Hydroxylation of S-tryptophan [66] thus leads to 5-S-hydroxytryptophan
[67], which on decarboxylation affords serotonin [68].[70] Tryptamine does

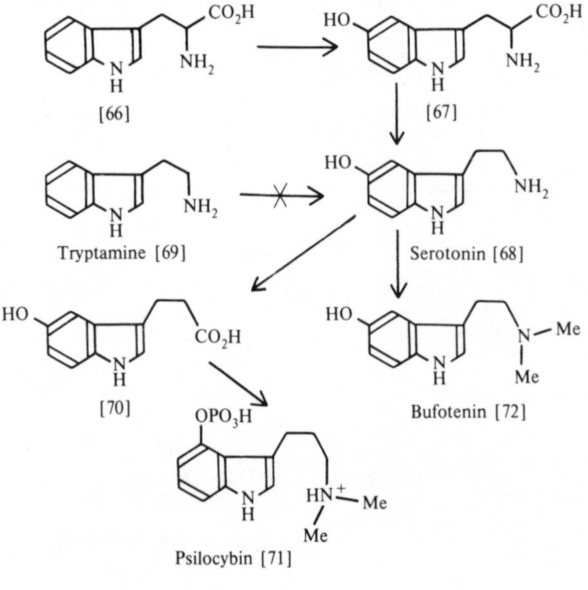

[66]   [67]

Tryptamine [69]   Serotonin [68]

[70]   Bufotenin [72]

Psilocybin [71]

SCHEME 6.10

not serve as a precursor to tryptophan, indicating that hydroxylation of tryptophan precedes its decarboxylation. Serotonin is found in bananas and stinging nettle. It has been shown that serotonin is converted to 5-hydroxyindole-3-acetic acid [70] in mammals (Scheme 6.10).[71] In *Psilocybe semperviva* psilocybin [71] is formed from tryptophan by hydroxylation at the 4-position,[72] followed by phosphorylation. It is not known whether 4-hydroxylation precedes or follows the decarboxylation of tryptophan. Bufotenin [72] probably arises by the direct methylation of tryptophan. Both psilocybin and bufotenin are hallucinogens.

The biosynthesis of gramine [79][73] has been studied by radio-labelling techniques and it has been established[74, 75] that in gramine the methylene attached to the indole β-position is retained. Conclusive evidence towards this was obtained by feeding [β-^{14}C]-tryptophan and [β-^{3}H]-tryptophan and demonstrating that the tritium–[14]C ratio was maintained during the experiment.[74] Evidence in support of such a pathway (Scheme 6.11) has been

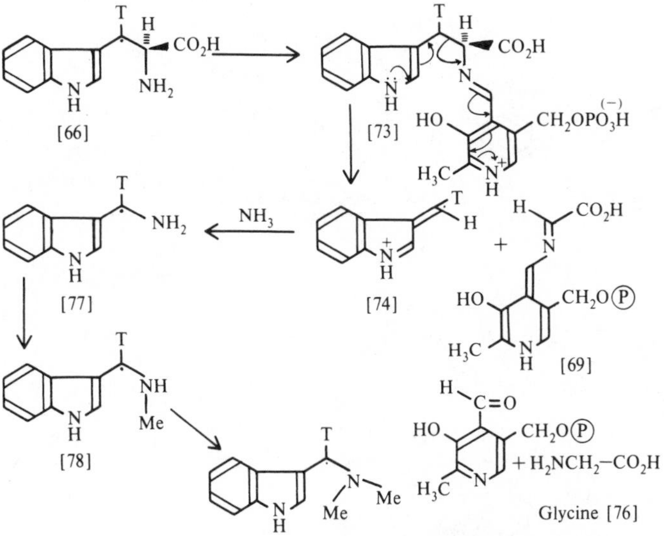

Gramine [79]   Pyridoxalphosphate [75]   SCHEME 6.11

provided by the isolation of [77] and [78] from barley and their conversion to gramine [79] on incubation with S-adenosylmethionine and a barley extract.[76] However, it has been shown that the terminal amino group of gramine is also tryptophan-derived, so that the later part of Scheme 6.11 is incorrect. An alternative pathway to gramine, which explains both the retention of β-carbon as well as the nitrogen of tryptophan, is illustrated in Scheme 6.12.

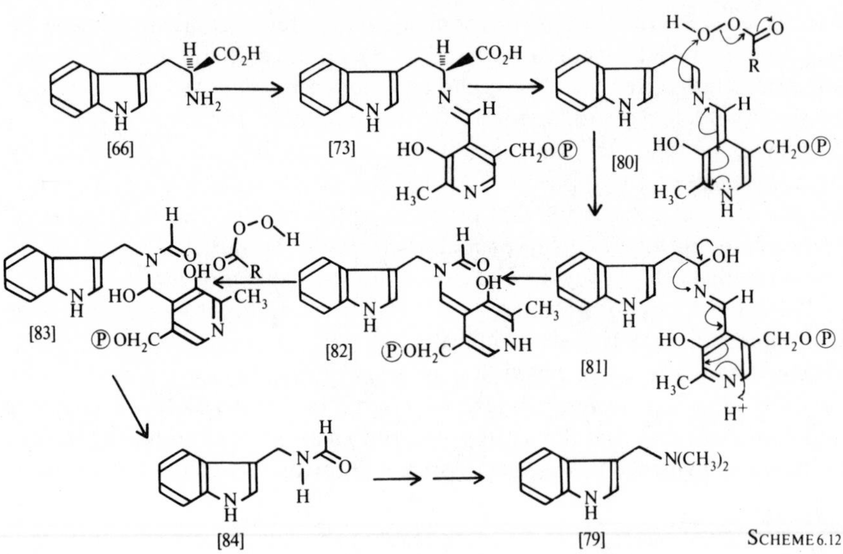

[66] [73] [80]

[83] [82] [81]

[84] [79] SCHEME 6.12

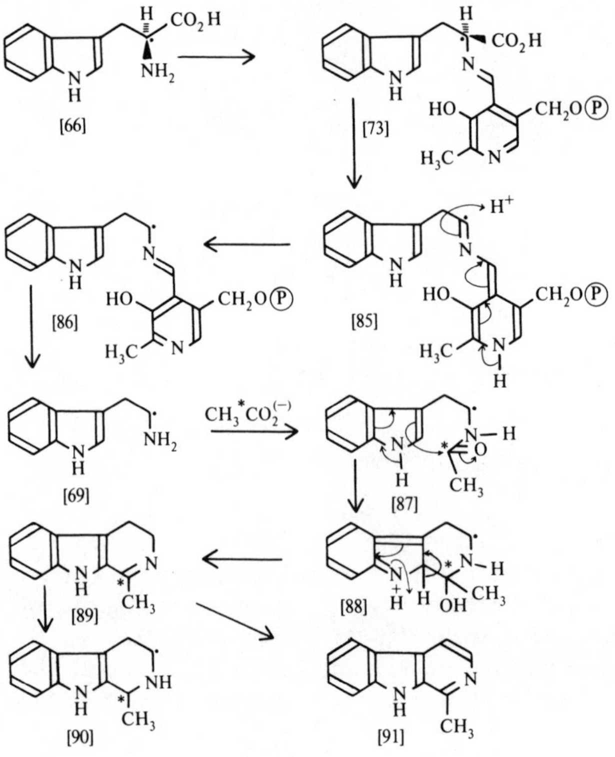

[66] [73]

[86] [85]

[69] [87]

[89] [88]

[90] [91] SCHEME 6.13

In a number of simple β-carboline alkaloids one, two, three, five, or more carbons have undergone condensation with the tryptophan moiety to afford appropriately substituted tri-, tetra-, or penta-cyclic indoles. The alkaloids of the harmaline group, e.g. eleagnine [90], serve as the simplest examples. As shown in Scheme 6.13, a pyridoxyl phosphate-catalysed decarboxylation of tryptophan [66] can afford tryptamine [69] which condenses with a C-2 providing species such as acetylcoenzyme A to ultimately afford eleagnine [90]. There are about two dozen alkaloids of this type, in which a two-carbon unit is seen to be condensed with tryptophan, at varying degrees of oxidation. The administration of labelled tryptophan and radioactive acetate to *Eleagnus augustifolia*[77] has established that the radioactivity was present at the corresponding carbon atoms in eleagnine [90] (Scheme 6.13).

The incorporation of $[\beta$-^{14}C-^{15}N]-tryptamine into harmine [91] in *Peganum harmala*, with retention of isotopic ratio, suggested that tryptophan undergoes decarboxylation before condensing with the C-2 unit. It has been found that sodium-$[1$-^{14}C]-acetate affords harmine containing majority of the label at C_1. The incorporation of $[2$- and 3-^{14}C]-pyruvate into harmine [91] has also been demonstrated, suggesting that it is a 3-carbon pyruvate unit that undergoes condensation with tryptamine and the species [92] then undergoes decarboxylation to afford the β-carboline [93][78] (Scheme 6.14). The issue is complicated to some extent by the known interconversions of acetate and pyruvate.[79] To clarify the pathway, labelled N-acetyltryptophan, N-acetyltryptamine, harmalan, and tetrahydroharman have been fed to *Passiflora edulis* and incorporation into harman observed. The results suggest that tryptophan → tryptamine → N-acetyl tryptamine → harmalan → harman route was operative (Scheme 6.13).[80] In eleagnine [90], however, acetyltryptamine [87] was not found to be incorporated.[81] This suggests that two different routes may be operative to β-carbolines, one to alkaloids such as harmalan [89] and harman [91] via N-acetyl tryptamine [87] (Scheme 6.13) and the other to tetrahydro-β-carboline [94] via pyruvate (Scheme 6.14).

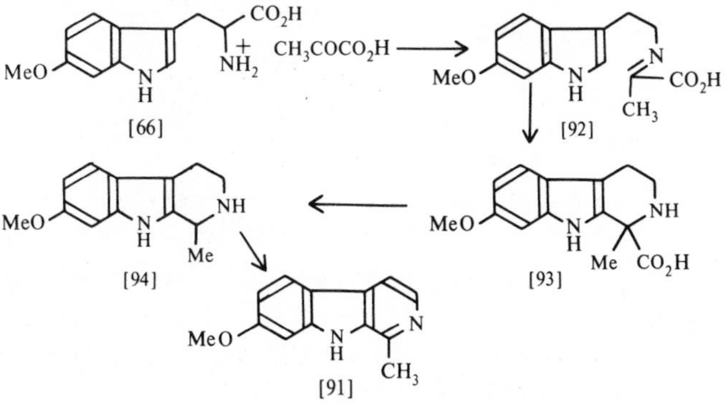

SCHEME 6.14

It has been shown recently that feeding of labelled tryptophan to *P. edulis* results in its incorporation into 1-methyl-1,2,3,4-tetrahydro-β-carboline-1-carboxylic acid, which in turn acts as an efficient precursor for harman.[226]

Several β-carboline alkaloids have been isolated in which the tryptophan carboxyl group is intact. These include cordifoline, adifoline, and harman-3-carboxylic acid. It therefore appears that while in most cases decarboxylation of tryptophan precedes condensation with an aliphatic moiety, in certain plant systems this need not happen and it is, therefore, possible to isolate condensation products with tryptophan. Recently the first tryptophan-derived indolo pyridoquinazoline alkaloid was isolated from the fruits of *Evodia rutaecarpa*.[82]

Some other closely-related groups such as the perlolyrine group, brevicolline group, etc. comprise a substituted harman system. Perlolyrine [96] can arise by the condensation of a furan aldehyde [95] with tryptamine [69] (Scheme 6.15). Similar rationalizations can be made for other

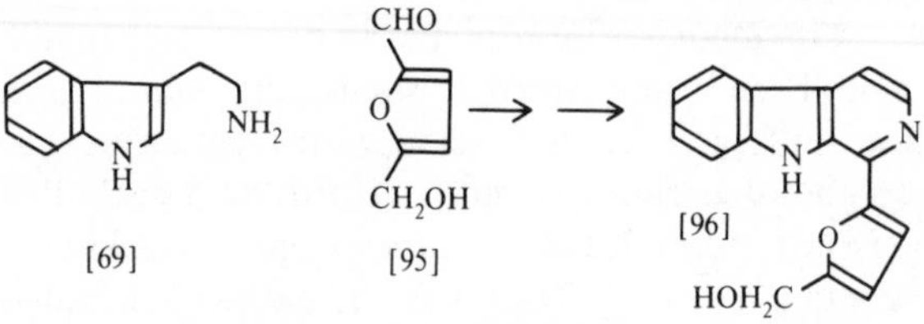

SCHEME 6.15

groups in this sub-class, but experimental evidence is lacking. It has been suggested that ornithine [99] may be involved in the biosynthesis of elaeocarpidine [98].[83] The antibiotic indolmycin [100] has been shown to

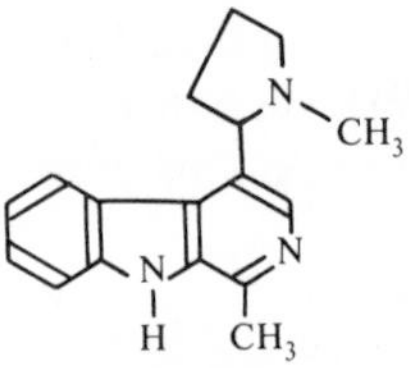

Brevicolline [97]

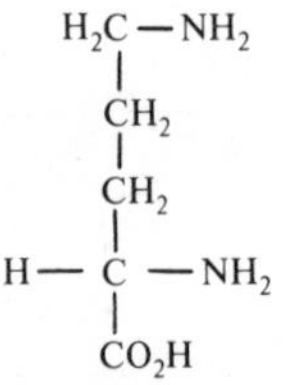

Ornithine [99]

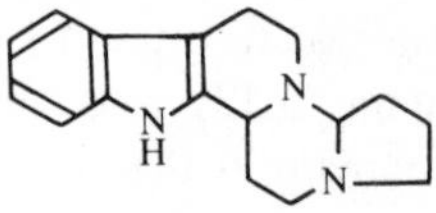

Elaeocarpidine [98]

be derived from indolmycenic acid [101]. It has been established[84] that tryptophan [66] supplies the C-3 side chain, methionine [102] supplies the two methyl groups while the guanidino group of arginine [103] supplies the carbon and nitrogens of the oxazoline ring (Scheme 6.16).

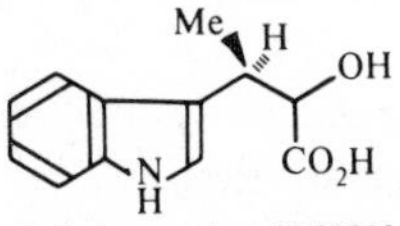

Indolmycenic acid [101]

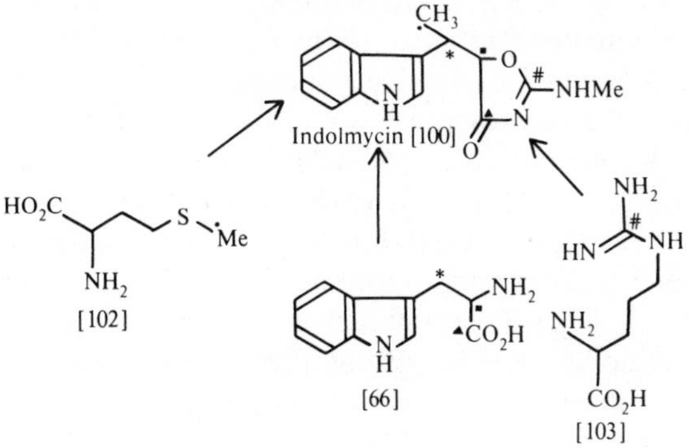

Indolmycin [100]

SCHEME 6.16

[102] [66] [103]

The closely-related physostigmine group of alkaloids can be considered to arise by the attack of the tryptamine nitrogen on the 3-methylated compound (Scheme 6.17). Physostigmine [106] is converted to geneserine [107] on treatment with hydrogen peroxide and is reduced back to physostigmine [106] on reduction with zinc-acid.[85]

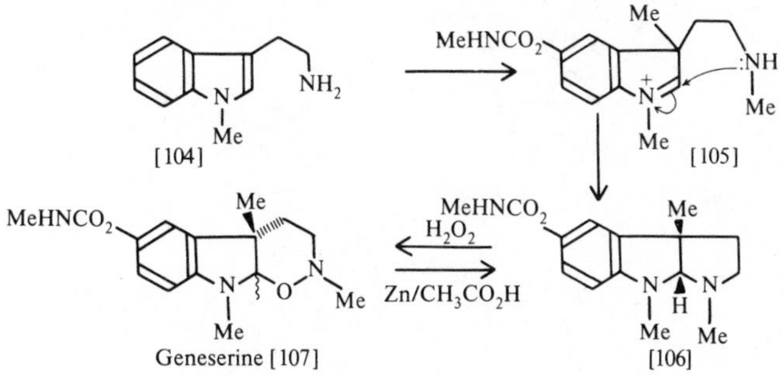

[104] [105]

Geneserine [107] [106]

SCHEME 6.17

6.3 Fungal indole alkaloids

The ergot alkaloids belong to a biogenetically distinct class of indole alkaloids. The toxic effects of the ergot alkaloids have been responsible for mass poisoning in man and animals, while the psychomimetic activity associated with lysergic acid diethylamide (LSD) has led to the development of a new field of knowledge, psychopharmacology. Several reviews on the biosynthesis of ergot alkaloids have been written.[87-95]

The participation of tryptophan [66] in the biosynthesis of ergot alkaloids was established by labelling experiments using (+)-[2-^{14}C]-tryptophan in rye plants.[86] The ergot alkaloids isolated were shown to have radioactivity in the lysergic acid portion. Saprophytic cultures of an ergot strain were found to assimilate tryptophan to the extent of 10–39 per cent. When tryptophan-^{14}CO$_2$H was fed, no incorporation of radioactivity was observed, thus indicating that the carboxyl group of tryptophan was not incorporated.[96] Extensive assimilation of (±)-[2-^{14}C]-tryptophan was observed in a *Claviceps* strain.[97] L-Tryptophan was found to be incorporated into elymoclavine with loss of carboxyl group.[98] Since labelled (+)-tryptophan also afforded positive incorporations, it proved that the (+)-form was also utilized.[99] [2-^{14}C]-Tryptamine was not found to be incorporated into the ergot alkaloids, thus suggesting that the decarboxylation of tryptophan occurs at a stage after it has been incorporated into the ergot alkaloids. [2-^{14}C]-Indole also afforded positive incorporation. Harley-Mason[100] and van Tamelen[101] independently proposed 5-hydroxytryptophan as an intermediate in the biosynthesis of ergot alkaloids. However, both 5- and 6-deuterated tryptophans were found to be incorporated without loss of deutereum. When deutereum was located at

Lysergic acid [108]   Festuclavine [109]   Chanoclavine I [110]

Pyroclavine [111]   Elymoclavine [112]   Isochanoclavine I [113]

Penniclavine [114]   Agroclavine [115]   Chanoclavine II [116]

the 4-position, it was found to be lost. These experiments[102] disproved the earlier hypotheses. $(-)$-$[^{14}CH_3]$-Methionine was found to be assimilated to the extent of 1.35 per cent, indicating that the N-methyl group of ergot alkaloids was derived from methionine by transmethylation.[103]

Various hypotheses have been advanced to explain the origin of the non-tryptophan portion of the ergot alkaloids.[86, 100-102, 104-107] Mothes was the first to postulate[86] the participation of a five-carbon isoprenoidal compound in the ergoline structure. $[2-^{14}C]$-mevalonic acid and 2-or 4-tritiated mevalonic acids afforded positive incorporations into the ergot alkaloids.[108-110] Acetate was also shown to be incorporated into ergot alkaloids.[110-111] Degradation experiments established the positions of incorporation of the label in the ergot alkaloids, and showed that C-2 of mevalonate becomes C-7 of the ergolines.[110, 112, 113] These incorporation results with labelled tryptophan, mevalonate, and methionine are shown in Scheme 6.18. The incorporation of $[2-^{14}C]$-mevalonic acid was lowered by

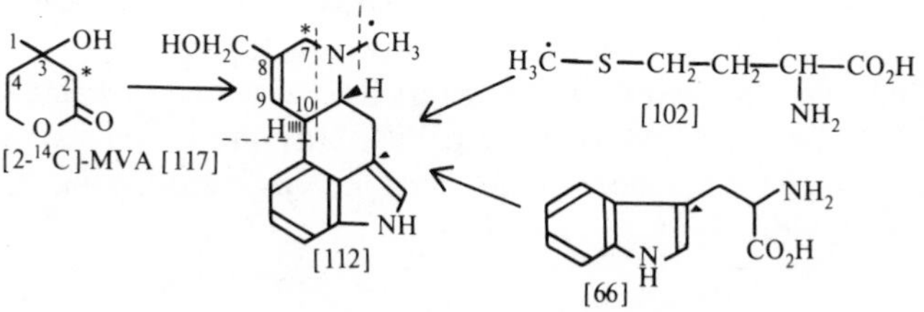

SCHEME 6.18

addition of isopentenyldiphosphate or $\gamma\gamma$-dimethylallyldiphosphate, suggesting that mevalonic acid was incorporated via the intermediacy of these 5-carbon isoprenoids[1] $[1-^3H]$-dimethylallylalcohol was not found to be incorporated.[115] It was shown by feeding labelled $\gamma\gamma$-dimethylallyltryptophan that the 4-E-methyl group of this substance appears at the C-methyl group of chanoclavine and at C-17 of agroclavine, elymoclavine, and lysergic acid.[205]

It therefore appears certain that the initial step in the biosynthesis of ergot alkaloids is an electrophilic substitution reaction of tryptophan with $\gamma\gamma$-dimethylallyldiphosphate [118] to afford the 4-substituted tryptophan derivative [119].[116] Added support to this is provided by the isolation of 4-$\gamma\gamma$-dimethylallyltryptophan [119] under conditions in which the normal biogenetic pathway to the ergot alkaloids is inhibited.[117-119] An enzyme catalysing the formation of dimethylallyl-L-tryptophan from L-tryptophan and $\gamma\gamma$-dimethylallylpyrophosphate has been isolated from *Claviceps*.[120] The incorporation of [119] into ergot alkaloids has been demonstrated.[114]

The mechanism of prenylation of tryptophan remains to be properly clarified, in spite of efforts of several groups.[116, 120, 121] It appears that the enzyme controlling the reaction is responsible for the site specificity, holding together the two substrates in such a geometrical manner that reaction at 4-position of the indole nucleus occurs preferentially. Furthermore, it has been shown that while N-methyl-4-($\gamma\gamma$-dimethylallyl)-tryptophan is incorporated into elymoclavine, the corresponding tryptamine derivative is not incorporated, suggesting that N-methylation of the basic nitrogen is the second pathway-specific step in ergot alkaloid biosynthesis[122] and that methylation of amino nitrogen precedes C-ring closure.

An examination of the structure of agroclavine [115], a typical ergot alkaloid, shows that the following main processes must occur for the conversion of the 4-$\gamma\gamma$-dimethylallyl tryptophan [119] to the ergot series: (a) loss of the tryptophan carboxyl group; (b) cyclization to afford a six-membered ring C; and (c) a second cyclization with one of the two methyl groups (at a suitably oxidized level) with the basic nitrogen to afford a six-membered piperidine ring.

One of the first questions to be answered was which of the two cyclizations occurs first. Under conditions of inhibition of biosynthesis an interesting metabolite, claviciptic acid [124], was isolated indicating that the cyclization of the basic nitrogen with one of the two methyl groups may occur first. This, however, proved to be incorrect as it was demonstrated that claviciptic acid [124] is not a precursor.[117] It was shown by using labelled alkaloids that the sequence of conversion was agroclavine → elymoclavine → penniclavine.[125, 126]

Claviciptic acid [124]

The occurrence of chanoclavine-I [110], chanoclavine-II [116], and isochanoclavine I [113] indicated that ring C is formed prior to the second cyclization to afford ring D. However, deoxychanoclavine I [122] and deoxy-nor-chanoclavine I [123] were shown not to be precursors.[123, 124] An explanation for this could be that one of the two methyl groups of the dimethylallyl group was hydroxylated *before* the formation of ring C, i.e. the alcohol [121] is formed first which is then cyclized to afford chanoclavine I [110], isochanoclavine I [113], and chanoclavine II [116] (Scheme 6.19). Ring D can then be readily considered to be formed from isochanoclavine I [113] by cyclization of the basic nitrogen with the sterically correctly -disposed hydroxymethyl function. Support for such a pathway has been provided by the demonstration that the label at C-2 of mevalonate ended up in all the isomeric chanoclavines in the C-methyl group.[125, 126]

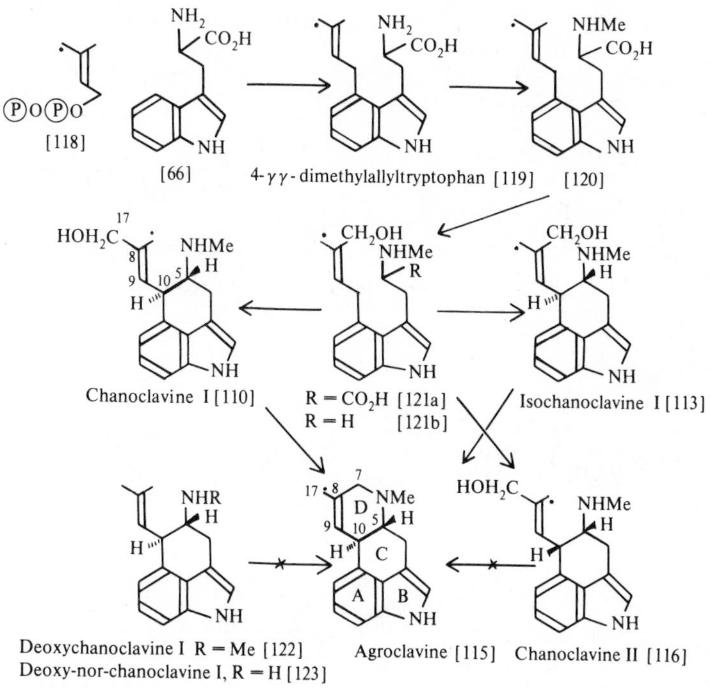

[118]

[66]   4-γγ-dimethylallyltryptophan [119]   [120]

Chanoclavine I [110]   R = CO₂H [121a]   R = H [121b]   Isochanoclavine I [113]

Deoxychanoclavine I R = Me [122]   Agroclavine [115]   Chanoclavine II [116]
Deoxy-nor-chanoclavine I, R = H [123]

SCHEME 6.19

That the formation of ring D was much more complicated than a simple cyclization in isochanoclavine was demonstrated by three groups of workers independently when it was shown that isochanoclavine I [113] and chanoclavine II [116] were *not* incorporated into agroclavine [115].[126, 129] On the other hand, chanoclavine I [110] in which the hydroxymethyl group was in the wrong (*trans*) steric disposition (for cyclization) was surprisingly found to afford positive incorporation into agroclavine [115]. Thus cyclization of chanoclavine I [110] to afford ring D of agroclavine [115] must involve a *cis–trans* isomerization at the allylic double bond.

A further element of intricacy was introduced when it was shown that tritium from 4R-tritiated mevalonate (but not from 4S-tritiated mevalonate) is incorporated into chanocalvine I [110] and elymoclavine [112]. It had previously been shown[130] with mevalonic acid [125] tritiated at C-4 that in the isomerization of isopentenyldiphosphate [128] (the initial decarboxylative elimination reaction product from mevalonic acid diphosphate [127]) to dimethylallyldiphosphate [118] it is the *S*-tritium which is lost (Scheme 6.20). The *S*-hydrogen is also lost in the condensation of isoprene units with the formation of *trans*-double bonds. In the biosynthesis of rubber, however, the *R*-hydrogen is lost, leading to *cis*-double bonds.[130] This indicated that label at C-2 of mevalonic acid ends up in the

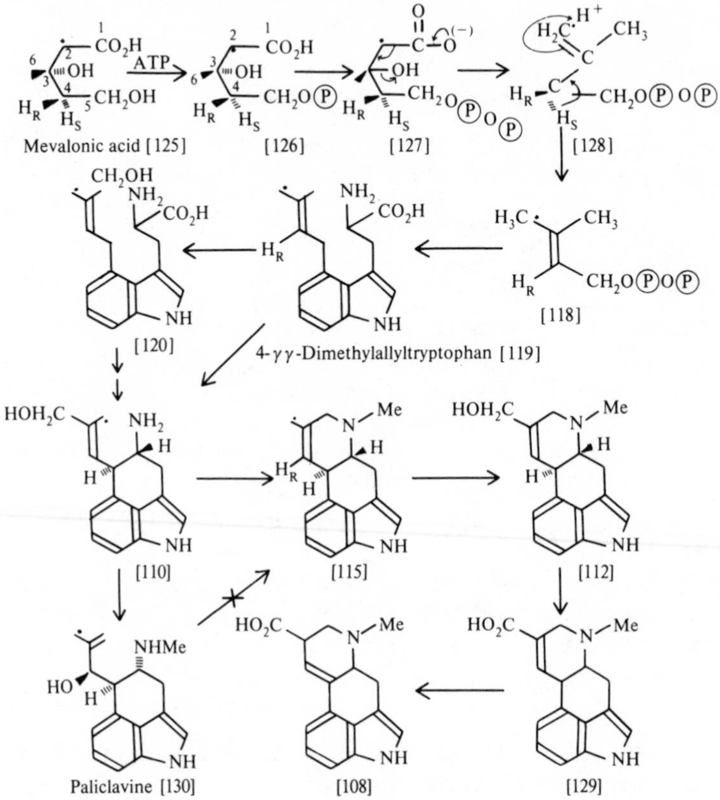

Mevalonic acid [125]   [126]   [127]   [128]

4-γγ-Dimethylallyltryptophan [119]   [118]

[120]

[110]   [115]   [112]

Paliclavine [130]   [108]   [129]

SCHEME 6.20

trans-methyl group[129, 131] in 4-γγ-dimethylallyltryptophan [119]. However, in chanoclavine I [110], the precursor to the tetracyclic ergot alkaloids, the label was found to be located in the *cis*-methyl group.[125, 126, 132] On cyclization to agroclavine [115], this labelled carbon then becomes *trans* oriented. Clearly the conversion of [120] to chanoclavine [110] must involve a *cis–trans* isomerization at the allylic double bond. Summarizing the sequence of events, 4-γγ-dimethylallyltryptophan [119] first undergoes hydroxylation at the *cis*-methyl group to afford [120]. This then undergoes cyclization and isomerization so that the hydroxymethyl group is found in a *trans* disposition, while the label is located in the *cis*-methyl group in chanoclavine I. A second isomerization at the double bond occurs in the cyclization of chanoclavine I [110], to agroclavine [115] as the label is now found to be located in the *trans*-methyl group while the hydroxy methyl group has cyclized after isomerization to the *cis*-disposition (Scheme 6.20). The formation of the tetracyclic ergot alkaloids thus involves two *cis–trans* isomerizations at the allylic double bond. Interestingly, an *in vitro* enzymatic oxidative ring closure of ring D of the ergolene nucleus of chanoclavine I [110] to elymoclavine [112] without the intermediacy of agroclavine [115] has been demonstrated.[133]

It has been shown that paliclavine-N-[14]CH$_3$ [130] is not utilized significantly by fungus strains, suggesting that it is not a precursor of ergolene carboxylic acids or tri- and tetracyclic clavines.[134]

In order to acquire a deeper understanding of the mechanisms of these conversions and isomerization, incorporations with double-labelled molecules have been carried out. Thus with a double-labelled compound it was shown that chanoclavine I [110] cyclized with complete retention of the C-10 proton.[129-135] However, when the 9-tritiated chanoclavine was used, 30–70 per cent of the tritium label was found to be lost. These results suggest that the C-9 proton is transferred intramolecularly. Further, when 17-tritiated chanoclavine I [110] was administered to *Claviceps*, the elymoclavine [112] isolated showed a 53 per cent retention of tritium. This suggested that one of the methylene hydrogens at C-17 is lost in the conversion of chanoclavine I to elymoclavine.[136, 137]

Double-labelling experiments by Floss and co-workers indicate that the cyclization of chanoclavine I to the tetracyclic ergolines involves an intermolecular transfer of the hydrogen at C-9 into the same position of a new molecule.[138] During this transfer process, one of the methylene hydrogens at C-17 of the precursor is also exchanged, the newly introduced hydrogen bearing the pro-R position at C-7. This and the efficient specific conversion of chanoclavine I aldehyde [131] into elymoclavine by the ergot fungus support the view that cyclization occurs through aldehyde intermediates. A

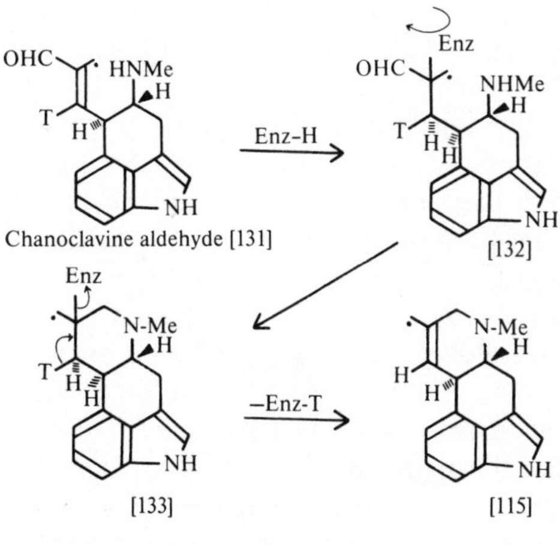

Chanoclavine aldehyde [131] [132] [133] [115]

SCHEME 6.21

mechanism suggested for the inversion of chanoclavine I aldehyde [131] is indicated in Scheme 6.21. Floss and co-workers have proposed a more detailed mechanism to account for all these observations (Scheme 6.22)

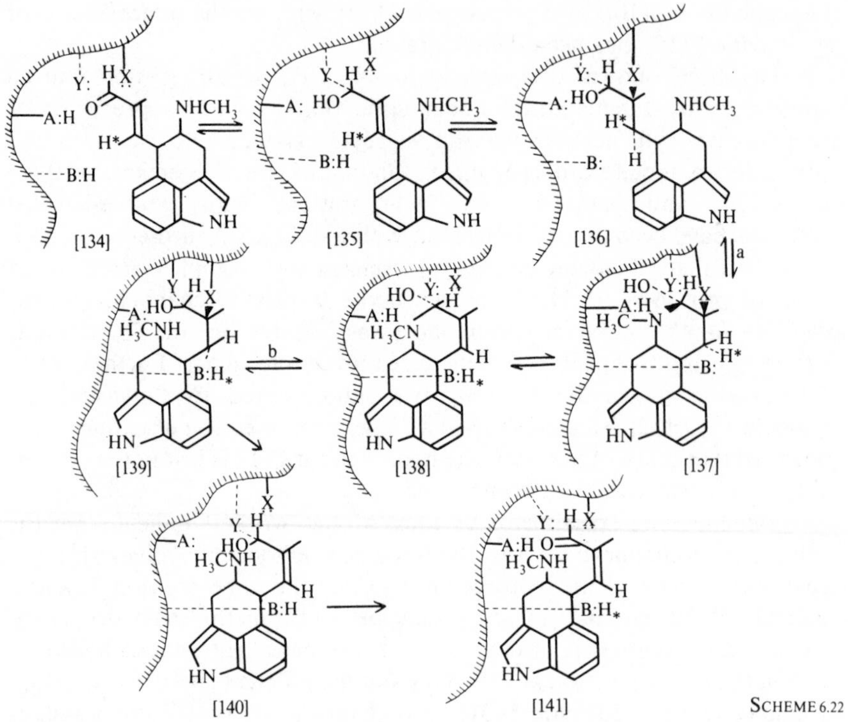

SCHEME 6.22

and it appears to overcome the deficiencies of the earlier mechanisms.[132, 133, 139–142]

The above discussion throws some light on the second *cis–trans* isomerization step of the olefin between chanoclavine I [110] and agroclavine [115]. As mentioned previously there is an earlier isomerization between 4-$\gamma\gamma$-dimethylallyltryptophan [119] and chanoclavine [110], the label in the *trans*-methyl group in [119] appearing in the *cis*-methyl group in chanoclavine I [110]. It has been suggested that this isomerization occurs during the formation of ring C.[129] One mechanism offered for this isomerization suggests that C-10 of the intermediate [134] is activated for ring C formation by hydroxylation, followed by an allylic rearrangement.[143] Ring closure then occurs by a concerted process by the attack of the electron pair of the C_5=N double bond (which may be formed through the participation of pyridoxylphosphate as suggested by Mothes and Weygand groups[96]) on C-10 accompanied by the elimination of the —OR group at C-8. Since the reaction involved is an S_N2' reaction, the stereochemistry of the initial hydroxylation would control the stereochemistry of the product of allylic rearrangement, and thereby determine the stereochemistry of attack at C-10 (Scheme 6.23). Thus if the initial hydroxylation afforded the

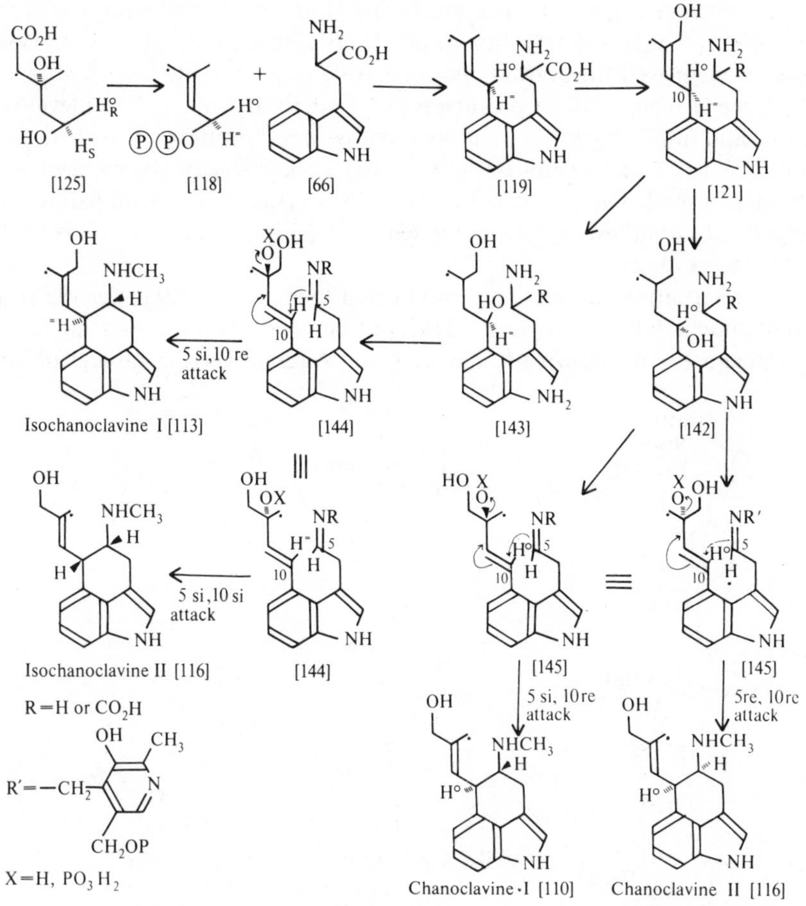

[125] [118] [66] [119] [121]

Isochanoclavine I [113] [144] [143] [142]

5 si,10 re attack

Isochanoclavine II [116] [144]

5 si,10 si attack

R = H or CO₂H

R' = —CH₂—

X = H, PO₃H₂

[145] [145]

5 si, 10re attack 5re, 10re attack

Chanoclavine ·I [110] Chanoclavine II [116]

SCHEME 6.23

10-R hydroxy derivative, allylic rearrangement would afford the 8-R hydroxylated product. On this only a 5 si attack is possible[144] to afford either isochanoclavine I [113] or chanoclavine II [116], attack at C-10 occurring in re and si manners respectively. In the former case, the geometry of the double bond remains unchanged while in the latter case cis–trans isomerization occurs at the allylic double bond. The formation of chanoclavine I and chanoclavine II can be explained on a similar basis.[145] 17-³H-N-Demethylchanoclavine I and N-demethylchanoclavine II were not significantly incorporated into elymoclavine by an ergot fungus, suggesting that these demethylchanoclavines are not precursors of chanoclavine but probably arise in the fungus by biological demethylation.[146]

Experiments using (3R, 5R) and (3R,5S) mevalonate-5-T have established that in a *Claviceps* strain, elymoclavine, agroclavine, chanoclavine I, isochanoclavine I, and (−)-chanoclavine II are all formed with loss of the pro-5R hydrogen and retention of pro-5S hydrogen of mevalonate.[147] These results disproved the above hypothesis which involved the loss of different hydrogens from C-10 of precursor [121] (derived from C-5 of mevalonic acid) and implied a correlation between the steric position of the hydrogen lost from C-5 of mevalonate and the stereochemistry of the chanoclavine isomer formed. These results, however, do not rule out the main pathway of ergoline biosynthesis involving the sequence [66] → [119] → [121] → [110] → [115] → [112].

An alternative explanation could be that the alcohol [121], derived from 4-γγ-dimethylallyltryptophan [119], could isomerize via two allylic rearrangements as shown in Scheme 6.24. Alternatively, a radical or car-

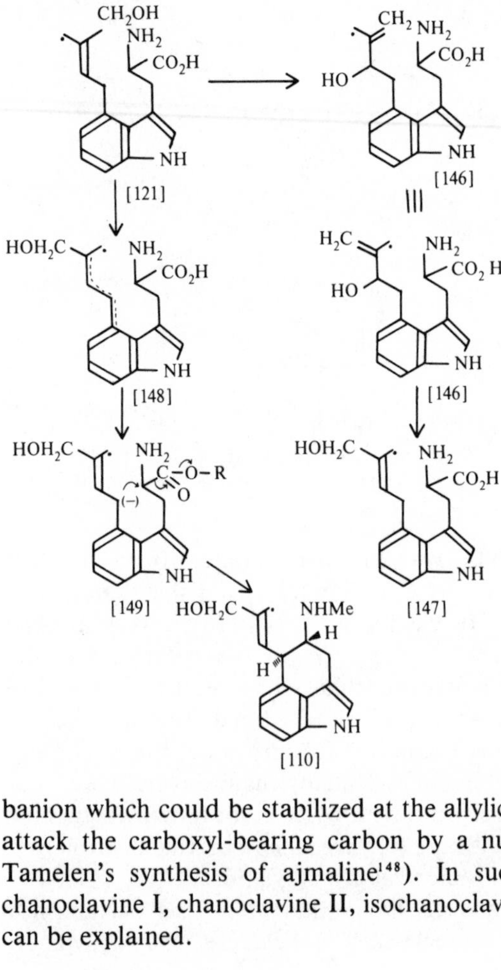

SCHEME 6.24

banion which could be stabilized at the allylic/benzylic carbon atom could attack the carboxyl-bearing carbon by a nucleophilic process. (c.f. van Tamelen's synthesis of ajmaline[148]). In such a case the formation of chanoclavine I, chanoclavine II, isochanoclavine I, and isochanoclavine II, can be explained.

It has been suggested that the isoprenylation of tryptophan could proceed with inversion of configuration at the allylic carbon atom, in which case the observed stereochemistry would be in accordance with the above sequence.[146, 130] The formation of chanoclavine II and isochanoclavine I would then have to be explained by an alternative process. It has been observed that the formation of chanoclavine I, chanoclavine II, and isochanoclavine I proceeds with complete retention of the pro-5R hydrogen of mevalonate.[146] The complete mechanism for these reactions still remains to be clarified.

The conversion of agroclavine [115] to lysergic acid [108] and other ergot alkaloids involves an initial hydroxylation of the C-methyl group to afford elymoclavine [112]. Elymoclavine is then converted to lysergic acid [108], α-hydroxyethylamide of lysergic acid [150], lysergylalanine [151], and ergotamine [152] as shown in Scheme 6.25.[124] It has been shown that

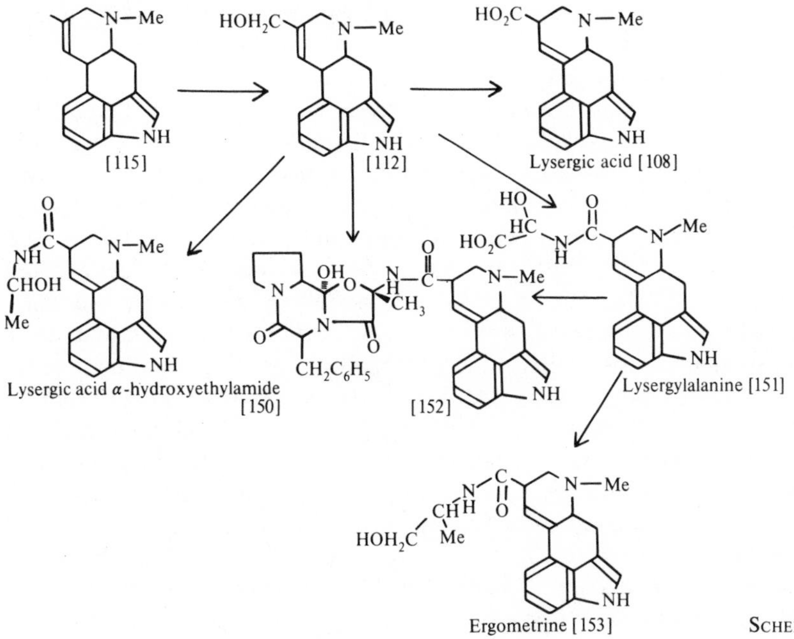

Lysergic acid α-hydroxyethylamide [150]

[152]

Lysergylalanine [151]

Ergometrine [153]   SCHEME 6.25

alanine is a precursor of the alaninol side chain of ergometrine [153], whereas alaninol and α-methyl serine are not incorporated.[149] Lysergic acid [108] and alanine can combine to afford lysergylalanine which can, by reduction, lead to ergometrine [153]. The conversion of chanoclavine I into elymoclavine in a cell-free system derived from *Claviceps* has been demonstrated in good yields.[150] Agroclavine [115] was not detectable as an intermediate in the system examined. Moreover, agroclavine was not found to act as an efficient precursor for elymoclavine [112]. The cell-free system studied required free oxygen and NADPH for the conversion of chanoclavine I to elymoclavine, suggesting the involvement of a mono-oxygenase in ergot alkaloid biosynthesis.[151] An alternative explanation could be that chanoclavine I [110] is converted by a dehydrogenase enzyme system to chanoclavine I aldehyde [131], which then undergoes a reductive condensation with the basic nitrogen. The observed loss of proton at C-9 during the *in vitro* transformation of chanoclavine I to elymoclavine is explained by the mechanism of isomerization of chanoclavine I aldehyde suggested in Scheme 6.21.[129]

There is now overwhelming evidence[126-128, 152, 153] that the production of lysergic acid derivatives in organisms proceeds via the intermediacy of the less-oxidized clavine alkaloids. Chanoclavine I [110] is one of the earlier products formed by the condensation of tryptophan with isoprenoid precursors. However, it is a minor metabolite. Larger amounts of agroclavine and elymoclavine are excreted by several strains. These are usually accompanied by smaller amounts of penniclavine, setoclavine, and other related alkaloids not lying on the main biosynthetic pathway to lysergic acid. It has been shown by radiotracer studies that many of the 'clavine' alkaloids are derived from elymoclavine and agroclavine.[124] Many organisms and plant tissues have been found to convert $\Delta^{8, 9}$-ergoline bases (e.g. elymoclavine) to the corresponding 8-hydroxylated compounds, e.g. penniclavine and isopenniclavine.[157][154-158] It has been shown by Taylor and co-workers[158, 159] that the reaction is catalysed by peroxidase and there appears to be a correlation between the production of peroxidase and the ability of an organism to effect these hydroxylations. The initial hydroxylation appears to occur at C-10 and the 8-hydroxy compounds then result by an allylic rearrangment.[159, 160] Thus Lin and co-workers have isolated 10-hydroxyelymoclavine by the oxidation of elymoclavine in the presence of horse-radish peroxidase and found that this substance readily rearranged under slightly acid conditions, to a mixture of penniclavine and isopenniclavine [157].[160] Therefore it probably represents one of the initial products of the peroxidase reaction (Scheme 6.26). The same reaction also results in the formation of a variety of oxidized clavine alkaloids formed by transformation of agroclavine and elymoclavine. The corresponding epoxides have also been isolated (Scheme 6.27).

It has been found that hydroxylation of agroclavine [115] to elymoclavine [112] occurs in cultures and mycelial homogenates of those *Claviceps* strains which produce large quantities of elymoclavine.[124, 161] This reaction has also been reported by Abe and co-workers to be effected by cultures of *Aspergillus* and *Penicillium*.[162] In general, it may be said that only those strains producing lysergic-acid derivatives and end products have

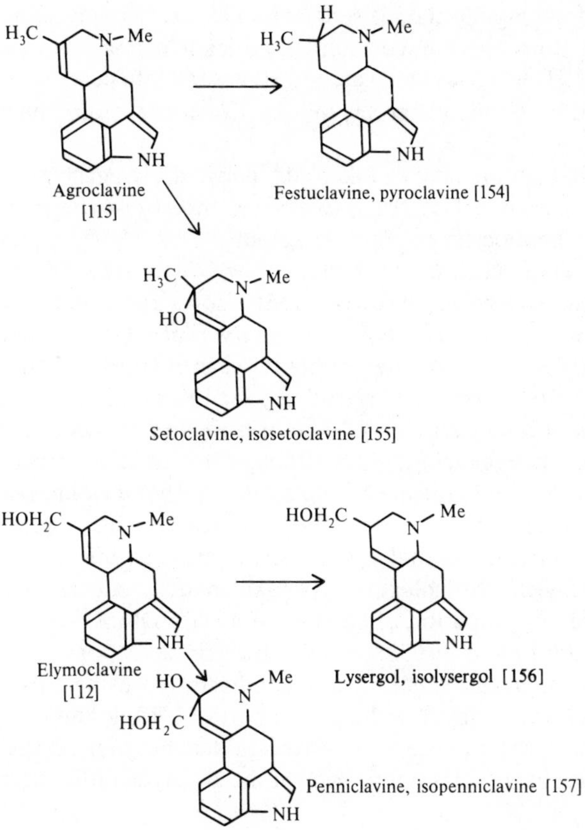
H$_3$C N-Me
NH
Agroclavine
[115]
H
H$_3$C N-Me
NH
Festuclavine, pyroclavine [154]
H$_3$C N-Me
HO
NH
Setoclavine, isosetoclavine [155]
HOH$_2$C N-Me
NH
Elymoclavine
[112]
HOH$_2$C N-Me
NH
Lysergol, isolysergol [156]
HO N-Me
HOH$_2$C
NH
Penniclavine, isopenniclavine [157]

SCHEME 6.26

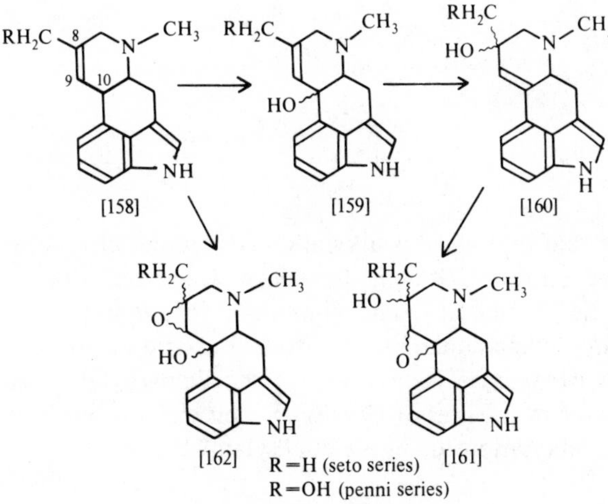
RH$_2$C 8 N CH$_3$
9 10
NH
[158]
RH$_2$C N CH$_3$
HO
NH
[159]
RH$_2$C
HO N CH$_3$
N H
[160]
RH$_2$C N CH$_3$
O
HO
NH
[162]
RH$_2$C
HO N CH$_3$
O
NH
[161]
R=H (seto series)
R=OH (penni series)

SCHEME 6.27

the capability of transforming the clavine series of alkaloids to the lysergic-acid series. Voigt and Bornschein have found significant increases in the yield of ergotamine [152], when ergometrine [153] was fed into cultures.[163] Reduction of $\Delta^{8, 9}$-double bond of agroclavine by *Claviceps* strains have also been observed.[124]

While it is known that simple clavine alkaloids, particularly agroclavine and elymoclavine, are precursors of the lysergic-acid moiety of the more complex amide and peptide-type ergot alkaloids,[164, 153, 165, 128, 166] the sequence of steps from elymoclavine [112] to the lysergic acid stage still remains to be elaborated. Lysergene, lysergol [156], isolysergol, and penniclavine [157] are apparently not precursors of lysergic-acid derivatives.[153, 165] Therefore, shift of the double bond into the 9, 10 positions cannot be the first step. 6-Methyl-8-ergolene-8-carboxylic acid ($\Delta^{8, 9}$-lysergic acid), a natural constituent of certain ergot strains was found to be incorporated into lysergic acid amides although not so efficiently as lysergic acid.[167] While this could indicate biological double-bond isomerization at the lysergic acid stage, the fact that the same reaction also occurs spontaneously at a measurable rate makes the interpretation ambiguous.[168]

Methyl-8-acetoxymethylene-9-ergolene   (lysergaldehyde   enolacetate) [163] has been prepared by oxidation of elymoclavine with DMSO–Ac$_2$O. The biological incorporation of [163] into the lysergic acid portion of ergotoxine by a *Claviceps purpurea* strain was demonstrated using [163] obtained from elymoclavine tritiated in the indole portion. While these experiments do not establish the intermediacy of lysergaldehyde, they do suggest the possibility of double bond isomerization at the aldehyde rather than the acid stage.[163]

[163]   [164]

There is some evidence that biological dealkylations of amines also occur in cultures of *Claviceps* strains. Thus it has been suggested that in *Claviceps*, tryptophan is formed from N-methyltryptophan[169] and chanoclavine I from N-methylchanoclavine I.[170] Ramstad and co-workers have suggested[171] that a peroxidase is involved in the demethylation of setoclavine to norsetoclavine in cultures of *Claviceps*. This enzyme has been reported to catalyse the dealkylation of simple N-alkylamines.[172]

The ability to convert agroclavine or elymoclavine to lysergic acid derivatives is of limited distribution. The conversion of elymoclavine to lysergic acid series has been demonstrated only for lysergic acid α-hydroxyethyl amide [150] and ergotamine [152].[137, 139] The role of lysergylalanine [151] in these conversions has been shown only for the biosynthesis of ergometrine [153]. Ergometrine itself does not appear to play any part in the biosynthesis of ergotamine [152].[173, 174] Lysergylalanine could also not be incorporated into lysergic acid hydroxyethylamide [150][173, 174] Lysergic acid, however, afforded positive incorporations into ergotamine [152] but the corresponding amide failed to act as a precursor.[174] The roles of L-alanine, pyruvate, and L-alaninol remain to be clarified.[149, 174] The biosynthesis of lysergic-acid amide derivatives also needs further clarification of pathways.[155] It has been shown that the biosynthetic pathway for dihydroergosine [164] formation in *Sphacelia sorghi* involves sequential formation of festuclavine [154], dihydroelymoclavine, and dihydrolysergic acid and that 9,10-unsaturated alkaloids were not involved.[175]

The biosynthesis of ergocryptine and ergocryptinine has been studied, and it has been found that D-lysergyl-L-valine-^{3}H as well as L-leucyl-L-proline-^{3}H were incorporated into the two ergolenes in the cell-free system.[176] Positive specific incorporations of L-tryptophan-[U-^{14}C], DL-mevalonic acid lactone-[2-^{14}C], sodium acetate-[2-^{14}C], sodium formate-[^{14}C], and L-methionine-[Me-^{14}C] have been observed into the ergolene moiety of ergotamine I in *C. purpurea*.[177]

The peptidic ergot alkaloids are derived by combination of three appropriate α-amino acids with lysergic acid. The order in which the various amino acids become linked to one another is however obscure. There is some evidence that a cyclic dipeptide combines with a lysergyl amino acid[178-179] but the bulk of the evidence is against such combination.[180-184]. Alternatively lysergic acid could link to a tripeptide. However L-valyl-L-leucyl-L-proline and L-valyl-L-valyl-L-proline were not incorporated directly into ergocornine [175] and ergocryptine [176] but only after hydrolysis into the individual amino acids.[184, 185] This has led to the suggestion that the peptide chain could be assembled in a concerted fashion on the surface of a multienzyme complex.[185] A possible way in which this might happen is shown in Scheme 6.28. This may involve the growth of the peptide chain from the lysergic-acid end or from the proline end. The lysergyl tripeptide thus produced would be initially bound to the enzyme, for example through a thio linkage at the proline carboxyl group. An internal lactam formation accompanied by displacement of the enzyme could follow.[92] Subsequent α-hydroxylation could take place either directly or by a dehydrogenation/hydration sequence via the 2,3-dehydro amino

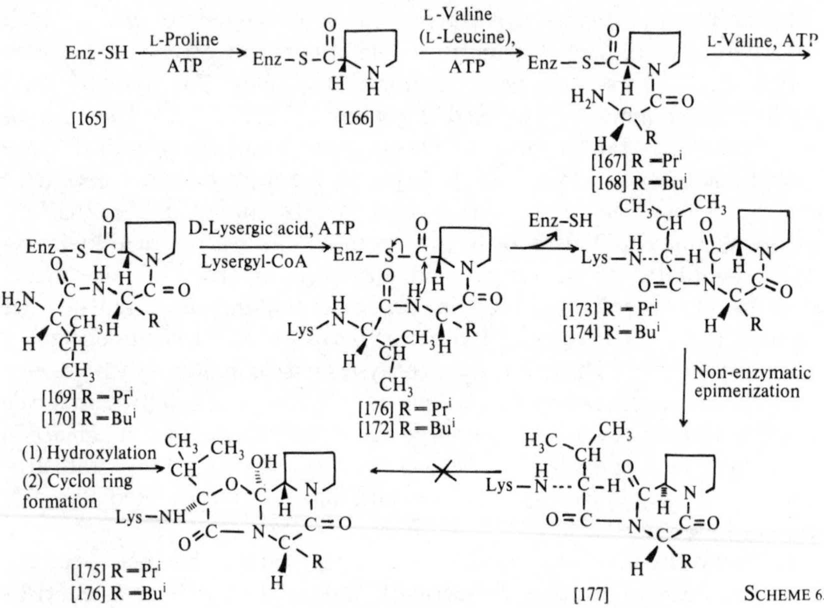

Enz-SH $\xrightarrow[\text{ATP}]{\text{L-Proline}}$ Enz–S–C ... N [165] ... [166] $\xrightarrow[\text{ATP}]{\substack{\text{L-Valine} \\ (\text{L-Leucine}),}}$ Enz—S–C ... N

[167] R = Pr^i
[168] R = Bu^i

L-Valine, ATP $\rightarrow$

Enz–S–C ...
[169] R = Pr^i
[170] R = Bu^i

$\xrightarrow[\text{Lysergyl-CoA}]{\text{D-Lysergic acid, ATP}}$ Enz–S–C ...

[176] R = Pr^i
[172] R = Bu^i

Enz–SH $\rightarrow$

[173] R = Pr^i
[174] R = Bu^i

Non-enzymatic epimerization

(1) Hydroxylation
(2) Cyclol ring formation

[175] R = Pr^i
[176] R = Bu^i

$\times$

[177]

SCHEME 6.28

acid or the imino acid and the α-hydroxy compound thus produced on cyclol formation would give rise to the ergocornine system.

The interesting tremergen group of alkaloids includes paspalin [185],[187–190] paspalicin [186],[187, 190, 191] paspalinin [187][191, 192] from *C. paspali* and aflavanine [188][193] and aflatrem [189][194] from *Aspergillus flavus*. Incorporation experiments with [1-^{13}C]-, [2-^{13}C]- and [1,2-^{13}C]-sodium acetate, [2-^{13}C]-mevalonic acid, 3-(1-[1-^{3}H]-geranylgeranyl)-indole and [17-^{3}H]-secoanhydropaspaline have given results in agreement with the biosynthetic pathway shown in Scheme 6.29 for paspalin [185], paspalicin [186], and paspalinin [187] | [195] The isolation of secoanhydropaspalin [180] from *Claviceps paspali* adds support to such a scheme.[192, 195]

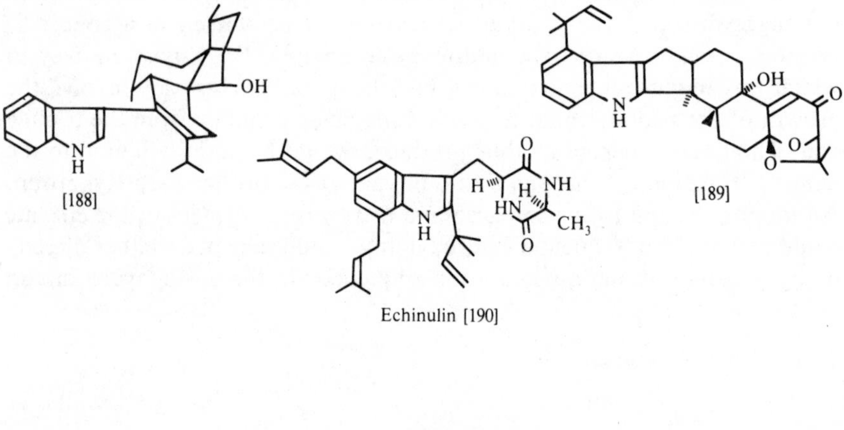

[188]

[189]

Echinulin [190]

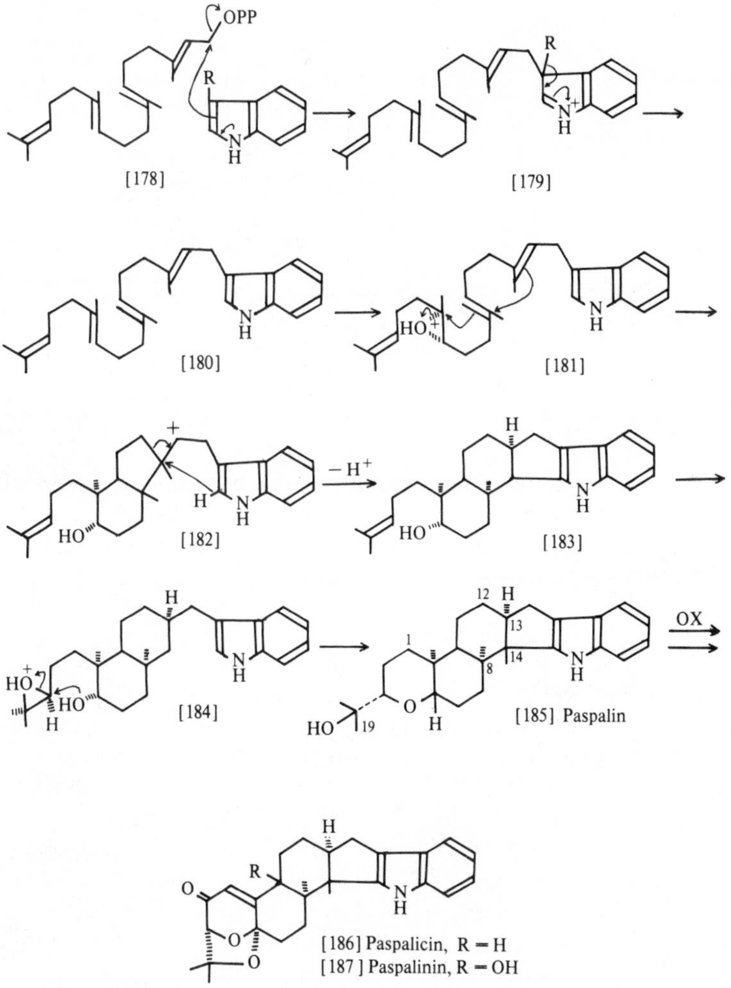

[178] → [179] →

[180] → [181] →

[182] $-H^+$ → [183] →

[184] → [185] Paspalin $\xrightarrow{OX}$

[186] Paspalicin, R = H
[187] Paspalinin, R = OH

SCHEME 6.29

A number of indole alkaloids have been isolated in the past few years in which one or more isoprene units attached to the indole system can be recognized. These alkaloids include 4-isopentenyltryptophan, a precursor to the ergot alkaloids discussed above. Echinulin [190] was shown by labelling experiments to be derived from tryptophan, alanine, and mevalonic acid.[190-202] The isoprenoid substituent at C-2 is believed to arise by initial isoprenylation at N-1 followed by allylic rearrangement to the 2-positon.[202] A plausible route to echinulin is presented in Scheme 6.30. A number of closely related derivatives have been isolated from *Aspergillus amstelodami* the major component being neoechinulin [193]. Later studies led to the

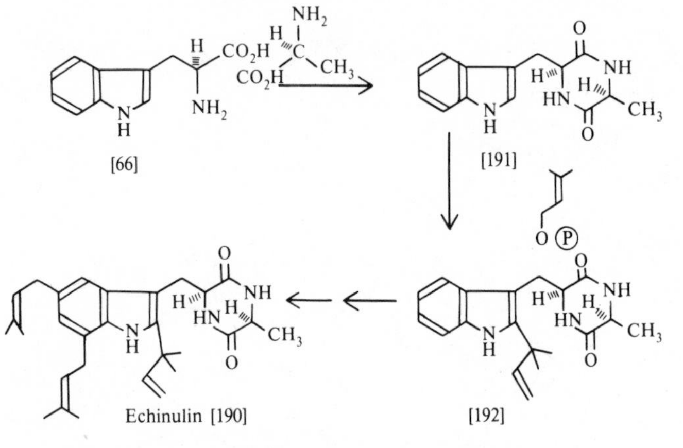

Echinulin [190]   [192]   SCHEME 6.30

isolation of four minor metabolites which are shown to be the precursors of neoechinulin.[203-204] Neoechinulin A [194] and B (196) do not have the isopentenyl group at C-6 and may well be precursors to neoechinulins D [195] and C[197] respectively.

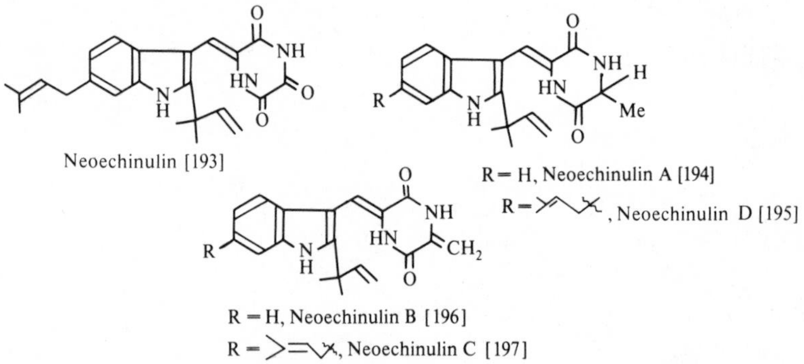

Neoechinulin [193]

R = H, Neoechinulin A [194]

R = ⅄⌣⤬ , Neoechinulin D [195]

R = H, Neoechinulin B [196]

R = ⟩=⤬ , Neoechinulin C [197]

Another closely related alkaloid cyclopiazonic acid [200] is found in *Penicillium cyclopium* and has been shown by labelling experiments to be derived from tryptophan, acetate, and mevalonate.[206, 207] The currently understood mode of biogenesis of cyclopiazonic acid, as elaborated by labelling experiments, is presented in Scheme 6.31.[207-213]

Another related alkaloid, roquefortine [204], has been isolated from cultures of *Penicillium roquefortii* and shown to be neurotoxic.[214, 215] Feeding experiments have established that labelled mevalonic acid, tryptophan, and histidine are incorporated. It has also been shown that the hydrogen at the 2-position of the indole ring is lost during this conversion supporting the view that the 'reverse' substituted dimethylallyl grouping at

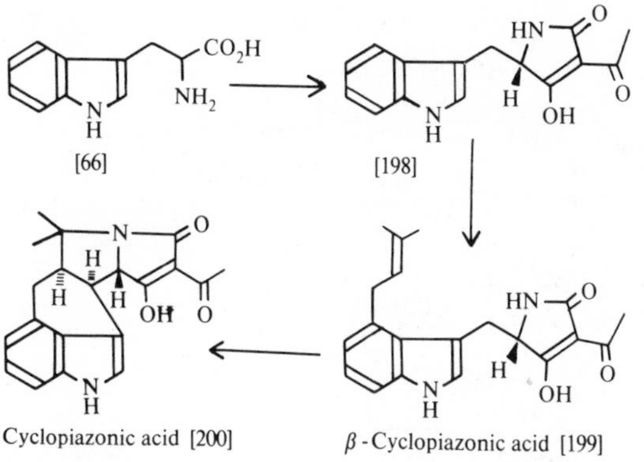

[66]   [198]

Cyclopiazonic acid [200]   β-Cyclopiazonic acid [199]   SCHEME 6.31

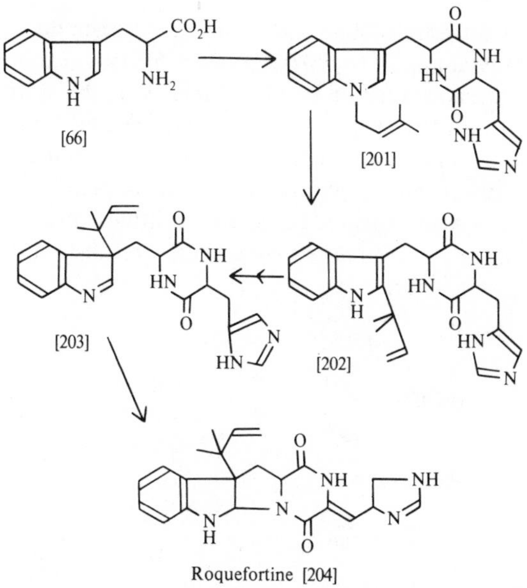

[66]   [201]

[203]   [202]

Roquefortine [204]   SCHEME 6.32

position-two is involved.[216] A possible biogenetic route to this alkaloid is presented in Scheme 6.32. The rearrangement of N-dimethylallyl grouping to the 2-position has been achieved *in vitro*.[217]

Besides echinulin and roquefortine, a number of other diketopiperazine derivatives of tryptophan and another amino acid modified by isoprenylation have been isolated.[218-222] These include the brevianamides, the austamides, lanosulin, and oxaline. All these substances are characterized by the presence of the 'reverse' substituted dimethylallyl grouping at the two-position of the indole ring. In brevianamide A [209] and E [207] tryptophan, proline, and an isoprene unit have reacted together to afford the respective ring systems. A precursor to these may be dioxopiperazine

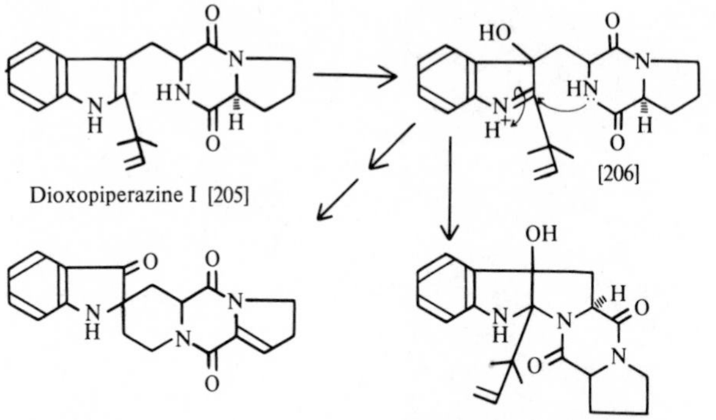

Dioxopiperazine I [205]

[206]

[208]

Brevianamide E [207]

SCHEME 6.33

[205][223] which could, by an intramolecular attack of the corresponding hydroxyindolenine, afford brevianamide E [207] (Scheme 6.33). Brevianamide B was found to be a stereoisomer of brevianamide A at the spiro centre. Brevianamide C [210] and D [211] were found to be *cis*-and *trans*-isomers at the olefinic double bond and are interconvertible by irradiation. They may well be artefacts produced by photolysis of brevianamide A. Brevianamide F [212] has been demonstrated to be a biosynthetic precursor of brevianamide A.[224] Other closely related substances are austamide

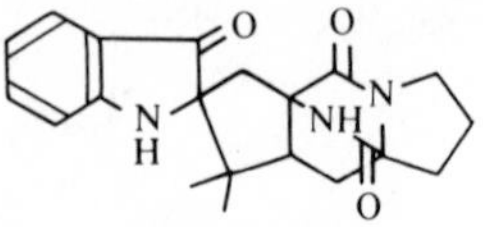

Brevianamide A [209]

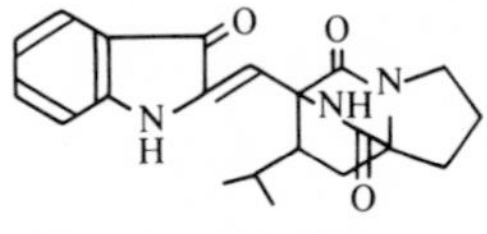

Brevianamide C [210]

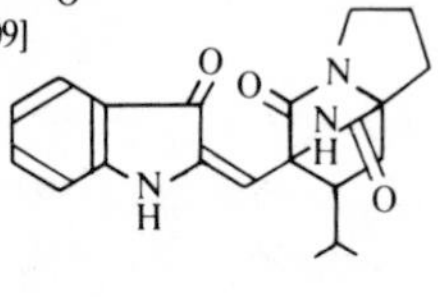

Brevianamide D [211]

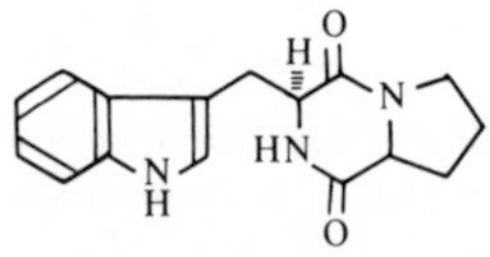

Brevianamide F [212]

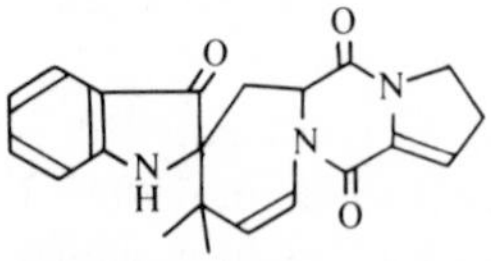

Austamide [213]

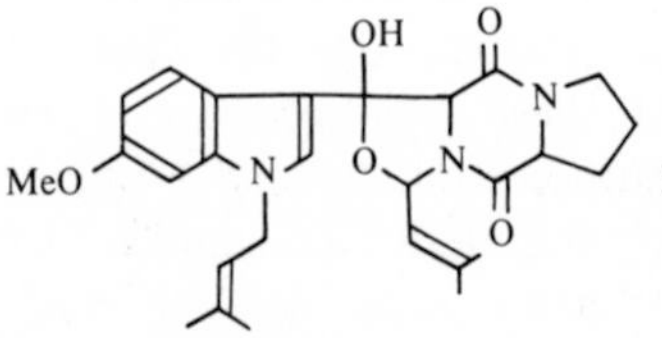

Lanosulin [214]

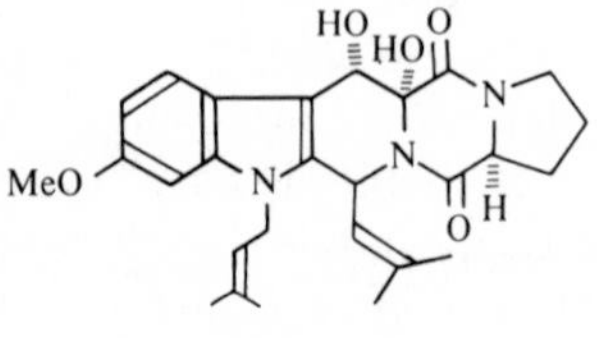

Fumitremorgin B [215]

[213] and lanosulin [214] . Fumitremorgin B [215] is a tremorgenic toxin isolated from *Aspergillus fumigatus*.[225] Interestingly, the isoprene unit is attached in the 'normal' sense to the amide nitrogen rather than in the 'reverse' sense as in echinulin and the brevianamides. The dioxopiperazine III group [208] probably arises from the dioxopiperazine I [205] system by side-chain migration of the corresponding hydroxyindolenine [206] and cyclization [Scheme 6.33] .

6.4 Summary

While the precise pathways to the various carbazole alkaloids have yet to be established, there is evidence to suggest that anthranilic acid is a precursor and that a mevalonic acid derivative is involved in the biosynthesis of ring C (Schemes 6.1–6.4).

Another group of alkaloids are those derived from tryptophan, but not by combination with isoprenoid units. These include the simple indoles such as gramine, harmaline, and serotonin. Two different routes appear to operate, one to alkaloids such as harmalan and harman via N-acetyltryptamine (Scheme 6.13) and the other to tetrahydro-β-carboline via pyruvate (Scheme 6.14).

The biosynthesis of the ergot alkaloids involves the reaction of tryptophan with $\gamma\gamma$-dimethylallyldiphosphate to afford 4-$\gamma\gamma$-dimethylallyltryptophan. This is hydroxylated at the *cis*-methyl group to afford the corresponding alcohol [121] . This then undergoes cyclization to afford a new 6-membered ring C, and an isomerization so that the hydroxymethyl group becomes *trans*-disposed in chanoclavine I [110] . A second isomerization at the olefinic bond occurs in the cyclization of chanoclavine I to agroclavine [115] which is subsequently oxidized to elymoclavine and lysergic acid. There appears to be an independent biosynthetic pathway from chanoclavine I directly to elymoclavine, as has been shown in cell-free preparations. The hydrogens at C-9 and C-10 of the tetracyclic ergolenes arise from the pro R-4H and pro S-5H, respectively, of the mevalonate moiety. The simple clavine alkaloids, e.g. agroclavine and elymoclavine act as precursors of the lysergic acid moiety of the more complex amide and peptide ergot alkaloids. The biosynthetic routes to a number of other isoprenoid tryptophan alkaloids are discussed.

References

1. CHAKRABORTY, D. P. *Tetrahedron Lett.* 661 (1966); *Phytochemistry* **8**, 769 (1969).
2. CHAKRABORTY, D. P. *J. Indian Chem. Soc.* **46**, 177 (1969).
3. CHAKRABORTY, D. P. and DAS, K. C. *Chem. Commun.* 967 (1968).
4. KUREEL, S. P., KAPIL, R. S., and POPLI, S. P. *Experientia* **25**, 790 (1969).

5. ERDTMAN, H. *Perspectives in phytochemistry,* (eds. J. B. HARBORNE, and T. SWAIN, p.107. Academic Press, New York (1969).
6. KAPIL, R. S. *The alkaloids,* Vol. XIII (ed. R. H. F. MANSKE, p. 299. Academic Press, New York (1971).
7. NARASIMHAN, N. S., PARADKAR, M. V., CHITGUPPI, V. P., and KELKAR, S. L. *Indian J. Chem.* **13**, 993 (1975).
8. KAPIL, R. S. *The alkaloids,* Vol. XIII (ed. R. H. F. MANSKE), p. 290 (1971).
9. KUREEL, S. P., KAPIL, S. S., and POPLI, S. P. *Tetrahedron Lett.* 3857 (1969).
10. SUHADOLNIK, R. J. and WINISTEAD, J. A. *J. Am. Chem. Soc.* **82**, 1644 (1960).
11. SUHADOLNIK, R. J., FISCHER, A., and WILSON, J. *Fedn. Am. Soc. exp. Biol.* **19**, 8. (1960).
12. SUHADOLNIK, R. J. and CHENOWETH, R. G. *J. Am. Chem. Soc.* **80**, 4391 (1958).
13. BOSE, A. K., DAS, K. G., FUNKE, P. T., KUGADHOVSKI, L., SUKLA, O. P., KANCHANDANI, K. S., and SUHADOLNIK, R. J. *J. Am. Chem. Soc.* **90**, 1038 (1968).
14. BU'LOCK, J. D. and RYLES, A. P. *J. Chem. Soc. Chem. Commun.* 1404 (1970).
15. JOHNS, N. and KIRBY, G. W. *J. Chem. Soc. Chem. Commun.* 163 (1971).
16. BU'LOCK, J. D., RYLES, A. P., KIRBY, G. W., and JOHNS, N. *J. Chem. Soc. Chem. Commun.* 100 (1972).
17. BU'LOCK, J. D. and LEIGH, C. *J. Chem. Soc. Chem. Commun.* 628 (1975).
18. KIRBY, G. W. and ROBINS, D. J. *J. Chem. Soc. Chem. Commun.* 354 (1976).
19. JEFFS, P. W., ARCHIE, W. C., HAWKS, R. L., and FARRIER, D. S. *J. Am. Chem. Soc.* **93**, 3752 (1971).
20. JEFFS, P. W., CAMBELL, H. F., FARRIER, D. S., GANGULI, G., MARTIN, N. H., and MOLINA, G. *Phytochemistry* **13**, 933 (1974).
21. JEFFS, P. W., JOHNSON, D. B., MARTIN, N. H., and RAUCKMAN, B. S. *J. Chem. Soc. Chem. Commun.* 82 (1976).
22. JEFFS, P. W., CAPPS, T., JOHNSON, D. B., KARLE, J. M., and RAUCKMAN, B. S. *J. org. Chem.* **39**, 2703 (1974).
23. JEFFS, P. W. and KARLE, J. M. *J. Chem. Soc. Chem. Commun.* 60 (1977).
24. DAVIS, B. D. *J. biol. Chem.* **191**, 315 (1951).
25. SRINIVASAN, P. R. and SPRINSON, D. B. *J. biol. Chem.* **234**, 716 (1959).
26. HARRISON, N. H., BOVER, P. D., and FALCONE, A. B. *J. biol. Chem.* **215**, 303 (1955).
27. MARUYAMA, H., EASTERDAY, R. L., CHANG, H. C., and LANE, M. D. *J. biol. Chem.* **241**, 2405 (1966).
28. CHANG, H. C. and LANE, M. D. *J. biol. Chem.* **241**, 2413, 2421 (1966).
29. CLELAND, W. W. *Biochim. biophys. Acta* **67**, 104 (1963).
30. STAUB, M. and DENES, G. *Biochim. biophys. Acta* **178**, 588, 599 (1969).
31. MOLDOVANYI, J. S. and DENES, G. *Acta. Biochem. Biophys Acad. Sci. Hung.* **3**, 259 (1968).
32. NAGANO, H. and ZALKIN, H. *Arch. Biochem. Biophys.* **138**, 58 (1970).
33. DELEO, A. B. and SPRINSON, D. B. *Biochem. Biophys. Res. Commun.* **32**, 373 (1968).
34. LEVIN, J. G. and SPRINSON, D. B. *J. biol. Chem.* **239**, 1142 (1964).
35. DELEO, A. B., DAYAN, J., and SPRINSON, D. B. *J. biol. Chem.* **248**, 2344 (1973).
36. FLOSS, H. G., ONDERKA, D. K., and CARROLL, M. *J. biol. Chem.* **247**, 736 (1972).

37. ONDERKA, D. K. and FLOSS, H. G. *J. Am. Chem. Soc.* **91**, 5894 (1969).
38. COHN, M., PEARSON, J. E., O'CONNELL, E. L., and ROSE, I. A. *J. Am. Chem. Soc.* **92**, 4095 (1970).
39. SCHONER, R. and HERRMANN, K. M. *J. biol. Chem.* **251**, 5440 (1976).
40. CORNFORTH, J. W. *Angew. Chem. Int. Ed.* **7**, 903 (1968).
41. HUISMAN, O. C. and KOSUGE, T. *J. biol. Chem.* **249**, 6842 (1974).
42. DOY, C. H. *Biochim. biophys. Acta* **132**, 528 (1967).
43. SPRINSON, D. B., ROTSCHILD, J., and SPRECHER, M. *J. biol. Chem.* **238**, 3170 (1963).
44. ALDERSBERG, M. and SPRINSON, D. B. *Biochemistry* **3**, 1855 (1964).
45. SMITH, B. W., TURNER, M. J., and HASLAM, E. *J. Chem. Soc. Chem. Commun.* 842 (1970).
46. TURNER, M. J., SMITH, B. W., and HASLAM, E. *J. Chem. Soc. Perkin Trans. I* 52 (1975).
47. HANSON, K. R. and ROSE, I. A. *Proc. natl. Acad. Sci. U.S.A.* **50**, 981 (1963).
48. BUTLER, R. J., ALWORTH, W. L., and NUGNET, M. J. *J. Am. Chem. Soc.* **96**, 1617 (1974).
49. GUNETILEKE, K. G. and ANWAR, R. A. *J. biol. Chem.* **243**, 5770 (1968).
50. LEVIN, J. G. and SPRINSON, D. B. *Biochem. Biophys. Res. Commun.* **3**, 157 (1960).
51. LEVIN, J. G. and SPRINSON, D. B. *J. biol. Chem.* **239**, 1142 (1964).
52. BONDINELL, W. E., YNEK, J., KNOWLES, P. F., SPRECHER, M., and SPRINSON, D. B. *J. biol. Chem.* **246**, 6191 (1971).
53. ZEMELL, R. I. and ANWAR, R. A. *J. biol. Chem.* **250**, 3185 (1975).
54. ZEMELL, R. I. and ANWAR, R. A. *J. biol. Chem.* **250**, 4959 (1975).
55. CASSIDY, P. J. and KAHAN, F. M. *Biochemistry* **12**, 1364 (1973).
56. HILL, R. K. and NEWKOME, G. R. *J. Am. Chem. Soc.* **91**, 5893 (1969).
57. ONDERKA, D. K. and FLOSS, H. G. *J. Am. Chem. Soc.* **91**, 5894 (1969).
58. ONDERKA, D. K. and FLOSS, H. G. *J. biol. Chem.* **247**, 736 (1972).
59. FUKUI, K. *Tetrahedron Lett.* 2427 (1965).
60. ANH, N. T. *J. Chem. Soc. Chem. Commun.* 1089 (1968).
61. ORLOFF, H. D. and KOLKA, A. J. *J. Am. Chem. Soc.* **76**, 5484 (1964).
62. CRITOL, S. J., BARASCH, W., and TIEMAN, C. H. *J. Am. Chem. Soc.* **77**, 583 (1955).
63. DE LAMARE, P. B. D., KOENIGSBERGER, R., and LOMAS, J. S. *J. Chem. Soc. B.,* 834 (1966).
64. HWANG, L. H. and ZALKIN, H. *J. biol. Chem.* **246**, 6889 (1971).
65. ROBB, F., HUTCHINSON, M. A., and BELSER, W. L. *J. biol. Chem.* **246**, 6908 (1971).
66. LEVIN, J. G. and SPRINSON, D. B. *J. biol. Chem.* **239**, 1142 (1964).
67. MCCORMICK, J. R. D., REICHENTHAL, J., KIRSCH, U., and SJOLANDER, N. O. *J. Am. Chem. Soc.* **84**, 3711 (1962).
68. RATLEDGE, C. *Nature* (Lond.) **203**, 428 (1964).
69. GANEM, B. *Tetrahedron* **34**, 3353 (1978).
70. MEISTER, A. *Biochemistry of the amino acids*, Vol. 2, 2nd edn., Ch. 6. Academic Press, New York (1965).
71. DALY, J. W. and WITKOP, B. *Angew. Chem. Int. Ed.* **2**, 421 (1963).
72. BRACK, A., HOFMANN, A., KALBERER, E., KOBEL, H., and RUTSCHMANN, J. *Arch. Pharm.* **294**, 230 (1961).
73. SPENSER, I. D. *Comprehensive biochemistry* (eds. M. Florkin and E. H. Stotz) Vol. 20, Ch. 6. Elsevier, Amsterdam (1968).

74. O'DONOVAN, D. G. and LEETE, E. *J. Am. Chem. Soc.* **85**, 461 (1963).
75. LEETE, E. and MINICH, M. L. *Phytochemistry* **16**, 149 (1977).
76. MUDD, S. H. *Nature* (Lond.) **189**, 489 (1961).
77. O'DONOVAN, D. G. and KENNEALLY, M. F. *J. Chem. Soc., C.* 1109 (1967).
78. STOLLE, K. and GROGER, D. *Arch. Pharm.* **301**, 561 (1968).
79. ROBINSON, R. *J. Chem. Soc.* 1079 (1936).
80. SLAYTOR, M. and MCFARLANE, I. J. *Phytochemistry* **7**, 605 (1968).
81. MCFARLANE, I. J. and SLAYTOR, M. *Phytochemistry* **11**, 229 (1972).
82. DANIELI, B., LESMA, G., and PALMISANO, G. *Experientia* **35**, 156 (1979).
83. SAXTON, J. E. *The alkaloids, Specialist periodical reports*, vol. I, p. 81 Chemical Society, London (1971).
84. HORNEMANN, U., HURLEY, L. H., SPEEDLE, M. K., and FLOSS, H. G. *Chem. Commun.*, 245 (1970); *Tetrahedron Lett.* 2255 (1971); *J. Am. Chem. Soc.* **93**, 3028 (1971).
85. HOOTELE, C. *Tetrahedron Lett.*, 2713 (1969).
86. MOTHES, K., WEYGAND, F., GROGER, D., and GRISEBACH, H. *Z. Naturforsch.* **13b**, 41 (1958).
87. TYLER, V. E. *J. Pharm. Sci.* **50**, 629 (1961).
88. WINKLER, K. and GROGER, D. *Pharmazie* **17**, 658 (1962).
89. WEYGAND, F. and FLOSS, H. G. *Angew. Chem. Int. Ed.* **2**, 243 (1963).
90. AGURELL, S. *Acta. Pharm. Suec.* **3**, 71 (1966).
91. VOIGT, R. *Pharmazie* **23**, 285, 353, 419 (1968).
92. RAMSTAD, E. *Lloydia* **31**, 327 (1968).
93. GROGER, D. *Biosynthese der alkaloide* (eds. K. Mothes and H. R. Schutte, p. 486. Deutsch Verlag Wiss, Berlin (1969).
94. THOMAS, R. and BASSETT, R. A. *Progr. Phytochem.* **3**, 47 (1972).
95. FLOSS, H. G. *Tetrahedron* **32**, 873 (1976).
96. GROGER, D., WENDT, H. J., MOTHES, K., and WEYGAND, F. *Z. Naturforsch.*, **14b**, 355 (1958).
97. TABER, W. A. and VINING, L. C. *Chem. Ind.* 1218 (1959).
98. WYGAND, F. and FLOSS, H. G. *Angew. Chem.* **75**, 783 (1963).
99. GROGER, D., MOTHES, K., SIMON, H., FLOSS, H. G., and WEYGAND, F. *Z. Naturforsch.* **16b**, 432 (1961).
100. HARLEY-MASON, J. *Chem. Ind.* 251 (1954).
101. VAN TAMELEN, E. E. *Experientia* **9**, 457 (1953).
102. PLIENINGER, H., FISCHER, R., KEILICH, G., and ORTH, H. D. *Ann.* **642**, 214 (1961).
103. BAXTER, R. M., KANDEL, S. I., and OKANY, A. *Chem. Ind.* 1453 (1961).
104. WENDLER, N. L. *Experientia* **10**, 338 (1954).
105. FELDSTEIN, A. *Experientia* **12**, 475 (1956).
106. ROBINSON, R. *The structural relation of natural products*, p. 106. Clarendon Press, Oxford (1955).
107. BIRCH, A. J. *Amino acids and peptides with antimetabolic activity* (eds. G. E.W. Westenholme, and C. M. O'Connor p. 247. Churchill, London (1958).
108. GROGER, D., MOTHES, K., SIMON, R., FLOSS, G. H., and WEYGAND, F. *Z. Naturforsch.* **15b**, 141 (1960).
109. TAYLOR, E. H. and RAMSTAD, E. *Nature (Lond.)*, **188**, 494 (1960).
110. BIRCH, A. J., MCLOUGHLIN, B. J., and SMITH, H. *Tetrahedron Lett.*, 1 (1960).
111. GROGER, D., WENDT, H. J., MOTHES, K., and WEYGAND, F. *Z. Naturforsch.*, **14b**, 355 (1959).

112. BHATTACHARJI, S., BIRCH, A. J., BRACK, A., HOFMANN, A., KOBEL, H., SMITH, D. C. C., SMITH, H., and WINTER, J. *J. Chem. Soc.* 421 (1962).
113. BAXTER, R.M., KANDEL, S.I., and OKANY, A. *Tetrahedron Lett.* 596 (1961).
114. PLIENINGER, H., IMMEL, H., and VOLKL, L. *Justus Leibig's Ann. Chem.*, **706**, 223 (1967).
115. SCOTT, A. I. *M.T.P. Int. Rev. Sci.* (Alkaloids) **9**, (1973).
116. PLIENINGER, H., FISCHER, R., and LIEDE, U. *Justus Liebig's Ann. Chem.*, **672**, 223 (1964).
117. ROBERRS, J. E. and FLOSS, H. G. *Tetrahedron Lett.* 1857 (1969).
118. AGURELL, S. and LINDGREN, J. E. *Tetrahedron Lett.* 5127 (1968).
119. ROBERRS, J. E. and FLOSS, H. G. *Arch. Biochem. Biophys.* **126**, 967 (1968).
120. HEINSTEIN, P. F., LEE, S.-L., and FLOSS, H. G. *Biochem. Biophys. Res. Commun.* **44**, 1244 (1971).
121. WENKERT, E. and SLIWA, H. *Bio-org. Chem.* **6**, 443 (1977).
122. OTSUKA, H., ANDERSON, J. A., and FLOSS, H. G. *J. Chem. Soc. Chem. Commun.*, 660 (1979).
123. AGURELL, S. and RAMSTAD, E. *Tetrahedron Lett.* 501 (1961).
124. AGURELL, S. and RAMSTAD, E. *Arch. Biochem. Biophys.* **98**, 457 (1962).
125. ACKLIN, W., FEHR, T., and ARIGONI, D. *Chem. Commun.* 799 (1966).
126. FEHR, T., ACKLIN, W., and ARIGONI, D. *Chem. Commun.* 801 (1966).
127. GROGER, D., ERGE, D., and FLOSS, H. G. *Z. Naturforsch.* **21b**, 827 (1966).
128. VOIGT, R., BORNSCHEIN, M., and RABITZSCH, G. *Pharmazie* **22**, 326 (1967).
129. FLOSS, H. G., HORNEMANN, U., SCHILLING, N., GROGER, D., and ERGE, D. *J. Am. Chem. Soc.* **90**, 6500 (1968).
130. POPJAK, G. and CORNFORTH, J. W. *Biochem. J.* **101**, 553 (1966).
131. FLOSS, H. G. *Chem. Commun.* 804 (1967).
132. FLOSS, H. G., HORNEMANN, U., SCHILLING, N., GROGER, D., and ERGE, D. *Chem. Commun.* 105 (1967).
133. OGUNLANA, E. O., WILSON, B. J., TYLER, E. V., MARRO, Jr. E., and RAMSTAD, E. *J. Chem. Soc., D* 775 (1970).
134. ACKLIN, W., FEHR, T., and STADLER, P. A. *Helv. Chim. Acta* **58**, 2492 (1975).
135. BAXTER, R. M., KANDEL, S. I., OKANY, A., and TAM, K. L. *J. Am. Chem. Soc.* **84**, 4350 (1962).
136. BATTERSBY, A. R. *Specialist periodical reports, The alkaloids*, Vol. I, pp. 39–40. Chemical Society, London.
137. SCOTT, A. I. *Acc. Chem. Res.* **3**, 151 (1970).
138. FLOSS, H. G., TSHEN-LIN, M., NAIDOO, B., CHING-JERCHANG, BLAIR, G. E., ABOU-CHAAR, C. I., and CASSADY, J. M. *J. Am. Chem. Soc.* **96**, 1896 (1974).
139. AGURELL, S. *Acta. Pharm. Suec.* **3**, 71 (1965).
140. VOIGT, R. *Pharmazie* **23**, 285, 353, 419 (1968).
141. RAMSTAD, E. *Lloydia* **31**, 327 (1968).
142. FLOSS, H. G. *Abh. Deut. Akad. Wiss., Berl., Kl. Chem. Geol. Biol.*, 395 (1972).
143. STAUFFACHER, D. and TSCHERTER, H. *Helv. Chim. Acta* **47**, 2186 (1964).
144. HANSON, K. R. *J. Am. Chem. Soc.* **88**, 2731 (1966).
145. SEILER, M. P. *Ph. D. Dissertation*, ETH, Zurich (1970).
146. CASSADY, J. M., ABOU-CHAAR, C. I., and FLOSS, H. G. *Lloydia* **36** (4), 390 (1973).
147. ABOU-CHAAR, C. I. GUENTHER, H. F., MANUEL, M. F., ROBERRS, J. E., and FLOSS, H. G. *Lloydia* **35**, 272 (1972).
148. VAN TAMELEN, E. E. and OLIVER, L. K. *J. Am. Chem. Soc.* **92**, 2136 (1970).
149. NELSON, V. and AGURELL, S. *Acta Chem. scand.* **23**, 3393 (1969).

150. OGUNLANA, E. O., WILSON, B. J., TYLER, V.E., and RAMSTAD, E. *Chem. Commun.* 775 (1970).
151. OGUNLANA, E. O., RAMSTAD, E., and TYLER, V. E. *J. Pharm. Sci.* **58**, 143 (1969).
152. GROGER, D., SCHUTTE, H. R., and STOLLE, K. *Z. Naturforsch.* **18b** 850 (1963).
153. AGURELL, S. and JOHANSON, *Acta Chem. scand.* **18**, 2285 (1964).
154. BRACK, A., BRUNNER, R., and KOBEL, H. *Helv. Chim. Acta* **45**, 276 (1962).
155. YAMATODANI, S., KOZU, Y., YAMADA, S., and ABE, M. *Annu. Reps., Takeda Research Lab.* **21**, 88 (1962).
156. GROGER, D. *Planta Med.* **4**, 444 (1963).
157. BELIVEAU, J. and RAMSTAD, E. *Lloydia* **29**, 234 (1966).
158. TAYLOR, E. H., GOLDNER, K. J., PONG, S. F., and SHOUGH, H. R. *Lloydia* **29**, 239 (1966).
159. TAYLOR, E. H. and SHOUGH, H. R. *Lloydia* **30**, 197 (1967).
160. CHAN LIN, W. N., RAMSTAD, E., and TAYLOR, E. H. *Lloydia* **30**, 202 (1967).
161. TYLER, Jr., V. E., ERGE, D., and GROGER, D. *Planta Med.* **13**, 315 (1965).
162. ABE, M., YAMATODANI, S., and YAMANO, T. *Nippon Nogeikagaku Kaishi*, **41**, 68 (1967).
163. VOIGT, R. and BORNSCHEIN, M. *Pharmazie* **20**, 521 (1965).
164. MOTHES, K., WINKLER, K., GROGER, D., FLOSS, H. G., MOTHES, W., and WEYGAND, F. *Tetrahedron Lett.* 933 (1962).
165. FLOSS, H. G., GUNTER, H., GROGER, D., and ERGE, D. *Z. Naturforsch.* **21b**, 128 (1966).
166. OHASHI, T., AOKI, S., and ABE, M. *J. agr. Chem. Soc. Jpn.* **44**, 527 (1970).
167. AGURELL, S. *Acta Pharm. Suec.* **3**, 65 (1966).
168. KOBEL, H., SCHREIER, E., and RUTSCHMANN, J. *Helv. Chim. Acta* **47**, 1052 (1964); LIN, C. C. L., BLAIR, G. E., CASSADY, J. M., GROGER, D., MAIER, W., and FLOSS, H. G. *J. org. Chem.* **38**, 2249 (1973).
169. FLOSS, H. G. and GROGER, D. *Z. Naturforsch.* **19**, 393 (1964).
170. VOIGT, R. and BORNSCHEIN, M. *Pharmazie* **22**, 258 (1967).
171. RAMSTAD, E., CHAN LIN, W. N., SHOUGH, H. R., GOLDNER, K. J., PARIKH, R. P., and TAYLOR, E. H. *Lloydia* **30**, 441 (1967).
172. GILLETTEE, J. R., DINGALL, J. V., and BORDIE, B. B. *Nature* (Lond.) **181**, 898 (1958).
173. BASMADJIAN, G., FLOSS, H. G., GROGER, D., and ERGE, D. *Chem. Commun.*, 418 (1969).
174. MINGHETTI, A., and ARCAMONE, F. *Experientia* **25**, 926 (1969).
175. BARROW, K. D., MANTTE, P. G., and QUIGLY, F. R. *Tetrahedron Lett.*, 1557 (1974).
176. OHASHI, T., TAKABIASHI, H., and ABE, M. *Nippon Nagoe Kagaku Kaishi*, **46**, 435 (1972).
177. BASSETT, R. A., CHAIN, E. B., and CORBETT, K. *Biochem. J.* **134**, 1 (1973).
178. ABE, M., OHASHI, T., OHMOTO, S., and TABUCHI, T. *Agric. biol. Chem. (Jpn)* **35A** 1 (1971).
179. OHASHI, T., TAKAHASHI, H., and ABE, M. *Nippon Nogei Kagaku Kaishi* **46**, 535 (1972).
180. FLOSS, H. G., BASMADJIAN, G. P., TCHENG, M., SPALLA, C., and MINGHETTI, A. *Lloydia* **34**, 442 (1971).
181. FLOSS, H. G., BASMADJIAN, G. P., TCHENG, M., GROGER, D., and ERGE, D. *Lloydia* **34**, 444 (1971).

182. GROGER, D. and JOHNE, S. *Experientia* **28**, 241 (1972).
183. MAIER, W., ERGE, D., and GROGER, D. *Biochem. Physiol. Pflanzen.* **165**, 479 (1974).
184. GROGER, D., JOHNE, S., and HARTLING, S. *Biochem. Physiol. Pflanzen.* **166**, 33 (1974).
185. FLOSS, H. G., TCHENG-LIN, M., KOBEL, H., and STADLER, P. *Experientia* **30**, 1369 (1974).
186. BYCROFT, B. W. *Nature* (Lond.) **224**, 595 (1969).
187. FEHR, T. and ACKLIN, W. *Helv. Chim. Acta* **49**, 1907 (1966).
188. STAMM, G. Diss. ETH Zurich Nr. 4418 (1969).
189. GYSI, P. Diss. ETH Zurich Nr. 4990 (1973).
190. SPRINGER, J. P., CLARDY, J., WELLS, J. M., COLE, R. J., and KIRKSEY, J. W. *Tetrahedron Lett.* 2531 (1975).
191. LEUTWILER, A. Diss. ETH Zurich Nr. 5163 (1973).
192. WEIBEL, F. Diss. ETH Zurich Nr. 6314 (1979).
193. GALLAGHER, R. T., MCCABE, T., HIROTSU, K., CLARDY, J., NICHOLSON, J., and WILSON, B. J. *Tetrahedron Lett.* 243 (1980).
194. GALLAGHER, R. T., CLARDY, J., and WILSON, B. J. *Tetrahedron Lett.* 243 (1980).
195. ACKLIN, W., WEIBEL, F., and ARIGONI, D. *Chimia* **31**, 2 (1977).
196. BIRCH, A. J., BLANCE, G. E., DAVID, S., and SMITH, H. *J. Chem. Soc.* 3128 (1961).
197. MACDONALD, J. C. and SLAYTOR, G. P. *Can. J. Microbiol.* **12**, 455 (1966).
198. BIRCH, A. J. and FARRAR, K R. *J. Chem. Soc.*, 4277 (1963).
199. SLAYTOR, G. P., MACDONALD, J. C., and NAKASHIMA, R. *Biochemistry* **9**, 2886 (1970).
200. ALLEN, C. M. *Biochemistry* **11**, 2154 (1972).
201. ALLEN, C. M. *J. Am. Chem. Soc.* **95**, 2386 (1973).
202. CASNATI, G. and POCHINI, A. *Chem. Commun.* 1328 (1970).
203. DOSSENA, A., MARCHELLI, R., and POCHINI, A. *Chem. Commun.* 779 (1975).
204. DOSSENA, A., MARCHELLI, R., and POCHINI, A. *Chem. Commun.* 771 (1974).
205. PLIENINGER, H., MEYER, E., MAIER, W., and BROGER, D. *Liebigs Ann. Chim.* 813 (1978).
206. HOLZAPFEL, C. W. and WILKINS, D. C. *5th Int. Symp. on the Chemistry of natural products*, Abstract C. 65, London (1968).
207. HOLZAPFEL, C. W. *Tetrahedron* **24**, 2101 (1968).
208. HOLZAPFEL, C. W. and WILKINS, D. C. *Phytochemistry* **10**, 351 (1971).
209. MCGRATH, R. M., STEYN, P. S., and FERREIRA, N. P. *Chem. Commun.* 812 (1973).
210. SCHABORT, J. C., WILKINS, D. C., HOLZAPFEL, C. W., POTGIETER, D. J. J., and NEITZ, A. W. *Biochim. biophys. Acta* **250**, 311 (1971).
211. STEENKAMP, D. J., SCHABORT, J. C., and FERREIRA, N. P. *Biochim. biophys. Acta* **309**, 440 (1975).
212. STEENKAMP, D. J. and SCHABORT, J. C. *Eur. J. Biochem.* **40**, 163 (1973).
213. STEYN, P. S., VLEGGAAR, R., FERREIRA, N. P., KIRBY, G. W., and VARLEY, M. J. *Chem. Commun.* 465 (1975).
214. SCOTT, P. M., MERRIEN, M., and POLONSKY, J. *Experientia* **32**, 140 (1976).
215. SCOTT, P. M. and KENNEDY, B. P. C. *J. agric. Food Chem.* **24**, 865 (1976).
216. BARROW, K. D., COLLEY, P. W., and TRIBE, D. E. *J. Chem. Soc. Chem. Commun.* 225 (1979).

217. CASNATI, G., MARCHELLI, R., and POCHINI, A. *J. Chem. Soc. Perkin Trans. 1* 754 (1969).
218. BARBETTA, M., CASNATI, G., POCHINI, A., and SELVA, A. *Tetrahedron Lett.* 4457 (1974).
219. STEYN, P. S. *Tetrahedron Lett.* 3331 (1971).
220. BIRCH, A. J. and WRIGHT, J. J. *Tetrahedron* **26**, 2329 (1970).
221. DIX, D. T., MARTIN, J., and MOPPETT, C. E. *J. Chem. Soc. Chem. Commun.* 1168 (1972).
222. NAGEL, D. W., PACHLER, K. G. R., STEYN, P. S., WESSELS, P. L., GAFNER, G., and KRUGER, G. K. *J. Chem. Soc. Chem. Commun.* 1021 (1974).
223. STEYN, P. S. *Tetrahedron* **29**, 107 (1973).
224. BALDAS, J., BIRCH, A. J., and RUSSELL, R. A., *J. Chem. Soc. Perkin Trans.* **I**, 50 (1974).
225. YAMAZAKI, Y., SASAGO, K., and MIYAK, K. *Chem. Commun.* 408 (1974).
226. HERBERT, R. B. and MANN, J. *J. Chem. Soc. Chem. Commun.* 841 (1980).

BIOSYNTHESIS OF CLASS V ALKALOIDS

7.1 Binary indole alkaloids

This class includes over a hundred alkaloids which each comprise two halves
linked together to afford binary molecules. These halves may either be true
indoles or may be modified units such as indolines, N-acylindolines, indolenines, oxindoles, etc. Certain binary alkaloids included in this class
show indolic (or modified indolic) character only in one moiety. The terms
'bis-indole' alkaloids or 'dimeric' alkaloids have therefore been deliberately
avoided.

The classification of these alkaloids in Chapter 2 was made on the basis
of whether the individual moieties of the binary molecules belonged to class
I, II, III, or IV. The anti-leukaemic alkaloids vinblastine [1] and vincristine
[2], for instance, thus fall in the II–III category, since one of the moieties,

Vinblastine, R = Me [1]
Vincristine, R = CHO [2]

vindoline, is a class II (*Aspidosperma*) alkaloid while the other cleavamine
moiety shows a class III (*Iboga*) skeleton. While with this system of
classification there are several gaps due to the non-existence of several
possible class combinations, it does result in a rational system of grouping
with the flexibility of being able to accommodate any new structural types
that may be encountered in the future.

7.1.1 O–II type

The haplophytine type alkaloids can be considered to arise by attack of the
highly nucleophilic carbon atom *para* to the basic indoline nitrogen of the
Aspidosperma structure [3] (class II alkaloids) on an electrophilic centre in
the non-indolic (Type 0) compound [4] to afford haplophytine [5]

(Scheme 7.1).[1, 2] This type of structure is not a bis-indole system as only one of the two moieties of the binary molecule is derived from an indole.

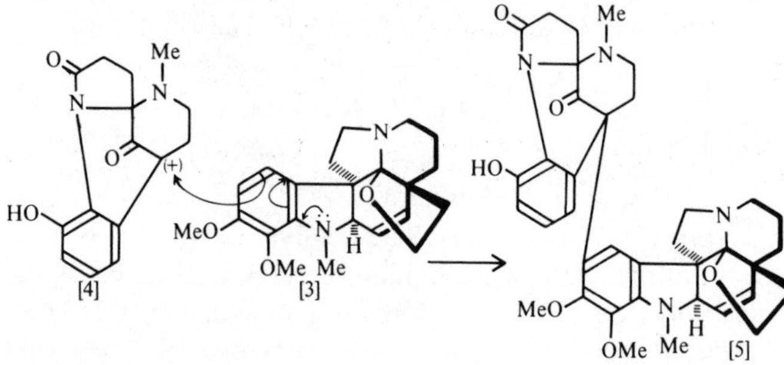

SCHEME 7.1

7.1.2 O–IV type

This type includes binary molecules formed by the combination of a class IV type indole alkaloid with a non-indolic moiety. This type of alkaloid has

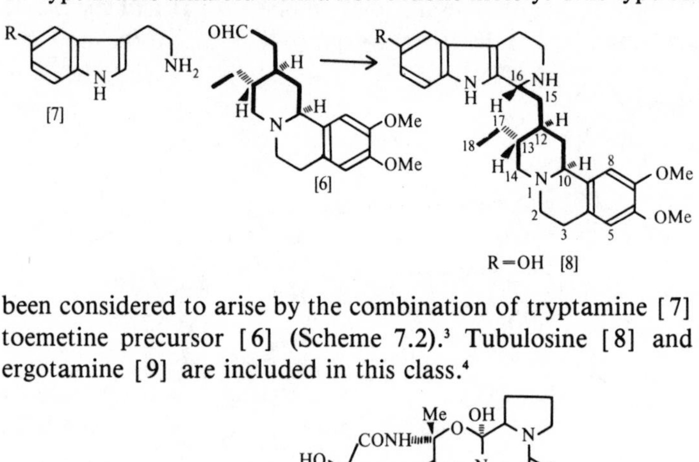

R=OH [8]   SCHEME 7.2

been considered to arise by the combination of tryptamine [7] with a pro-toemetine precursor [6] (Scheme 7.2).[3] Tubulosine [8] and 8-hydroxy-ergotamine [9] are included in this class.[4]

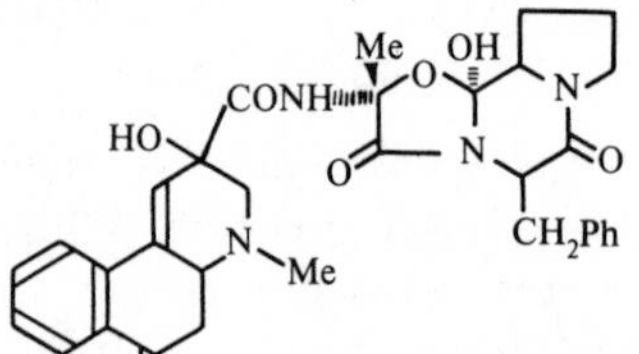

8-Hydroxyergotamine [9]

7.1.3 I–I type

These binary alkaloids are formed by the combination of two class I alkaloid units; the majority of binary indole alkaloids belong to this group. Serpentinine [12] can be considered to arise by the combination of serpentine [10] with a dihydrocorynantheine derivative [11]. This can be rationalized by envisaging an attack of the enamine system of serpentine [10] on [11] (Scheme 7.3).

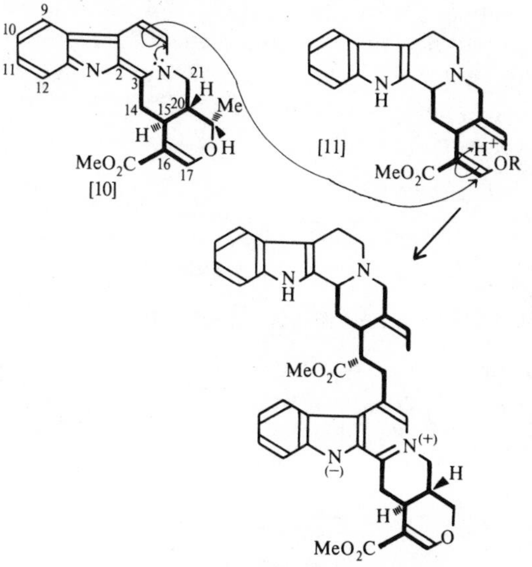

Serpentinine [12]

SCHEME 7.3

The binary alkaloids from *Calabash curare* fall in the same group. Over a dozen alkaloids of this type are known. They all contain a $C_{38}N_4$ skeleton and they are derived by the combination of Wieland–Gumlich aldehyde [13] and 18-desoxy Wieland–Gumlich aldehyde or their corresponding $N_{(b)}$-metho salts [14] and [15]. Wieland–Gumlich aldehyde [13] has been isolated from *Strychnos toxifera* but it is more readily accessible by the degradation of strychnine.[5-7] The biosynthetic route to the Wieland–Gumlich aldehyde has been described already. When Wieland–Gumlich aldehyde [13] is heated in acetic acid–sodium acetate, two molecules intermolecularly condense by combination of the potential aldehyde centres at C_{17} with the secondary basic nitrogen atoms to afford the dimeric substance [16] which deprotonates to afford C-toxiferine [17]

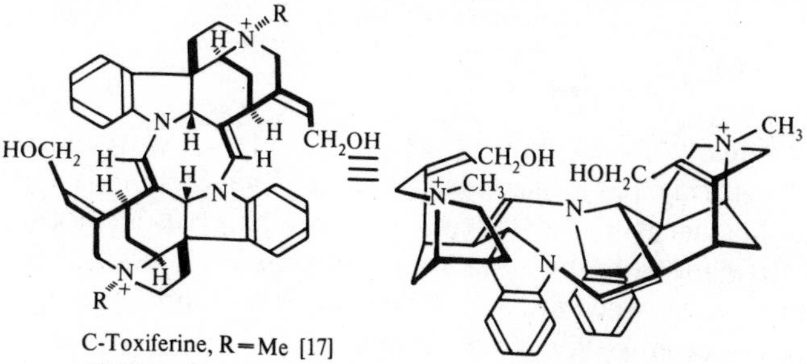

C-Toxiferine, R = Me [17]
Alloferine, R = CH$_2$—CH=CH$_2$ [18]

containing a biazacyclooctadiene ring. A similar enzyme-controlled reaction is likely to prevail in nature to afford the various *C. curare* alkaloids. Mild *in vitro* equilibrations between C-toxiferine [17] and caracurine V-dimetho salt [19] and between alloferine [18], tautoferine [22], and the $N_{(b)}N_{(b)}$-diallyl-caracurine V salt [20] are known.[8] The same equilibria may be invoked to explain the biogenetic origins of the various alkaloids of this type (Scheme 7.4). 12-Hydroxyisostrychnobiline [23], an alkaloid isolated

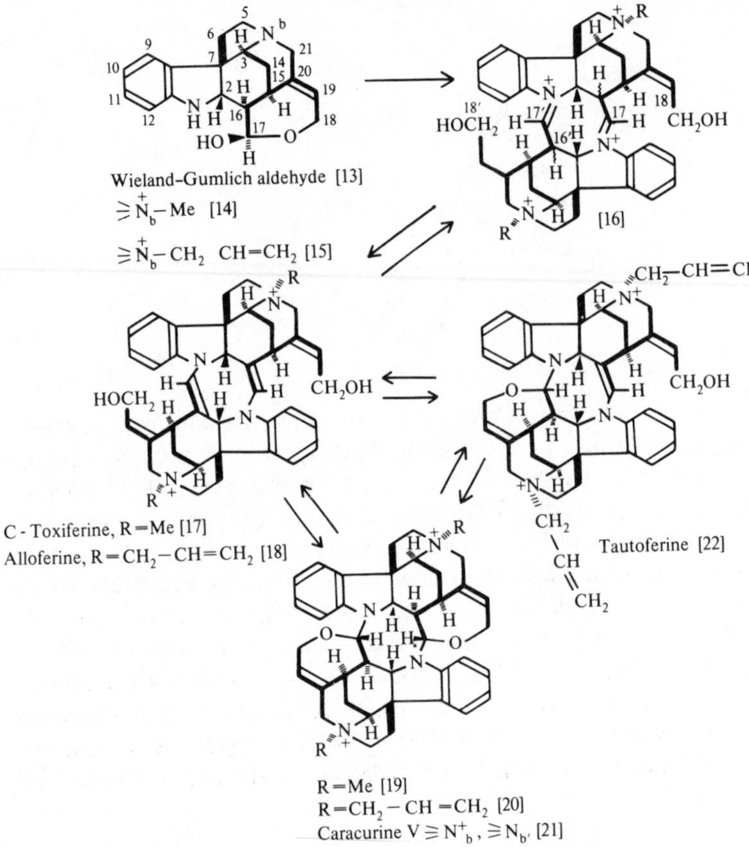

Wieland–Gumlich aldehyde [13]
$\geqq N_b^+ - Me$ [14]
$\geqq N_b^+ - CH_2\ CH=CH_2$ [15]

C - Toxiferine, R=Me [17]
Alloferine, R=CH₂−CH=CH₂ [18]

Tautoferine [22]

R=Me [19]
R=CH₂−CH=CH₂ [20]
Caracurine V ≧ N⁺_b , ≧ N_b' [21]

Scheme 7.4

from *Strychnos variabilis*, falls in the same group.[9] A different bis-indoline alkaloid sungucine [24] has been isolated from *Strychnos ijaca*[10] which shows bonding across $C_{23}-C_5$ between the two parts of the molecule, and as such it is different from the other binary *Strychnos* alkaloids, e.g. toxiferine and strychnobiline which are derived from the Wieland–Gumlich aldehyde.

The geissospermine-type binary indole alkaloid [25] shows a *Strychnos* structure (geissoschizoline) in combination with a *Corynanthe* unit

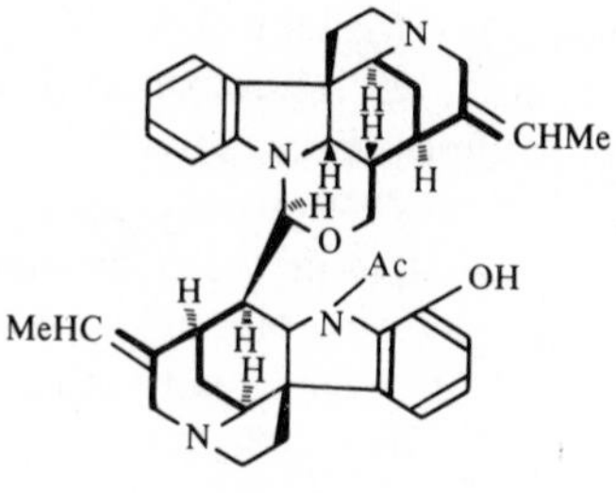

12-Hydroxyisostrychrobiline [23]

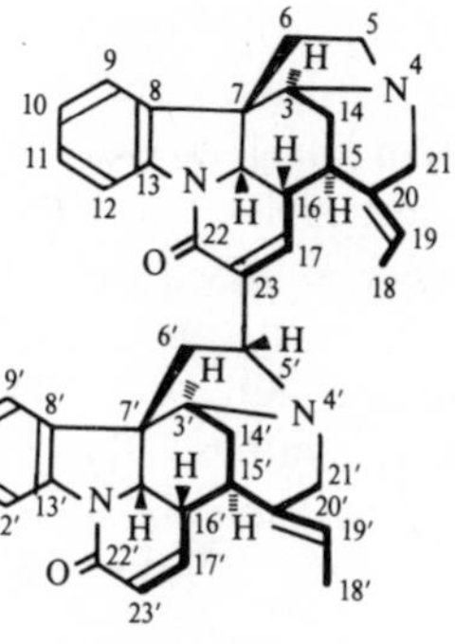

Sungucine [24]

(geissoschizine). A partial biomimetic synthesis of geissospermine has been achieved starting from these monomeric units in 10 per cent acetic acid.[11, 12]

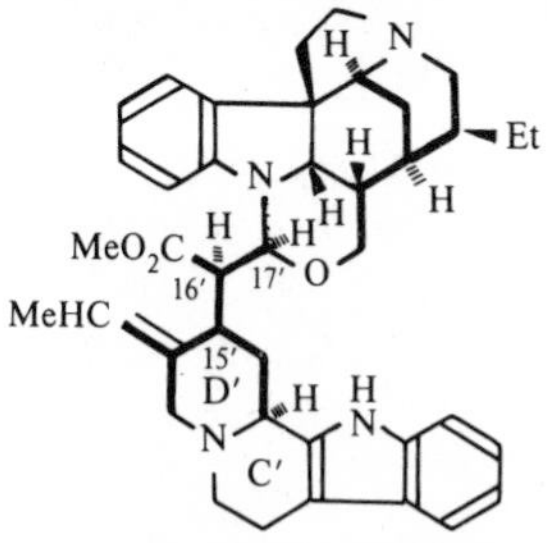

Geissospermine [25]

Some binary indole alkaloids have been isolated from *Alstonia* species. These contain the base macroline [27] in combination with another indole

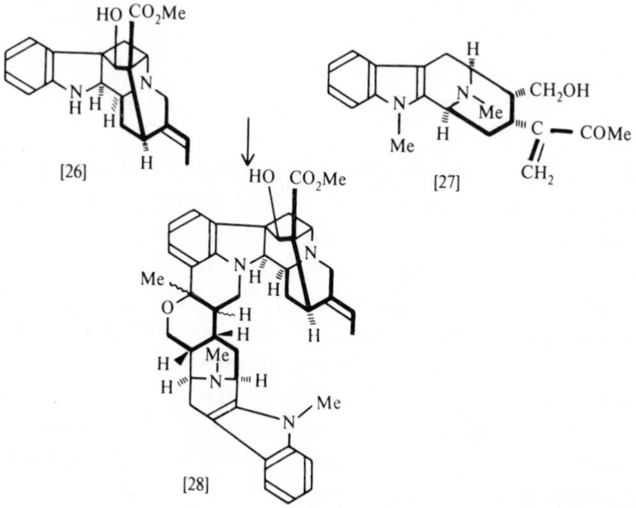

[26]

[27]

[28]

SCHEME 7.5

moiety. Biomimetic syntheses of many of these alkaloids have been achieved.[13-19] Alstonidine [28], for instance, has been prepared by reaction of quebrachidine [26] with macroline [27] (Scheme 7.5). Macroline has not itself been isolated as a natural product, but it has been considered as a likely biogenetic precursor to these bis-indoles, either directly or through a closely related analogue.

Villalstonine [30], another binary indole alkaloid from the same species, has also been synthesized by what is regarded as a biomimetic route by the same group by allowing pleiocarpamine [29] to react with macroline [27] in acid condition[19] [Scheme 6].

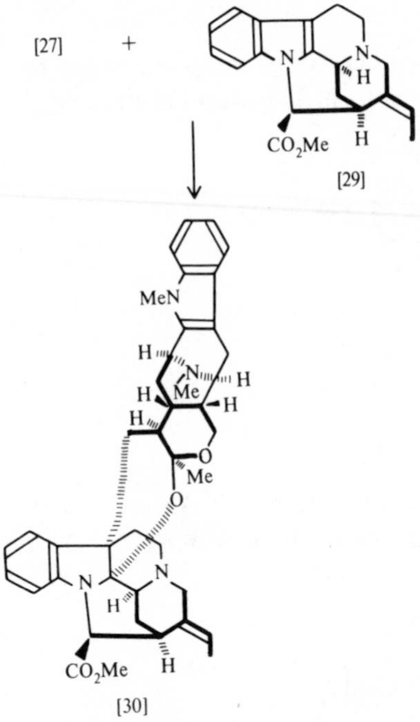

SCHEME 7.6

The presecamines are a group of bis-indole alkaloids isolated from *Rhazya stricta* and *Rhazya orientalis*.[20, 21] When heated in vacuum, presecamine [33] undergoes a retro Diels–Alder reaction to give secodine [31]. Tetrahydro-presecamine [34] on similar treatment affords dihydrosecodine [32] on standing. Secodine [31] and dihydrosecodine [32] dimerize to afford presecamine [33] and tetrahydropresecamine [34] respectively (Scheme 7.7). Acid-catalysed rearrangement of presecamines affords secamine [35] by a reaction which probably represents its biogenetic origin also. Tetrahydrosecamine has been isolated from *Amsonia elliptica*[22] while a diastereoisomer of (−)-5,6,5′,6′-tetrahydropresecamine has been found in *Pandaca minutiflora*.[23]

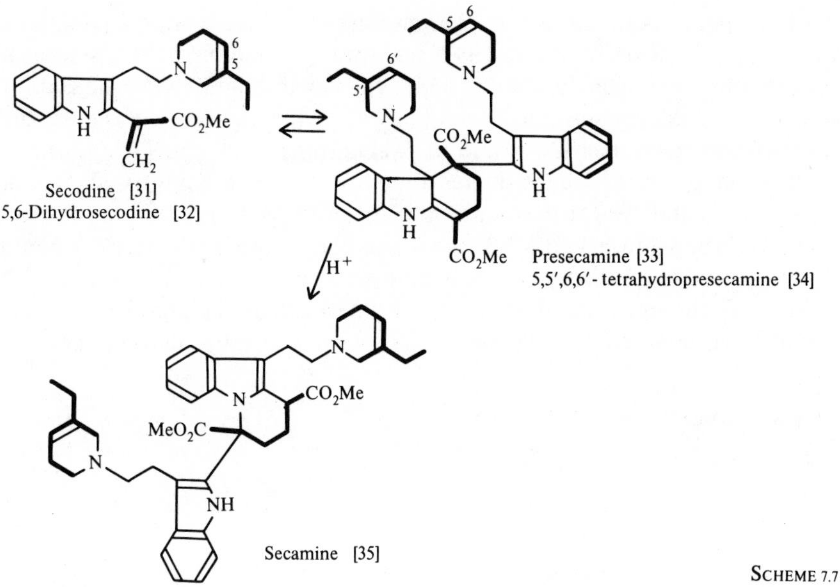

Secodine [31]
5,6-Dihydrosecodine [32]

Presecamine [33]
5,5′,6,6′-tetrahydropresecamine [34]

Secamine [35]

SCHEME 7.7

7.1.4 I–II type

Binary alkaloids formed by the combination of a class I alkaloid with a class II alkaloid are included in this section. Pycnanthine [39] and certain

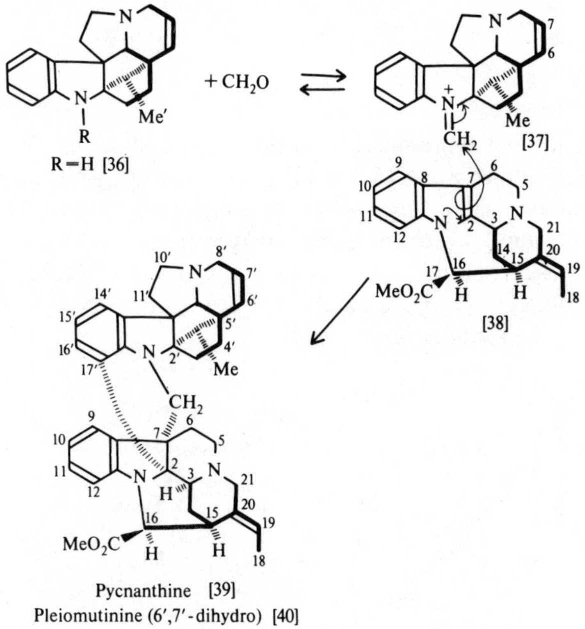

R=H [36]

[37]

[38]

Pycnanthine [39]
Pleiomutinine (6′,7′-dihydro) [40]

SCHEME 7.8

closely-related alkaloids such as pleiomutinine [40] fall among alkaloids of this category. In pycnanthine [39], an isotuboxenine unit [36] is seen in combination with pleiocarpamine [38]. In pycnanthinine [42], the class II moiety is 6,7-dehydro-aspidospermidine [41] while the class I moiety is the same pleiocarpamine [38] found in pycnanthine [39] and pleiomutinine [40]. The probable biosynthetic route to these alkaloids is shown in Schemes 7.8 and 7.9 for pycnanthine [39] and pycnanthinine [42] respectively. As shown in Scheme 7.8, the β-position of the indole nucleus could attack an electrophilic one-carbon unit attached to $N_{(a)}$ in isotuboxenine [37] and the resulting indoleninium intermediate is attacked by the aromatic carbon ortho to the basic nitrogen to afford pycnanthine [39].

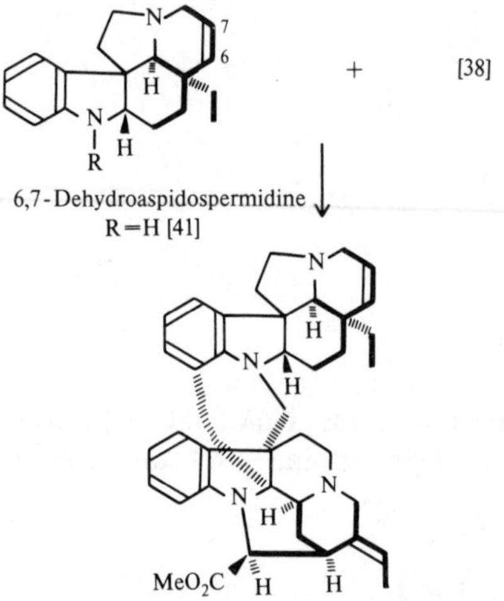

6,7-Dehydroaspidospermidine
R=H [41]

Pycnanthinine [42]   SCHEME 7.9

Umbellamine [43] is another representative of this type of alkaloid. One of the moieties in umbellamine is 14,15-dihydroeburnamenine while the other is pseudoakuammigine. Umbellamine can be considered to arise by the attack of the carbon atom *para* to the basic nitrogen in pseudoakuammigine on to a suitable electrophilic centre created at C-14 in the eburnamenine system. Ervafoline [44], an alkaloid isolated from *Stenosolen heterophyllus*, belongs to the same series.[24]

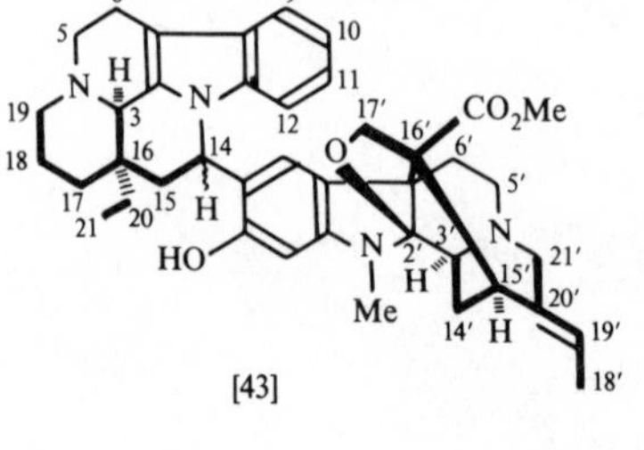

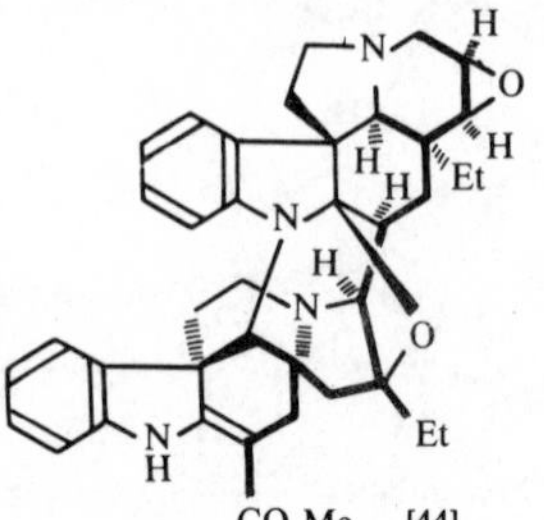

7.1.5 I–III type

Voacamine [49] and related binary alkaloids are included in this section. One of the moieties in voacamine is derived from the class I alkaloid vobasine [46] while the other is the *Iboga* (class III) alkaloid, voacangine [48]. Voacamine can be considered to arise by the generation of an electrophilic delocalized carbonium ion centre at the carbon α- to the indole ring in vobasinol which is then attacked by the activated carbon α- to the methoxyl group in voacangine [48] (Scheme 7.10). A biomimetic synthesis

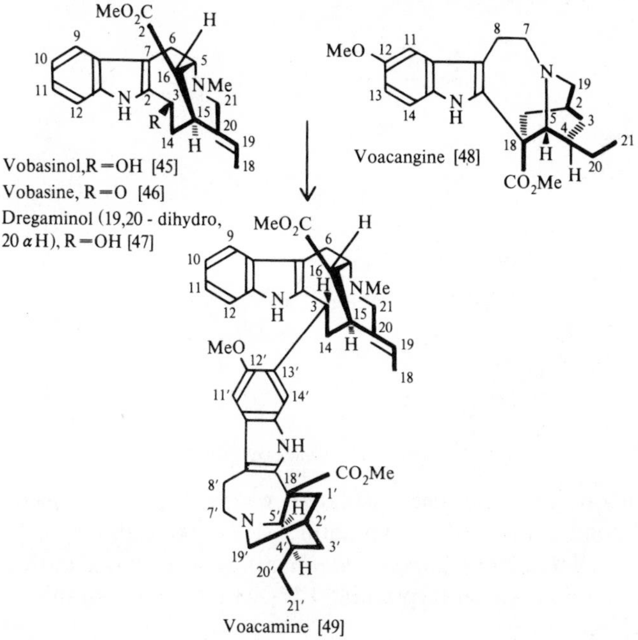

Vobasinol, R = OH [45]
Vobasine, R = O [46]
Dregaminol (19,20 - dihydro, 20 α H), R = OH [47]

Voacangine [48]

Voacamine [49]

Scheme 7.10

of 19,20-dihydrovoacamine from dregaminol [47] and voacangine [48] has been reported.[25] Tabernamine [50] also shows a sarpagine unit (class I) in combination with an ibogamine-type structure. The structure has been

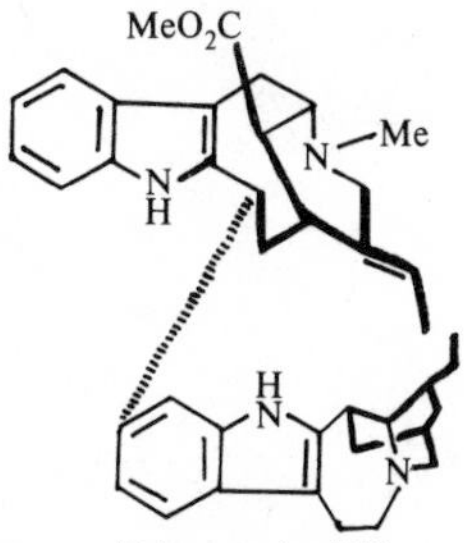

Tabernamine [50]

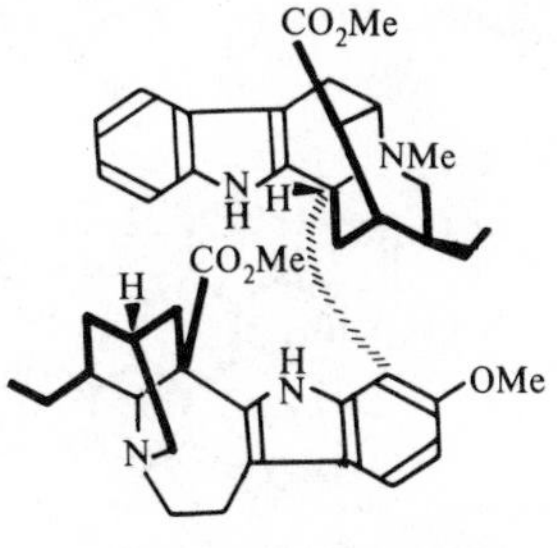

Tabernaelegantine A [51]

confirmed by its synthesis form ibogamine and vobasinol,[25] and other alkaloids of closely related structure have also been isolated, e.g. tabernaelegantine [51].[26]

7.1.6 I–IV type

Cinchophyllamine [54] is a binary alkaloid of I–IV type. It can be considered to arise from the precursor [52] which is closely related to quinamine [55], an alkaloid which co-occurs with cinchophyllamine in *Cinchona ledgeriana* (Scheme 7.11).

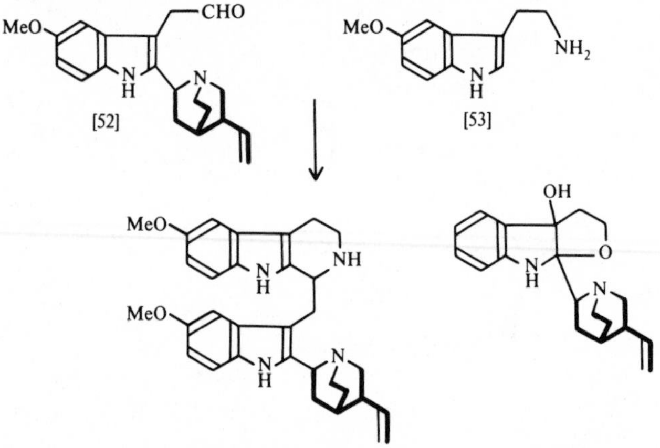

Cinchophyllamine [54]   Quinamine [55]   SCHEME 7.11

A number of alkaloids have been isolated which contain a seco-yohimbine unit in combination with tryptamine. The roxburghines, e.g. roxburghine D [58] isolated from *Uncaria gambier* may be included in this type. They arise by combination of tryptamine [57] with the corynantheine derivative [51] (Scheme 7.12).

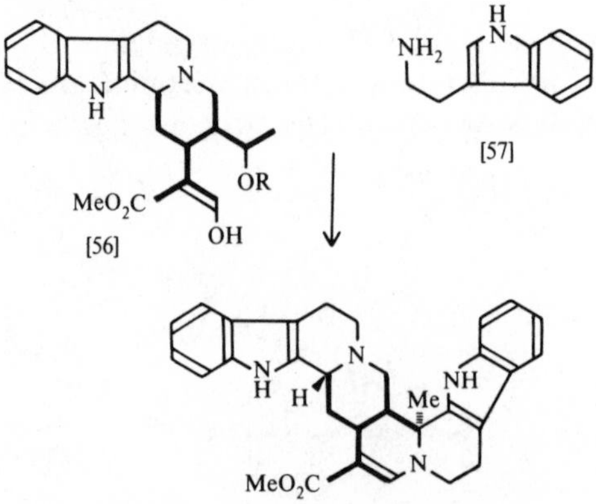

Roxburghine D [58]   SCHEME 7.12

Ochrolifuanine [60] also exhibits a I–IV type structure.[27] Such alkaloids can be considered to arise from tryptamine [57] and a decarboxylated *Corynanthe* derivative [59] as shown in Scheme 7.13.

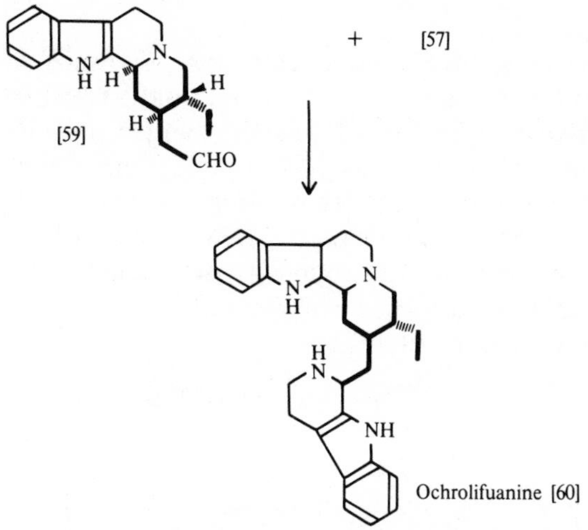

SCHEME 7.13

A number of closely related alkaloids have been isolated during the last few years. These include usambarensine [61],[28] 18,19-dihydrousambarine [62],[29] usambarine,[30] and nigritanine.[31] A biomimetic synthesis of usambarine by condensation of (±)-geissoschizol with N_b-methyltryptamine has been achieved.[30]

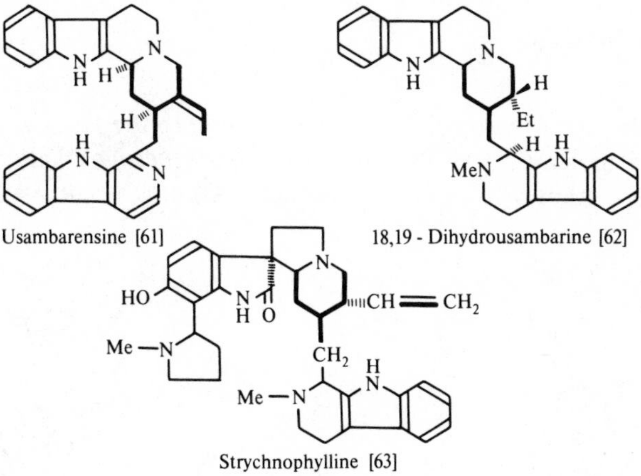

Usambarensine [61]

18,19 - Dihydrousambarine [62]

Strychnophylline [63]

A number of binary oxindole alkaloids of Class I type in combination with tryptamine have been isolated. Strychnophylline [63] belongs to this group of alkaloids,[32] and it can be considered to be formed from an usam-barensine derivative by rearrangement of the corresponding hydroxyin-dolenine.

7.1.7 II–II type

Pleiomutine [66] is included under this alkaloid type in which both moieties of the binary molecule consist of ring II alkaloids. One of the building blocks is the *Hunteria* alkaloid 14,15-dihydroeburnamenine while the other is pleiocarpinine [65]. A plausible route to this binary system is the attack of the carbon atom *para* to the indoline nitrogen in pleiocarpinine [65] on to the hydroxyl-bearing potentially electrophilic carbon in (−)-eburnamine [64] (Scheme 7.14). This coupling reaction has been suc-cessfully accomplished in the laboratory by reaction of (−)-pleiocarpinine with (−)-eburnamine under acid conditions.[33, 34]

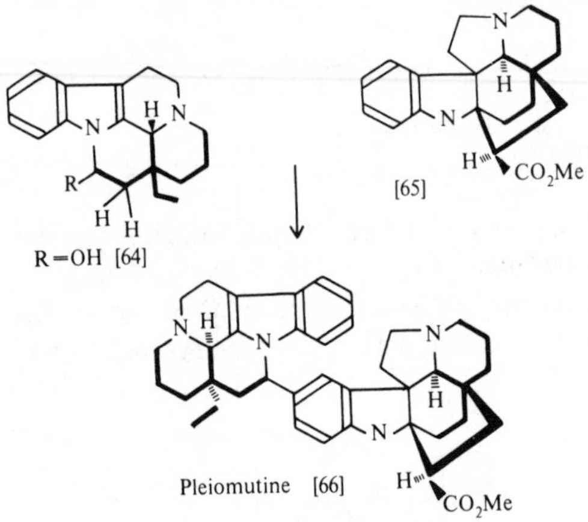

R=OH [64]

[65]

Pleiomutine   [66]

SCHEME 7.14

Alkaloids such as vobtusine [67] also belong to the II–II alkaloid type. One of the moieties in vobtusine is the *Aspidosperma* alkaloid beninine. The other moiety contains another *Aspidosperma* system with

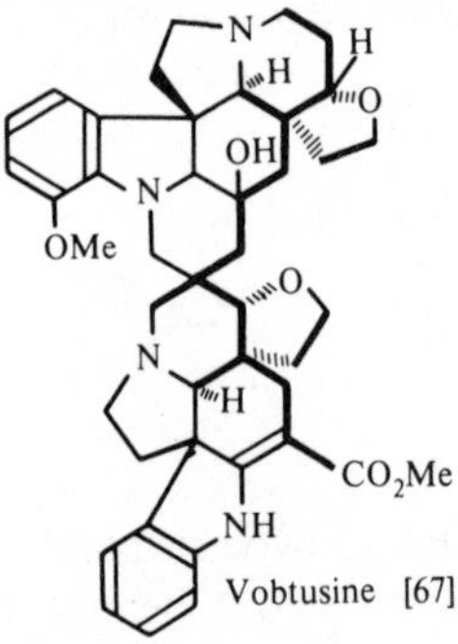

Vobtusine [67]

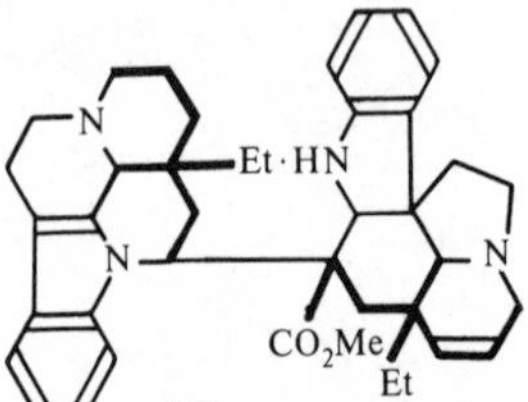

[68]

vincadifformine-like anilinoacrylate structure. The electrophilic nature of C-22′ and C-23′ in one moiety and the nucleophilic nature of C-7 and C-20 in the other readily explain the biogenetic origin of vobtusine and related alkaloids. Another II–II type alkaloid isolated form *Melodinus balansae* is shown in [68]. A *Hunteria* skeleton is seen in combination with an *Aspidosperma* unit in this alkaloid.

7.1.8 II–III type

The binary anti-tumour alkaloids vinblastine [1] and vincristine [2] occurring in *Catharanthus roseus* serve as examples of this type of alkaloid. It had generally been assumed that these alkaloids arise by combination of vindoline [69] with a suitably activated tetracyclic cleavamine unit such as

Vinblastine, R = CH₃ [1]
Vincristine, R = CHO [2]

[69]

[70]

[71]

[71]. All the earlier synthetic efforts to vinblastine were directed at synthesizing the cleavamine units and activating them for attack by vindoline.[34–39] In spite of intensive investigation of *C. roseus*, cleavamines have not been isolated from this plant. This led to a novel biogenetic hypothesis[40, 44] that vinblastine [1] or vincristine [2] may arise by the attack of vindoline directly at C-16 of an *Iboga* alkaloid such as a suitable derivative of catharanthine [70] with the concomitant cleavage of C_{16}–C_{21} bond to afford the desired nine-membered nitrogen containing 'cleavamine' moiety.

The possibility of preparing vinblastine analogues from catharanthine and vindoline was first demonstrated by the synthesis of 16-epi-anhydrovinblastine [74][40] from 16-carbomethoxycleavamine [72] prepared by reduction of catharanthine under acid conditions (Scheme 7.15). Subsequently, Potier[44, 48] and others[41–43, 45–47, 49–52] developed successful coupling procedures based on the modified Polonovski reactions

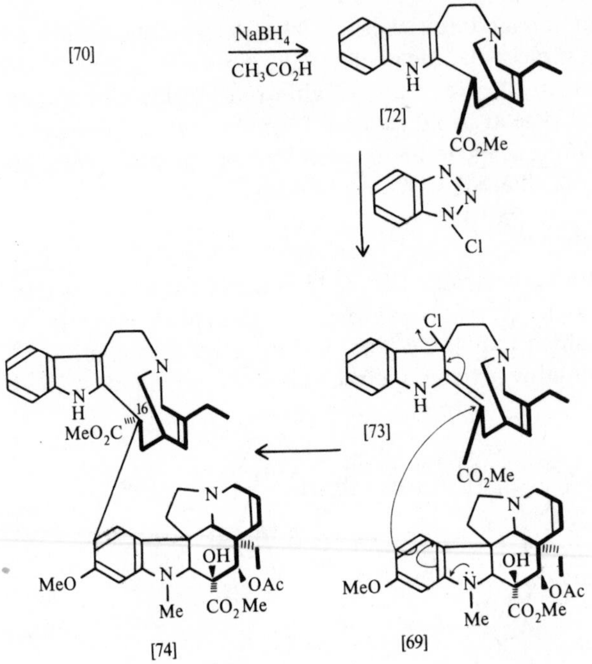

SCHEME 7.15

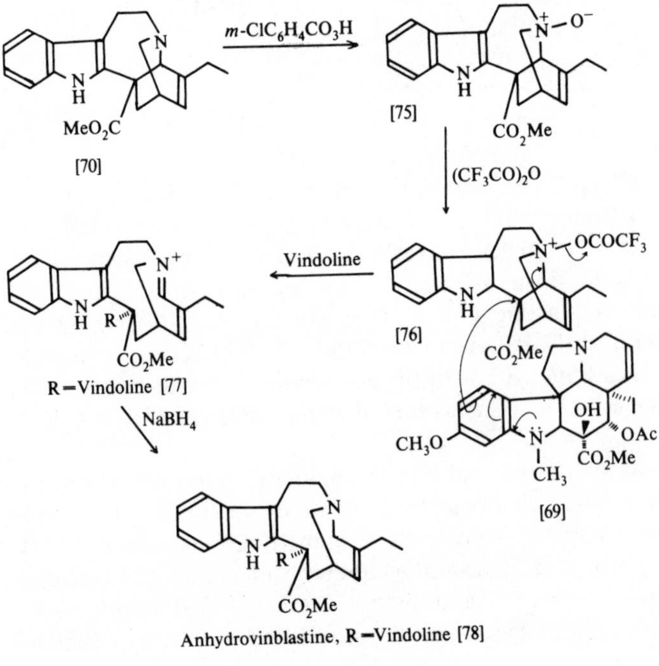

SCHEME 7.16

starting from vindoline and catharanthine (Scheme 7.16). A number of biogenetically-oriented partial syntheses have subsequently been reported to vinblastine[46-48] and its derivatives, all starting from catharanthine and vindoline.[41-52] Anhydrovinblastine [78] prepared from catharanthine and vindoline has been converted to vinblastine [1] and vinrosidine [82] in a biomimetic sequence of reactions[47] (Scheme 7.17) and a parallel approach has later been reported by the French group.[48] It has been shown that anhydrovinblastine can be enzymically converted to vinblastine.[67]

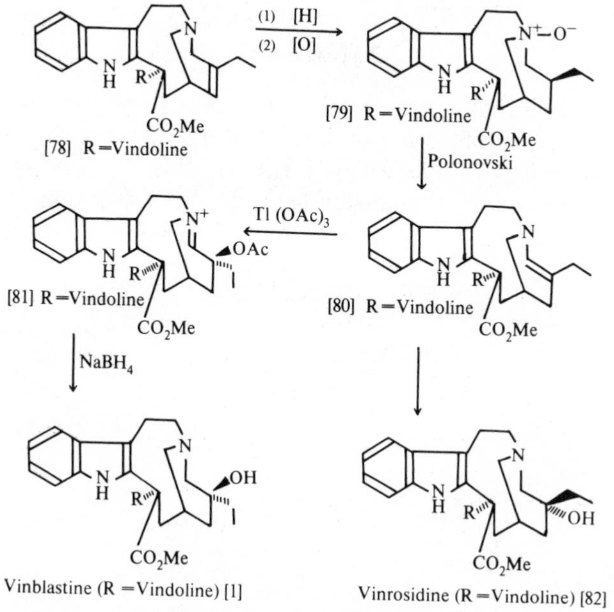

Vinblastine (R = Vindoline) [1]

Vinrosidine (R = Vindoline) [82]

A biomimetic synthesis of vinblastine[47]

SCHEME 7.17

Feeding of [³H-CO₂CH₃]-catharanthine and [¹⁴C-OCOCH₃]-vindoline to apical cuttings of 3–4 month-old *C. roseus* plants afforded low but definite incorporations of both alkaloids into vinblastine, demonstrating that these alkaloids are indeed precursors of vinblastine.[53] It has also been shown that anhydrovinblastine [78] obtained by the coupling of vindoline with catharanthine-N-oxide could be converted to the major binary antitumour alkaloids such as leurosine [83], catharine [86], vinblastine [1], leurosidine [82], and their deoxy derivatives [84] and [85] simply by aerial oxidation affected by stirring a solution of anhydrovinblastine in acetonitrile for a few hours.[54] This raises the interesting question of whether many of the binary alkaloids such as leurosine are 'natural artefacts'. While anhydrovinblastine [78] is not normally isolable from *C. roseus* extracts

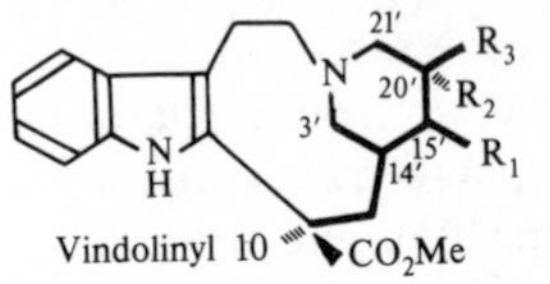

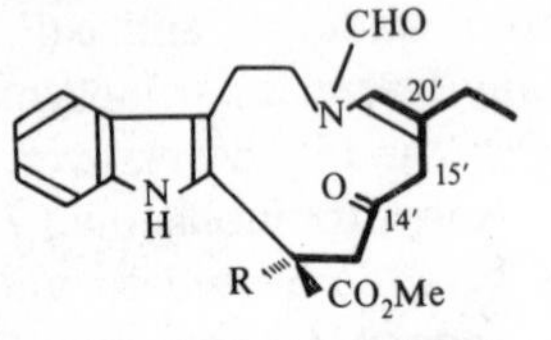

Catharine [86] (R = vindoline)

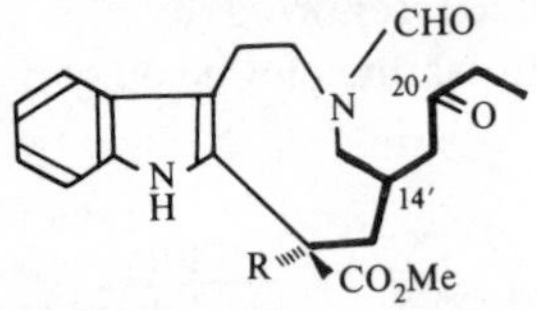

Catharinine [87] (R = vindoline)

	R_1	R_2	R_3
Anhydrovinblastine	$\Delta^{15'-20'}$		Et [78]
Leurosine	— O —		Et [83]
Vinblastine	H	Et	OH [1]
Leurosidine	H	OH	Et [82]
Deoxyvinblastine	H	Et	H [84]
Deoxyleurosidine	H	H	Et [85]

because of its relative instability,[55] Scott has demonstrated that anhydrovinblastine formed from catharanthine and vindoline can be isolated if the extraction is carried out swiftly.[56] There is growing evidence to suggest that the appropriate functionalizations in the piperidine ring of the indole moiety of vinblastine occur in the later biosynthetic stages.[57-59] A

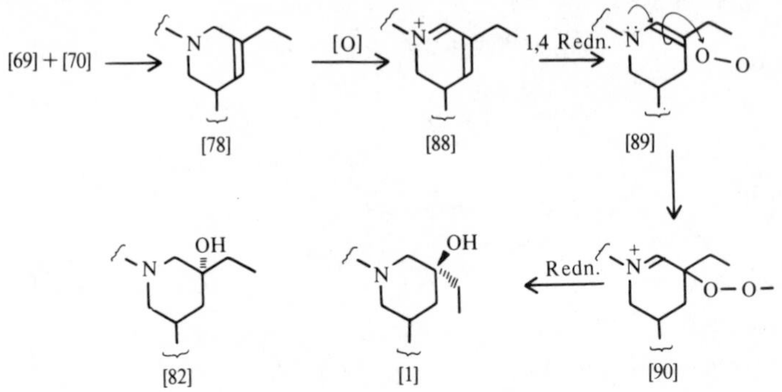

SCHEME 7.18

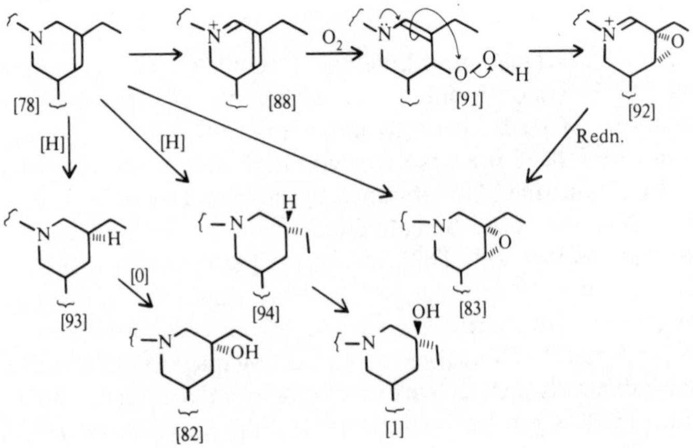

SCHEME 7.19

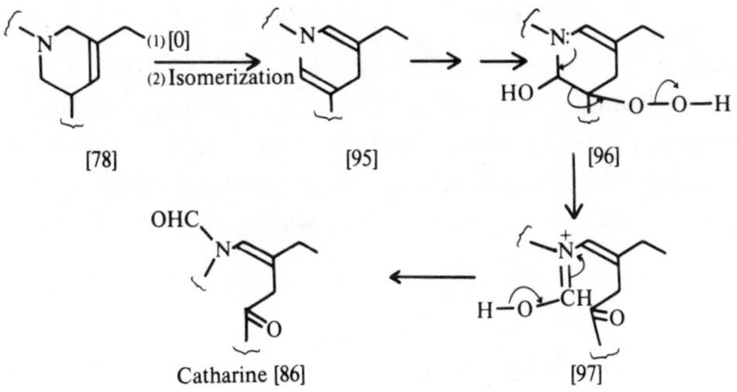

Catharine [86]   [97]   SCHEME 7.20

number of routes can be visualized from anhydrovinblastine to the various binary alkaloids of this type, and some of these are shown in Schemes 7.18, 7.19 and 7.20. The fact that labelled anhydrovinblastine was incorporated into vinblastine by cell-free preparations of *C. roseus*[59] strongly supports these or closely related pathways to the binary indole alkaloids of this type.

Anhydrovinblastine [78] has been found[55] to be rapidly oxidizable to leurosine [83] with various oxidants, including aerial oxidation. Yields of up to 40 per cent of leurosine by aerial oxidation of anhydrovinblastine have been reported which suggests that leurosine may not be a 'natural' alkaloid but formed by oxidation during the extraction process.

It is interesting that no catharanthine derivatives bearing oxygen functionalities at C-15 or C-20 have been isolated. This strongly suggests that the functionalization of the olefinic bond in the piperidine ring of the indole moiety of vinblastine takes place after the combination of catharanthine (as the N-oxide?) with vindoline.

7.1.9 III–IV type

The alkaloid bonafousine [98] has been isolated from *Bonafousia tetrastachya*.[60] The nucleophilic carbon *ortho* to the phenolic OH group in the isovoacangine unit can attack an electrophilic site *α*- to the indole nitrogen in the hydrocanthine moiety.

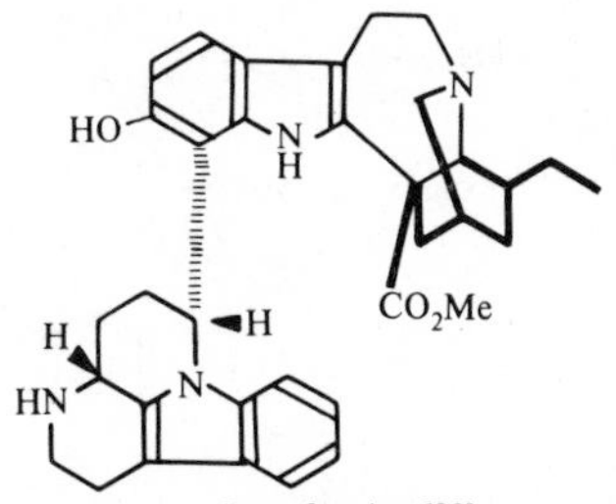

Bonafousine [98]

7.1.10 IV–IV type

Several binary alkaloids which occur in *Calycanthus* and *Chimonanthus* species are formed by combination of two class IV type structures, folicanthine [100] being a typical example. It has been demonstrated in *Calycanthus floridus* plants that [2-^{14}C]-tryptophan is specifically incorporated into folicanthine.[61] It thus seems very probable that folicanthine [100] arises

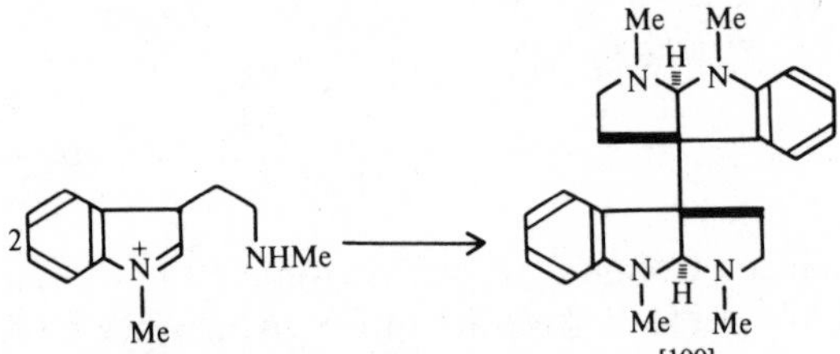

SCHEME 7.21

by the coupling of two indolyl radicals (Scheme 7.21). Other related alkaloids of the same series may also arise by radical coupling of N,N-dimethyltryptamine derivatives. The alkaloids chaetocin [101] and

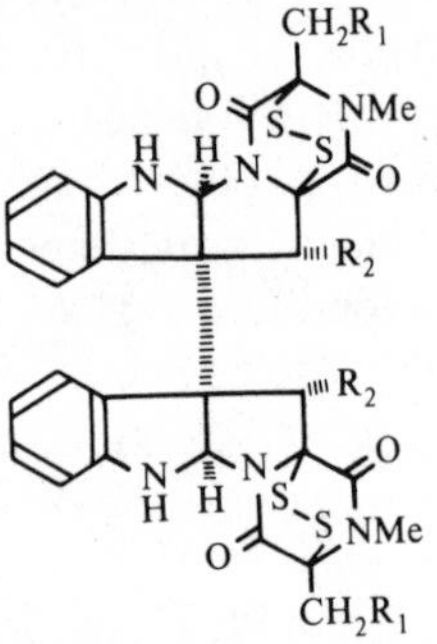

Chaetocin, R$_1$=OH, R$_2$=H [101]
Verticillin A, R$_1$=H, R$_2$=OH [102]

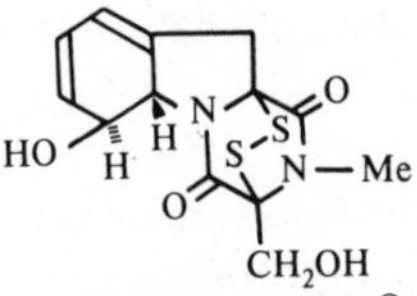

Gliotoxin [103]

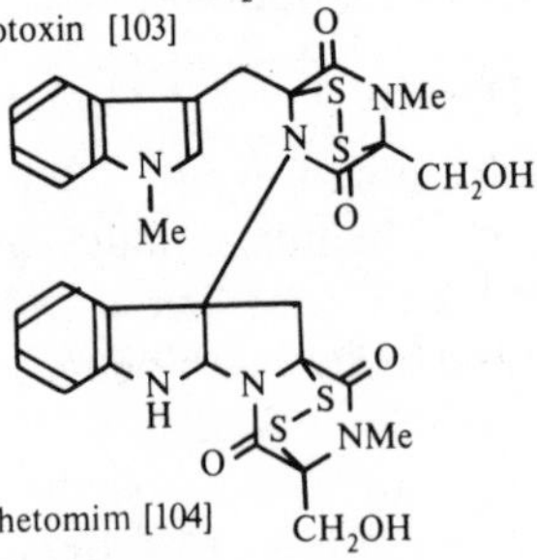

Chetomim [104]

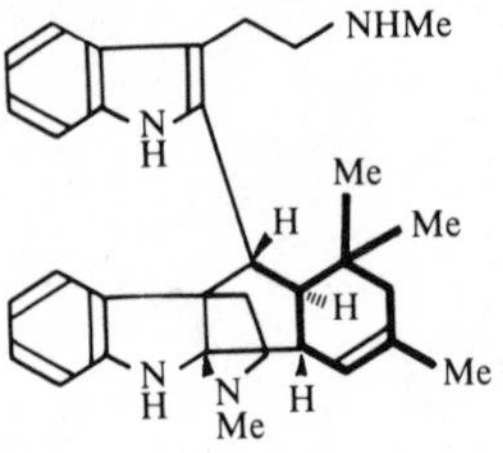

[105]

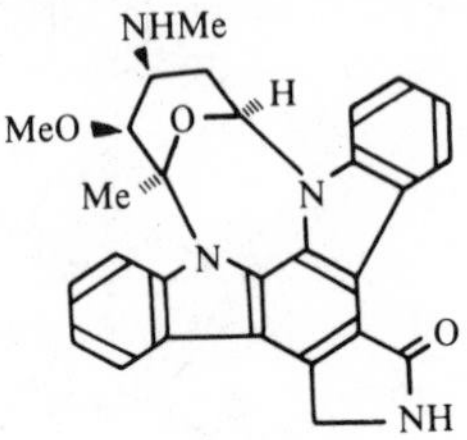

[106]

verticillin A [102] isolated from *Chaetomium minutum* consist of two gliotoxin [103]-type units in combination.[62, 63] A closely related alkaloid, chetomin [104], is a toxic metabolite of *Chaetomium cochiliodes* and *C. globusum*.

The bis-indole alkaloids [105] and [106] which have been isolated from *Flindersia fournier*[64] and *Streptomyces* sp.[65] show two class IV-type units in combination.

7.2 Trimeric and tetrameric indole alkaloids

Hodgkinsine [107], isolated from *Hodgkinsonia frutescens* exhibits a trimeric structure.[66] Such alkaloids probably arise in a similar way as folicanthine, described earlier. The third unit added is also an N-methyltryptamine derivative which has undergone radical coupling with the aromatic ring of the dimeric moiety.

A repetition of the same process is observed to afford tetramers in quadrigemine A [108] and quadrigemine B which co-occur with hodgkinsine in the same plant.

[107]   [108]

7.3 Summary

A number of binary indole alkaloids occur in nature in which two different alkaloidal units have become linked, often by C—C bonds. The individual moieties may be true indoles or may be found in modified forms such as indolines, N-acylindolines, indolenines, oxindoles, etc. The formation of these binary alkaloids usually occurs by the attack of an electron-rich centre in one molecule (which may be an enamine, enolate, activated aromatic ring, etc.) on to an electron-deficient carbon (or a carbon bearing a good leaving group) in the other molecule. Because of the structural complexity of these substances it is only during the last decade that the biosynthetic pathways to some of them have been established, and the biomimetic partial syntheses to many have been achieved. Some alkaloids with trimeric and tetrameric structures (e.g. hodgkinsine) [107] and quadrigemine A [108], respectively, have also been isolated.

References

1. YATES, P., MACLACHLAN, F. N., RAE, I. D., ROSENBERGER, M., SZABO, A. G., WILLIS, C. R., CAVA, M. P., BECHFOROUZ, M., LAKSHMIKANTHAM, M. V., and ZEIGER, W. *J. Am. Chem. Soc.* **95**, 7842 (1973).
2. CHENG, P. T., NYBURG, S. C., MACLACHLAN, F. M., and YATES, P. *Can. J. Chem.* **54**, 726 (1976).
3. BRAUCHLI, P., DEULOFEU, V., BUDZIKIEWICZ, H., and DJERASSI, C. *J. Am. Chem. Soc.* **86**, 1895 (1964).
4. KRAJICEK, A., TRTIK, B., SPACIL, J., SEDMERA, P., VOKOUN, J., and REHACEK, Z. *Coll. Czech. Chem. Commun.* **44**, 2255 (1979).
5. WIELAND, H. and KAZIRO, K. *Justus Liebigs Ann. Chem.* **506**, 60 (1933).
6. WIELAND, H. and GUMLICH, W. *Justus Liebigs Ann. Chem.* **494**, 191 (1932).
7. EDWARDS, P. N. and SMITH, G. F. *J. Chem. Soc.* 1952 (1961).
8. FURST, A., BOLLER, A., and ELS, H. unpublished results.
9. TITS, M., TAVERNIER, D., and ANGENOT, L. *Phytochemistry* **18**, 515 (1979).
10. LAMOTTE, J., DUPONT, L., DIDEBERG, O., KAMBU, K., and ANGENOT, L. *Tetrahedron Lett.* 4227 (1979).
11. PUISIEUX, F., GOUTAREL, R., JANOT, M. M., LEMEN, J., and LE HIR, A. *C. R. Acad. Sci., Paris* **250**, 1285 (1960).
12. CHIARONI, A. and RICHE, C. *Acta Crystallogr.* **B35**, 1820 (1979).
13. BURKE, D. E., COOK, J. M., and LEQUESNE, P. W. *Chem. Commun.* 697 (1972).
14. COOK, J. M., LEQUESNE, P. W., and ELDERFIELD, R. C. *J. Chem. Soc. Chem. Commun.* 1306 (1969).
15. BURKE, D. E., DEMARKEY, C. A., LEQUESNE, P. W., and COOK, J. M. *J. Chem. Soc. Chem. Commun.* 1346 (1972).
16. BURKE, D. E., COOK, J. M., and LEQUESNE, P. W. *J. Am. Chem. Soc.* **95**, 546 (1973).
17. GARNICK, R. L. and LEQUESNE, P. W. *Tetrahedron Lett.* 3249 (1976).
18. GARNICK, R. L. and LEQUESNE, P. W. *J. Am. Chem. Soc.* **100**, 4213 (1978).
19. BURKE, D. E. and LEQUESNE, P. W. *Chem. Commun.* 678 (1972).
20. CORDELL, G. A., SMITH, G. F., and SMITH, G. N. *Chem. Commun.* 191 (1970).
21. BROWN, R. T., SMITH, G. F., STAPLEFORD, K. S. J., and TAYLOR, D. A. *Chem. Commun.* 190 (1970).
22. SAKAI, S., AIMI, N., KATO, K., IDO, H., MASUDA, K., WATANABE, Y., and HAGINIWA, J. *Yakugaku Zasshi* **95**, 1152 (1975).
23. PETITFRERE, E., MORFAUX, A. M., DEBRAY, M. M., LEMEN-OLIVIER, L., and LEMEN, J. *Phytochemistry* **14**, 1648 (1975).
24. HENRIQUES, A., KAN-FAN, C:. AHOND, A., RICHE, C., and HUSSON, H. P., *Tetrahedron Lett.* 3707 (1976).

25. KINGSTON, D. G. I., GERHART, B. B., and IONESCU, F. *Tetrahedron Lett.* 649 (1976).
26. BOMBARDELLI, E., BONATI, A., GABETTA, B., MARTINELLI, E. M., MUSTICH, G., and DANIELI, B. *J. Chem. Soc. Perkin Trans.* I, 1432 (1976).
27. KOCH, M. C., PLAT, M. M., PREAUX, N., GOTTLIEB, H. E., HAGAMAN, E. W., SCHELL, F. M., and WENKERT, E. *J. Org. Chem.* **40**, 2836 (1975).
28. DIDEBERG, O., DUPONT, L., and ANGENOT, L. *Acta Crystallogr.* **B31** 1571 (1975).
29. ANGENOT, L., COUNE, C., and TITS, M. *J. Pharm. Belg.* **33**, 11 (1978).
30. ANGENOT, L., COUNE, C., TITS, M., and YAMADA, K. *Phytochemistry* **17**, 1687 (1978).
31. OGUAKWA, J. U., GALEFFI, C., MESSANA, I., LABUA, R., NICOLETTI, M., and MARINI BETTOLO, B. *Gazz. Chim. Ital.* **108**, 615 (1978).
32. ANGENOT, L. *Planta. Med. Phytother.* **12**, 123 (1978).
33. THOMAS, D. W., ACHENBACH, H., and BIEMANN, K. *J. Am. Chem. Soc.* **88**, 1537 (1966).
34. HARLEY-MASON, J., ATTA-UR-RAHMAN, and BEISLER, J. A. *J. Chem. Soc. Chem. Commun.* 743 (1966).
35. HARLEY-MASON, J. and ATTA-UR-RAHMAN *J. Chem. Soc. Chem. Commun.* 208 (1967).
36. HARLEY-MASON, J. and ATTA-UR-RAHMAN, *J. Chem. Soc. Chem. Commun.* 1048 (1967).
37. HARLEY-MASON, J. and ATTA-UR-RAHMAN, *Chem. Ind. (Lond.)* **52**, 1845 (1968).
38. ATTA-UR-RAHMAN and HARLEY-MASON, J. *Tetrahedron* **36**, 1063 (1980).
39. HARLEY-MASON, J. and ATTA-UR-RAHMAN, *Tetrahedron* **36**, 1057 (1980).
40. ATTA-UR-RAHMAN, *Pakistan J. Sci. Ind. Res.* **14**, 487 (1971).
41. ATTA-UR-RAHMAN, BASHA, A., WAHEED, N., and GHAZALA, M. *Z. Naturforsch* **31B**, 1416 (1977).
42. ATTA-UR-RAHMAN, *J. Chem. Soc. Pakistan* **1**, 81 (1979).
43. KUTNEY, J. P., GREGONIS, D. E., IMHOF, R., ITOH, I., JAHNGEN, E., SCOTT, A. I., and CHAN, W. K. *J. Am. Chem. Soc.* **97**, 5013 (1975).
44. POTIER, P., LANGLOIS, N., LANGLOIS, Y., and GUERITTE, F. *Chem. Commun.* 670 (1975); *Idem. J. Am. Chem. Soc.* **98**, 7017 (1976).
45. ATTA-UR-RAHMAN, R. WAHEED, N., and GHAZALA, M. *Z. Naturforsch.* **31B**, 264 (1976).
46. ATTA-UR-RAHMAN, BASHA, A., and GHAZALA, M. *Tetrahedron Lett.* 2351 (1976).
47. ATTA-UR-RAHMAN, Pakistan Patent No. 126852, 14 February 1978.
48. MANGENEY, P., ANDRIAMIALISOA, R. Z., LANGLOIS, N., LANGLOIS, Y., and POTIER, P. *J. Am. Chem. Soc.* **101**, 2243 (1979).
49. KUTNEY, J. P., BALSEVICH, J., BOKELMAN, G. H., HIBINO, T., ITOH, I., and RATCLIFFE, A. H. *Heterocycles* **4**, 997 (1976).
50. HONMA, Y. and BAN, Y. *Tetrahedron Lett.* 155 (1978).
51. KUTNEY, J. P., BALSEVICH, J., BOKELMAN, G. H., HIBINO, T., HONDA, T., ITOH, I., RATCLIFFE, A. H., and WORTH, B. R. *Can. J. Chem.* **56**, 62 (1978).
52. KUTNEY, J. P., JOSHUA, A. V., LIAO, P-H., and WORTH, B. R. *Can J. Chem.* **55**, 3233 (1977).
53. HASSAM, S. B. and HUTCHINSON, C. R. *Tetrahedron Lett.* 1681 (1978).
54. LANGLOIS, N. and POTIER, P. *J. Chem. Soc. Chem. Commun.* 582 (1979).
55. LANGLOIS, N. and POTIER, P. *J. Chem. Soc. Chem. Commun.* 102 (1978).

56. SCOTT, A. I., GUERITTE, F., and LEE, S. L. *J. Am. Chem. Soc.* **100**, 6253 (1978).
57. STUART, K. L., KUTNEY, J. P., and WORTH, B. R. *Heterocycles* **9**, 1015 (1978).
58. STUART, K. L., KUTNEY, J. P., HONDA, T., and WORTH, B. R. *Heterocycles* **9**, 1391, 1419 (1978).
59. BAXTER, R. L., DORSCHEL, C. A., LEE, S. L., and SCOTT, A. I. *J. Chem. Soc. Chem. Commun.* 257 (1979).
60. DAMAK, M., AHOND, H., DOUCERAIN, H., and RICHE, C. *J. Chem. Soc. Chem. Commun.* 510 (1976).
61. O'DONOVAN, D. G. and KOEGH, M. F. *J. Chem. Soc., (C)* 1570 (1966).
62. HAUSER, D., WEBER, H. P., and SIGG, H. P. *Helv. Chim. Acta* **53**, 1961 (1970).
63. MINATO, H., MATSUMOTO, M., and KATAYAMA, T. *Chem. Commun.* **44** (1971).
64. TILLEQUIN, F., KOCH, M., BERT, M., and SEVENET, T. *J. Nat. Prod.* **42**, 92 (1979).
65. FURUSAKI, A., HASHIBA, N., MATSUMOTO, T., HIRANO, A., IWAI, Y., and OMURA, S. *J. Chem. Soc. Chem. Commun.* 800 (1978).
66. ANET, E. F. L. J., HUGHES, G. K., and RITCHIE, E. *Aust. J. Chem.* **14**, 173 (1961).
67. SCOTT, A. I., LEE, S.-L., CULVER, M. G., WAN, W., HIRATA, T., GUÉRITTE, F., BAXTER, R. L., NORDLÖV, H., DORSCHEL, C. A., MIZUKAMI, H., and MACKENZIE, N. E. *Heterocyles* **15**, 1257 (1981).

8

TABLE OF INCORPORATION DATA

Compound	Precursor	Plant	Incorporation/further information.*	Reference
Agroclavine	Isochanoclavine I	*Claviceps*	+	1–4
	Chanoclavine II	*Claviceps*	+	
	Chanoclavine I	*Claviceps*	+	
	^{13}C-dimethylallyltryptophan	*Claviceps*	0.26	5, 6
	(3R, 4R)-mevalonic			
	2-^{14}C-4t-acid (PP = 5.00)	*Claviceps*	PP = 3.14	7
Ajmalicine	[1-^{14}C]-acetate	*Catharanthus roseus*	0.0003, T	8
	[1-^{14}C]-glycine	*C. roseus*	0	9
	[2-^{14}C]-glycine	*C. roseus*	0.4 U, A, T	9
	[2-^{14}C]-mevalonate	*C. roseus*	0.03	10
	[4-^{3}H, 2-^{14}C]-4R-mevalonate	*C. roseus*	E, V	11
	[4-^{3}H, 2-^{14}C]-4S-mevalonate	*C. roseus*	E, G	12
	[3-^{14}C]-mevalonate	*C. roseus*	E, I	12
	[6-^{14}C]-mevalonate	*C. roseus*	0.06	13
	[1-^{3}H$_2$]-geraniol	*C. roseus*	E, W	14
	[1-^{3}H, 2-^{14}C]-geraniol	*C. roseus*	0.05	11
	[2-^{3}H, 2-^{14}C]-geraniol	*C. roseus*	E, G	12
			E, I	12
	[2-^{14}C]-gernayl pyrophosphate	*C. roseus*	0.16	15
	[1-^{3}H$_2$]-nerol	*C. roseus*	0.13	11
	[2-^{3}H, 2-^{14}C]-nerol	*C. roseus*	E, I	12
	[1-^{3}H$_2$]-10-hydroxygeraniol	*C. roseus*	0.36, P	16
	[8-^{14}C]-10-hydroxygeraniol	*C. roseus*	0.17	14
	[1-^{3}H$_2$]-citronellol	*C. roseus*	0.14	14
			0.0004	
			0.002	
	[1-^{3}H$_2$]-10-hydroxy-citronellol	*C. roseus*	0.001	16
	[1-^{3}H$_2$]-10-oxo-citronellol	*C. roseus*	0.005	16
	[1-^{3}H]-citronellal	*C. roseus*	0	16
	[aryl-^{3}H]-geissoschizine	*C. roseus*	+	17
	[1-^{3}H$_2$]-linalool	*C. roseus*	0	16
	[8-^{14}C]-10-hydroxylinalool	*C. roseus*	0.0004	14
	[OMe-^{3}H]-deoxyloganin	*C. roseus*	0.10	18
	[OMe-^{3}H]-loganin	*C. roseus*	0.026	19, 20
	[9-^{14}C]-loganin	*C. roseus*	0.47	21
	[10-^{14}C]-loganin	*C. roseus*	0.1	22

Compound	Precursor	Plant	Incorporation/further information.*	Reference
	[7-³H]-loganin	C. roseus	0.2	23
	[5-³H, OMe-³H]-loganin	C. roseus	+	23
	[7-³H, OMe-³H]-loganin	C. roseus	0.38	24
	[6,8-³H, OMe-¹⁴C]-loganin	C. roseus	0.09	25
	[OMe-³H]-secologanin	C. roseus	0.55	26, 27
	[7-³H]-verbenalin	C. roseus	0	19, 20
	[OMe-¹⁴C]-monotropeine methyl ester	C. roseus	0	19, 20
	[U-¹⁴C]-shikimic acid	C. roseus	2.2	28–36
	[2-¹⁴C]-tryptophan	C. roseus	E	31
	[3-¹⁴C]-tryptophan	C. roseus	0.8	32, 33
	[2-¹⁴C]-tryptamine	C. roseus	+	17
	[ar-³H]-tryptamine	C. roseus	0.26	33–35
	[6-¹⁴C]-isovincoside	C. roseus	5.20	36
	[OMe-³H]-isovincoside	C. roseus	0.4	32
	[ar-³H]-vincoside	C. roseus	0.01	37
	[OMe-³H]-vincoside	C. roseus	0.51	24, 35
	[OMe-³H]-vincoside and isovincoside	C. roseus	0.47	34, 35
	[OMe-³H]-isovincoside	C. roseus	2.85	36
	[ar-³H]-vincoside and isovincoside	C. roseus	0.95	38
	[³H: ¹⁴C]-isovincoside (7.4:1)	C. roseus	1.53 (7.0:1)	36
	[³H: ¹⁴C]-isovincoside (13:3)	C. roseus	4.9 (12:8)	39
	[³H:¹⁴C]-vincoside (13:3)	C. roseus	0	39
	[ar-³H, OMe-³H]-vincoside and isovincoside	C. roseus	0.65	34, 35
	[ar-³H]-isovincoside	C. roseus	1.1	37
	[OMe-³H]-dihydrovincoside	C. roseus	0	24, 35
	[ar-³H]-geissoschizine	C. roseus	0.12	38
	[OMe-³H]-geissoschizine	C. roseus	0.22	40
	[ar-³H]-geissoschizine	C. roseus	0.12	41
	[ar-³H, OMe-³H]-geissoschizine	C. roseus	0.12	38
	[OMe-³H]-corynantheine aldehyde	C. roseus	0.001	12
	[ar-³H]-corynantheine aldehyde	C. roseus	0.001, J	12
	[ar-³H]-16,17-dihydro-secodine-17-ol	C. roseus	0.001	40, 42
	[ar-³H]-secodine	C. roseus	0.001	40
	[ar-³H]-tabersonine	C. roseus	0	32, 33
	[ar-³H]-15,20-dihydro-vincadine	C. roseus	0	32, 33
	[ar-³H]-carbomethoxy-cleavamine	C. roseus	0	32, 33
	[3-³H, 6-¹⁴C]-strictosidine	C. roseus	0.46	43
	[ar-³H]-strictosidine	C. roseus	0.2	44

Compound	Precursor	Plant	Incorporation/further information.*	Reference
19-Epiajmalicine	[³H: ¹⁴C]-isovincoside	*C. roseus*	1.3 (13:3)	39
Ajmaline	[¹⁴C]-methionine	*Rauwolfia serpentina*	0.029	35
	[2-¹⁴C]-alanine	*R. serpentina*	A, T	46
	[1-¹⁴C]-acetate	*R. serpentina*	0.2, F	46, 47
			0.01, T	48
			E, F	49
	[2-¹⁴C]-acetate	*R. serpentina*	E, T	50, 51
	[¹⁴C]-glycine	*R. serpentina*	E, F	50, 51
	[¹⁴C]-formate	*R. serpentina*	0.01, T	48
			0.01, T	52
			0.1, F	47
	[1,3-¹⁴C]-mevalonate	*R. serpentina*	0	46
	[2-¹⁴C]-mevalonate	*R. serpentina*	E	53
	[2-¹⁴C]-mevalonolactone	*R. serpentina*	0	54
	[1-³H]-loganin	*R. serpentina*	0.04	21
	[(3-³H, 6-¹⁴C]-strictosidine	*Rauwolfia vomitoria*	0.72	43
	[ar-³H)-5α-carboxystrictosidine	*R. vomitoria*	0	55
	[ar-³H]-5α-carboxyvincoside	*R. vomitoria*	0	55
	[2-¹⁴C]-tyrosine	*Rauwolfia serpentina*	0	46
	[2-¹⁴C]-tryptophan	*R. serpentina*	0.01	56, 57
	[3-¹⁴C]-tryptophan	*Rauwolfia verticillata*	0.28	35
	[N_b-Me¹⁴C]-methyltryptophan	*R. verticillata*	0	35
	[N_b-Me-¹⁴C]-methyltryptamine	*R. verticillata*	0	35
	[ar-³H]-deoxyajmaline	*Rauwolfia serpentina*	0.048	35
Akuammicine	[2-¹⁴C]-tryptophan	*C. roseus*	E, K	31
	[ar-³H]-vincoside and isovincoside	*C. roseus*	0.76	38
	[ar-³H]-geissoschizine	*C. roseus*	1.53	58
	[ar-³H]-geissoschizine	*C. roseus*	0.63	38
	[ar-³H, OMe-³H]-geissoschizine	*C. roseus*	2.0	38
	[ar-³H]-geissoschizine oxindole	*C. roseus*	0.55	40
Akuammidine	5α-Carboxystrictosidine	*Vallesia glabra*	0.01	55
	5α-Carboxyvincoside	*Vacanga africana*	0.002	55
Akuammigine	[ar-³H]-strictosidine	*C. roseus*	0.001	44
Apparicine	[ar-³H]-16,17-dihydrosecodine-17-ol	*C. roseus*	A	59, 42

Compound	Precursor	Plant	Incorporation/further information.*	Reference
Aricine	[7-³H]-secologanin	*Rauwolfia canescens*	0.02	43
	[2-¹⁴C]-tryptamine	*R. canescens*	PP = 10.49:1	
	[3-³H, 6-¹⁴C]-strictosidine	*R. canescens*	0.79 PP = 5.57:1	43
Aspidospermidine	[ar-³H]-quebrachamine	*Vinca minor*	0	32
	[ar-³H]-vincaminoreine	*V. minor*	0.008	32
Asperuloside	[10-³H]-deoxyloganic acid	*Daphniphyllum macropodum*	0.6	60–62
	[7-³H]-loganin	*D. macropodum*	0.05	61
	[10-³H]-loganin	*D. macropodum*	0.5 0.45	63 62, 63
	[7-³H]-epiloganin	*D. macropodum*	0	60, 61
	[10-³H]-epiloganin	*D. macropodum*	0.44	61
Acubin	[2-¹⁴C]-mevalonate	*Verbascum thapsus*	A, B	64
	[10-³H]-deoxyloganic acid	*Aucuba japonica*	0.5	60–62
Brevianamide A	[³H]-mevalonate	*Penicillium breviocompactum*	+	63
	[³H]-proline	*P. breviocompactum*	+	63
	[³H]-tryptophan	*P. breviocompactum*	+	63
	cyclo-L-[methylene-¹⁴C]-tryptophyl-L-[5-³H]-proline	*P. breviocompactum*	+	63
Brucine	[—CH₂-¹⁴C]-5-hydroxy tryptophan	*Strychnos nux vomica*	0.003	64
	[U-¹⁴C]-strychnine	*S. nux vomica*	E	64
Catharanthine	[1-¹⁴C]-acetate	*C. roseus*	0.41, D 0.85, Z 0.009 0.0009, T	65 29, 66 8 8
	[2-¹⁴C]-acetate	*C. roseus*	0.41, D	65
	[¹⁴C]-leucine	*C. roseus*	0.38, Z 0.003, D	29, 66 67
	[1,3-¹⁴C]-malonate	*C. roseus*	0.018, D	65
	[2-¹⁴C]-malonate	*C. roseus*	0.038, D	65
	[2-¹⁴C]-pyruvate	*C. roseus*	0.20, D	65
	[2-¹⁴C]-mevalonate	*C. roseus*	0.4	10
	[4-³H, 2-¹⁴C]-4R-mevalonate	*C. roseus*	E, G	12
	[4-³H, 2-¹⁴C]-4S-mevalonate	*C. roseus*	E, I	12
	[3-¹⁴C]-mevalonate	*C. roseus*	E, V	12
	[4-¹⁴C]-mevalonate	*C. roseus*	E, V	11
	[5-¹⁴C]-mevalonate	*C. roseus*	E	15

Compound	Precursor	Plant	Incorporation/further information.*	Reference
	[6-¹⁴C]-mevalonate	*C. roseus*	E, W	14
	[1-³H₂]-geraniol	*C. roseus*	0.25	11
	[1-³H₂, 2-¹⁴C]-geraniol	*C. roseus*	E, G	12
	[3-¹⁴C]-geraniol	*C. roseus*	0.41	68
	[2-¹⁴C]-geranyl pyrophosphate	*C. roseus*	0.2	15
	[1-³H₂]-nerol	*C. roseus*	0.21	11
	[2-³H, 2¹⁴C]-nerol	*C. roseus*	E, I	12
	[1-³H₂]-10-hydroxygeraniol	*C. roseus*	0.36, P	16
	[8-¹⁴C]-10-hydroxynerol/geraniol	*C. roseus*	0.5	14
	[8-¹⁴C]-10-hydroxynerol	*C. roseus*	0.9	14
	[1-³H₂]-citronellol	*C. roseus*	0.0004	14
	[1-³H₂]-10-hydroxycitronellol	*C. roseus*	0.02	16
	[8-¹⁴C]-10-hydroxycitronellol	*C. roseus*	0.002	14
	[1-³H₂]-10-oxocitronellol	*C. roseus*	0.011	16
	[1-³H₂]-linalool	*C. roseus*	0.0008	16
	[8-¹⁴C]-10-hydroxylinalool	*C. roseus*	0.002	14
	[OMe-³H]-deoxyloganin	*C. roseus*	0.29	18
	[OMe-³H]-loganin	*C. roseus*	0.8	19, 20
	[7-³H]-loganin	*C. roseus*	1.2	23
	[9-¹⁴C]-loganin	*C. roseus*	0.5	21
	[10-¹⁴C]-loganin	*C. roseus*	0.3	22
	[5-³H, OMe-³H]-loganin	*C. roseus*	+	23
	[7-³H, OMe-³H]-loganin	*C. roseus*	1.5	24
	[6,8-³H, OMe-¹⁴C]-loganin	*C. roseus*	0.65	25
	[OMe-³H]-secologanin	*C. roseus*	0.16	26, 27
	[7-³H]-verbenalin	*C. roseus*	0	19, 20
	[OMe-¹⁴C]-monotropeine methyl ester	*C. roseus*	0	19, 20
	[3-³H, 6-¹⁴C]-strictosidine	*C. roseus*	0.03	43
	[ar-³H]-strictosidine	*C. roseus*	0.25	44
	[2-¹⁴C]-tryptophan	*C. roseus*	E, K	31
	[3-¹⁴C]-tryptophan	*C. roseus*	1.5–3.3 0.05	29, 66
	[ar-³H]-tryptamine	*C. roseus*	0.61	32, 33
	[OMe-³H]-isovincoside	*C. roseus*	2.08	36
	[OMe-³H]-isovincoside	*C. roseus*	0.01	32, 33
	[OMe-³H]-vincoside	*C. roseus*	0.001	36
	[OMe-³H]-vincoside	*C. roseus*	0.89	24, 35
	[6-¹⁴C]-isovincoside	*C. roseus*	4.51	36
	[ar-³H]-vincoside and isovincoside	*C. roseus*	0.35	38
	[OMe-³H]-vincoside and isovincoside	*C. roseus*	0.84	34, 35
	[ar-³H]-vincoside	*C. roseus*	0.01	37
	[ar-³H, OMe-³H]-vincoside and isovincoside	*C. roseus*	0.41	34, 35
	[ar-³H]-isovincoside	*C. roseus*	1.5	37

TABLE OF INCORPORATION DATA

Compound	Precursor	Plant	Incorporation/further information.*	Reference
	[OMe-³H]-dihydrovincoside	C. roseus	0	24, 35
	[ar-³H]-geissoschizine	C. roseus	0.21	38
	[OMe-³H]-geissoschizine	C. roseus	0.41	40
	[ar-³H, OMe-³H]-geissoschizine	C. roseus	0.47	38
	[ar-³H]-corynantheine aldehyde	C. roseus	0.001, J	12
	[OMe-³H]-corynantheine aldehyde	C. roseus	0.001, J	12
	[ar-³H]-ajmalicine	C. roseus	0.3, K	69
	[OMe-³H]-stemmadenine	C. roseus	0.56, K	69
	[OMe-³H, 6-¹⁴C]-stemmadenine	C. roseus	0.3, K	69
	[ar-³H]-16,17-dihydro-secodin-17-ol	C. roseus	0.001	40, 59
	[ar-³H]-secodine	C. roseus	0.001	50
	[ar-³H]-tabersonine	C. roseus	0.05	32, 33
	[OMe-³H]-tabersonine	C. roseus	0.8, K	69
	[OMe-³H, 6-¹⁴C]-tabersonine	C. roseus	0.14, K	69
	[ar-³H]-carbomethoxy-cleavamine	C. roseus	0.05	32, 33
	[ar-³H]-coronaridine	C. roseus	0	31
Cephaeline	[1-¹⁴C]-acetate	Cephaelis acuminata	0.0001	51
	[1-¹⁴C]-acetate	Cephaelis ipecacuanha	0.40, T	53
	[2-¹⁴C]-acetate	C. acuminata	0.0008, T A, T	51 70
	[1-¹⁴C]-glycine	C. acuminata	0.0003	51
	[2-¹⁴C]-glycine	C. acuminata	0.02, S, T F, BB 0.005, F. S	51 70
	[2-¹⁴C]-glycine	C. ipecacuanha	0.01 F	71
	[1,3-¹⁴C₂]-glycerol	C. acuminata	0.0004	51
	[¹⁴C]-formate	C. ipecacuanha	0, 4, T	48, 53
	[6,7-³H, U-¹⁴C]-leucine	C. ipecacuanha	0	71
	[6,7-³H, U-¹⁴C]-leucine	C. acuminata	0	51
	[6,7-³H-²⁴C₂]-malonate	C. ipecacuanha	0.06, T	53
	[2,3-¹⁴C₂]-succinic acid	C. ipecacuanha	0.0006	51
	[1,4-¹⁴C₂]-succinic acid	C. acuminata	A	71
	[1,4-¹⁴C₂]-succinic acid	C. ipecacuanha	0	51
	[2-¹⁴C]-glyoxylate	C. acuminata	0.006, T	51
	[1-¹⁴C]-glycollate	C. acuminata	0.0025, T	51
	[1-¹⁴C]-glycollate	C. ipecacuanha	0.02, T	71
	[2-¹⁴C]-glycollate	C. acuminata	0.02	51
	[2-¹⁴C]-pyruvate	C. acuminata	0.0002	51
	[3-¹⁴C]-pyruvate	C. acuminata	0.0002	51
	[2-¹⁴C]-mevalonate	C. ipecacuanha	E	53
	[2-¹⁴C]-mevalonolactone	C. ipecacuanha	0	54
	[2-¹⁴C]-geraniol	C. ipecacuanha	0.015	72

Compound	Precursor	Plant	Incorporation/further information.*	Reference
	[2-^{3}H, 2-^{14}C]-geraniol	C. ipecacuanha	0.01, I	73
	[7,^{3}H]-loganin	C. ipecacuanha	1.1	73
	[OMe-^{3}H, 9-^{14}C]-loganin	C. ipecacuanha	0.97	72
	[OMe-^{3}H, 6-^{3}H$_2$]-secologanin	C. ipecacuanha	0.11	73
	[2-^{14}C]-phenylalanine	C. ipecacuanha	0.018	53
	[2-^{14}C]-tyrosine	C. ipecacuanha	0.8	53
	[3-^{14}C]-tyrosine	C. ipecacuanha	E	71
	[3-^{14}C]-tyrosine	C. acuminata	0.006	51
	[3$'$-^{14}C]-desacetylipecoside	C. ipecacuanha	0.34	73
	[3$'$-^{14}C]-desacetylisoipecoside	C. ipecacuanha	0.009	73
	[3$'$-^{14}C]-dihydrodesacetylipecoside	C. ipecacuanha	0.06	73
	$\gamma\gamma$-Dimethylallyltryptophan	Claviceps	0.61	5, 6
	Mevalonate-3$'$-d_3	Claviceps	SS = Do21%, D$_1$ 11.3%, D$_2$ 63%, D$_3$ 4%	7
	[3R, 4R]-mevalonate-2-^{14}C-4-t (PP = 5.00)	Claviceps	PP = 7.29	7
	[3R, 4R]-mevalonate-2-^{14}C-4-t (PP = 5.00)	Claviceps	PP = 5.25	7
Cinchonidine	[1-^{3}H$_2$, 1-^{14}C]-tryptamine	Cinchona ledgeriana	0.12, G	74
	[ar-^{3}H]-vincoside and isovincoside	C. ledgeriana	0.008	75
	[ar-^{3}H]-corynantheine aldehyde	C. ledgeriana	0	75
	[ar-^{3}H]-corynantheal	C. ledgeriana	0.04	75
	[11-^{3}H]-cinchonidinone	C. ledgeriana	+	74
	[18-^{3}H$_2$]-cinchonidinone	C. ledgeriana	0.03	74
	[ar-^{3}H]-cinchonamine	C. ledgeriana	0.0008	74
	[18-^{3}H$_2$]-cinchonidine	C. ledgeriana	0.06	74
Cinchonine	[7-^{3}H]-loganin	C. ledgeriana	0.06	76
	[2-^{14}C, 1-^{15}N]-tryptophan	Cinchona succirubra	E, BB	77
	[1-^{3}H^2, 1-^{14}C]-tryptamine	C. ledgeriana	0.2, G	74
	[ar-^{3}H]-vincoside and isovincoside	C. ledgeriana	0.07	75
	[ar-^{3}H]-corynantheine aldehyde	C. ledgeriana	0	75
	[ar-^{3}H]-dihydrocorynantheine	C. succirubra	0	77
	[ar-^{3}H]-corynantheal	C. ledgeriana	0.13	75
	[18-^{3}H$_2$]-cinchonamine	C. ledgeriana	0.14	74
	[ar-^{3}H]-cinchonamine	C. ledgeriana	0.001	74
	[ar-^{3}H]-cinchonamine	C. succirubra	A	77

TABLE OF INCORPORATION DATA

Compound	Precursor	Plant	Incorporation/further information.*	Reference
Coronaridine	[2-^{14}C]-tryptophan	*C. roseus*	0.08, K	58
	[ar-^{2}H]-geissoschizine	*C. roseus*	0.35, K	58
	[5-^{14}C]-tabersonine	*C. roseus*	E, K	31
	[ar-^{3}H]-catharanthine	*C. roseus*	0	78
Corynantheine	[OMe-^{3}H]-corynantheine aldehyde	*C. roseus*	13, K	69
β-Cyclopiazonic acid	[3-^{3}H]-tryptophan	*C. roseus*	+	79
Deacetyl vindoline	[5-^{14}C]-mevalonate	*C. roseus*	E	15
1,2-Dehydroaspidospermidine	[2-^{14}C]-mevalonate	*Rhazya stricta*	0.16	10
	[3-^{14}C]-mevalonate	*R. stricta*	0.15	10
Deoxyloganin	[1-^{3}H]-geraniol	*Menyanthes trifoliata*	C, E	18
Dihydrofoliamenthin	[2-^{14}C]-geraniol	*M. trifoliata*	S, O	182
16,17-Dihydrosecodin-17-ol	[OMe-^{3}H]-loganin	*Rhazya orientalis*	0.013, C	80
	[OMe-^{3}H]-loganin	*C. roseus*	A	80
4,γγ-dimethylallyltryptophan	[2-^{14}C]-mevalonate	*C. roseus*	+	4, 81
Echinulin	Cyclo-L-[U-^{14}C]-alanyl-L-[5,7-^{3}H$_2$] tryptophyl [5,7-^{3}H$_2$,	*Aspergillus amstelodami*	+	82, 83
	3-^{14}C]-tryptophan	*A. amstelodami*	+	84, 86
	[^{3}H]-alanine	*A. amstelodami*	+	87
	[^{3}H]-mevalonic acid	*A. amstelodami*	+	87
Elymoclavine	L-Tryptophan	*Claviceps*	+	88
	[^{14}C-17-^{3}H]-chanoclavine I-aldehyde	*Claviceps*	+	89, 90
	[17-^{3}H, ^{14}C]-chanoclavine	*Claviceps*	+	4, 90
	[17-^{3}H-N]-demethylchanoclavine I	Fungus strain	0	91
	N-demethylchanoclavine II	Fungus strain	0	91
	[3R, 4R]-mevalonic-2-^{14}C-4-t-acid (PP = 5.00)	*Claviceps*	PP = 2.61	7
	[4S-^{3}H]-mevalonate	*Claviceps*	+	92
	[3R, 4R]-mevalonate-2-^{14}C-4-t (PP = 2.50)	*Claviceps*	PP = 1.58, RR = 63	7
	[3R, 5R]-mevalonate 5-T	*Claviceps* strain	+	91

Compound	Precursor	Plant	Incorporation/further information.*	Reference
	Mevalonate-3'-d_3	*Claviceps* strain	SS = Do 28.2% D_1 7.9% D_2 3.9%	7
	[3R, 5S]-mevalonate-5-T	*Claviceps* strain	0.47	93
	^{13}C-$\gamma\gamma$-dimethylallyl-tryptophan	*Claviceps* strain	0.47	93
	N-methyl-4-($\gamma\gamma$-dimethyl-allyl-)-tryptophan	*Claviceps* strain	+	93
	[^{15}N-C^{2}H$_3$]-tryptamine	*Claviceps* strain	–	93
Ergolene carboxylic acid and tri- and tetra-cyclic clavines	Phaliclavine-N-^{14}CH$_3$	Fungus strain	+	94
Ergot alkaloids	[2-^{14}C]-tryptamine	Fungus strain	+	95
	[5 & 6]-deuterated tryptophan	Fungus strain	+	92
	[-^{14}CH$_3$]-methionine	Fungus strain	1.35%	96
	[2-^{14}C]-mevalonic acid	Fungus strain	+	96
	2- or 4-tritiated mevalonic acid	Fungus strain	+	96
	[1-^{3}H]-dimethylallyl alcohol	Fungus strain	0	97
	4-[$\gamma\gamma$-dimethylallylpyrophosphate]-tryptophan	Fungus strain	+	98
Ergometrine	Alanine	Fungus strain	+	99
	Alaninol and α-methyl serine	Fungus strain	+	99
	Lysergylalanine	*Claviceps paspali*	R	100, 101
	Lysergylalanine	*C. paspali*	+	100
	Lysergic acid	*C. paspali*	+	102
Ergotoxine	Lysergaldehyde enol acetate	*Claviceps purpurea*	+	103
	DL-[1-^{14}C]-valine	*Claviceps* strain	R	108
	[1-^{14}C]-L-alanine	*Claviceps* strain	R	108
	[^{14}C]-lysergyl-L-valine	*Claviceps* strain	R	108
Ergotamine	Ergometrinine	*C. purpurea*, *C. paspali*	A, T	104
	Ergometrine	*C. purpurea*, *C. paspali*	A, T	102, 104
	Lysergic acid	*C. purpurea*, *C. paspali*	+	105, 102
	Lysergic acid amide	*C. purpurea*, *C. paspali*	–	105
	[^{14}C]-lysergylalanine	*C. paspali*	A, T	104
	[U-^{14}C]-L-Tryptophan	*C. purpurea*	+	105

Compound	Precursor	Plant	Incorpora-tion/further informa-tion.*	Refer-ence
	DL-Mevalonic acid lactone-2-[14]C	C. purpurea	+	105
	Proline	C. purpurea	+	106
	[14C]-Sodium formate	C. purpurea	+	105
	[2-14C]-Sodium acetate	C. purpurea	+	105
	L-Methionine-Me[14]C	C. purpurea	+	105
Foliamenthin	[9-14C]-geraniol	M. trifoliata	2.5	107
Gardenoside	[10-3H]-deoxyloganic acid	Gardenia jasminoides	0.4	61, 62
Geissoschizine	[2-14C]-tryptophan	C. roseus	E, K	31
Gelsemine	[ar-3H]-5α-carboxy-strictosidine	Gelsemium sempervirens	0.001	55
	[ar-3H]-5α-carboxyvin-coside	G. sempervirens	0.001	55
Gliotoxin	Phenylalanine	Trichoderma viride	+	109
Geniposide	[10-3H]-deoxyloganic acid	G. jasminoides	0.05	61, 26
Gentianine	[1-14C]-acetate	Gentiana lutea	0.0026, D	110
	[2-14C]-glycine	Gentiana asclepiadea	E, F	111
Gentioflavine	[1-14C]-geraniol	G. asclepiadea	3.0	111
	[1-14C]-linalool	G. asclepiadea	3.0	111
Gentiopicroside	[2-14C]-acetate	Swertia caroliniensis	0.01, B	112
	[2-14C]-mevalonate	S. caroliniensis	0.12	113
			0.02–011	112
			0.04, F	114
	[2-3H2, 2-14C]-mevalonate	S. caroliniensis	0.02–0.11, G	115
	[2,3H, 2-14C]-2R-mevalonate	S. caroliniensis	0.1, H	115
	[2-3H, 2-14C]-2S-mevalonate	S. caroliniensis	0.1, H	115
	[4-3H, 2-14C]-4R-mevalonate	S. caroliniensis	0.1, E	116
	[4-3H, 2-14C]-4S-mevalonate	S. caroliniensis	0.04, I E, I	115
	[2-14C)-mevalonolactone	Gentiana triflora var. japonica	0.06	117, 118
	[5,9-3H, 3,7,11]- loganic acid	S. caroliniensis	1.3, G	115, 116

Compound	Precursor	Plant	Incorporation/further information.*	Reference
	[10-³H]-loganic acid	G. triflora var. japonica	4.5	60
	[2-¹⁴C]-loganin	Swertia petiolata	3.2	119
Harman	[2-acetyl-¹⁴C]-N-acetyltryptamine	Passiflora edulis	+	120
	Eleagnin	P. edulis	+	120
	Harmalan	P. edulis	+	120
	[10-¹⁴C]-sweroside	Gentiana scabra	40	121, 122
Gramine	[3-¹⁴C]-tryptophan	Phalaris arundinacea	8.4	44
	[2-¹⁴C, 2-³H]-tryptophan	P. arundinacea	+	123, 124
Ibogaine	[3-¹⁴C]-tryptophan	Tabernanthe iboga	6.7	125
Indolmycin	Indolmycenic acid	Streptomyces griseus	+	126
	[2-¹⁴C]-[RS]-tryptophan	S. griseus	+	127
	[¹⁴C]-S-aryinine	S. griseus	+	127
	[¹⁴C]-S-methionine	S. griseus	+	127
	[1-¹⁴C]-tryptophan	S. griseus	+	127
	[2-¹⁴C]-indole	S. griseus	+	127
	[Me-¹⁴C, Me-³H]-methionine	S. griseus	TT	127
	RS-[3-¹⁴C, 2-³H]-tryptophan	S. griseus	T	127
	RS-[3-¹⁴C, 3-³H]-tryptophan	S. griseus	UU = 52	127
Ipecoside	[2-¹⁴]-glycine	Cephaelis acuminata	A, S, T	51
	[2-¹⁴C]-glyoxylate	C. acuminata	0.008, T	51
	[2-¹⁴C-pyruvate	C. acuminata	0.0003	
	[2-¹⁴C]-geraniol	Cephaelis ipecacuanha	0.038	72
	[2-³H, 2-¹⁴C]-geraniol	C. ipecacuanha	0.03, L	73
	[OMe-³H]-loganin	C. ipecacuanha	1.7	72
	[OMe-³H, 2-¹⁴C]-loganin	C. ipecacuanha	1.9, L	72
	[7-³H]-loganin	C. ipecacuanha	2.2	73
	[OMe-³H, 6-³H]-secologanin	C. ipecacuanha	0.10	73
	[3′-¹⁴C]-desacetylipecoside	C. ipecacuanha	0.59	73
	[3′-¹⁴C]-desacetyldihydro-ipecoside	C. ipecacuanha	0.59	73
Isochanoclavine I	[3R, 4R]-mevalonic-2-¹⁴C-4-t-acid (PP = 5.00)	Claviceps	PP = 4.97	7

Compound	Precursor	Plant	Incorpora-tion/further informa-tion.*	Refer-ence
3-Isoajmalicine	Ajmalicine	*Mitragyna parvifolia*	CC	128, 129
3-Isomitraphylline	Ajmalicine	*M. parvifolia*	CC	128, 129
	3-Isoajmalicine	*M. parvifolia*	CC	128, 129
Isoreserpiline	[7-^{3}H]-secologanin	*Rauwolfia canescens*	0.05	
	[2-^{14}C]-tryptamine	*R. canescens*	PP = 14.82:1	43
	[3-^{3}H, 6-^{14}C]-strictosidine	*R. canescens*	1.46 PP = 5.67:1	43
Isoreserpinine	[3-^{3}H, 6-^{14}C]-strictosidine	*R. canescens*	0.16 PP = 0.43:1	43
Jasminine	[10-^{3}H]-deoxyloganic acid	*Jasminium primulum*	0.01	130
	[10-^{3}H]-loganin	*J. primulum*	0.03	130
Loganic acid	[2-^{14}C]-mevalonate	*S. caroliniensis*	1.16, J	131
			0.52, K	131
			0.08–0.3	113
			0.08–1.2	112
	[2-^{2}H$_2$, 2-^{14}C]-mevalonate	*S. caroliniensis*	0.08–1.2, G	112
	[4-^{3}H, 2-^{14}C]-4R-mevalonate	*Catharanthus roseus*	0.25, G	132
	[4-^{3}H, 2-^{14}C]-4S-mevalonate	*S. caroliniensis*	1.2, L E, L	115
	[I-^{14}C]-geranyl pyro-phosphate	*S. caroliniensis*	0.002	113
Loganin	[^{14}C]-methionine	*C. roseus*	E	133
	[2-^{14}C]-mevalonate	*M. trifoliata*	1.0 M 0	134 134
		C. roseus	0.01 C	131
	[2-^{3}H$_2$, 2-^{14}C]-mevalonate	*C. roseus*	0.02 G	132
	[2-^{3}H, 2-^{14}C]-2R-mevalonate	*S. caroliniensis*	0.8	115
	[2-^{3}H, 2-^{14}C]-2S-mevalonate	*S. caroliniensis*	0.1	115
	[4-^{3}H, 2-^{14}C]-4R-mevalonate	*C. roseus*	E, L E, L	115, 116 12
	[4-^{3}H, 2-^{14}C]-4S-mevalonate	*S. caroliniensis*	E, I	115, 116
		C. roseus	E, I	12
	[3'-^{14}C-]3-hydroxy-3-methylglutaric acid	*C. roseus*	0.01	131
	[1-^{3}H$_2$]-geraniol	*M. trifoliata*	0.02 0.2	135 135
	[2-^{14}C]-geraniol	*M. trifoliata*	0.25 E, N	136 21
	[1-^{3}H$_2$, 2-^{14}C]-geraniol	*C. roseus*	E, G	12
	[2-^{3}H, 2-^{14}C]-geraniol	*C. roseus*	E, L	136
	[5-^{3}H$_2$, 2-^{14}C]-geraniol	*C. roseus*	E, L	136

Compound	Precursor	Plant	Incorporation/further information.*	Reference
	[9-^{14}C]-geraniol	*M. trifoliata*	0.1	134
	[2-^{14}C]-geranyl pyrophosphate	*M. trifoliata*	0.41	119
	[2-^{3}H, 2^{14}C]-nerol	*C. roseus*	E, L	12
	[1-^{3}H$_2$]-10-hydroxygeraniol	*C. roseus*	0.31, P	16
	[8-^{14}C]-10-hydroxygeraniol	*C. roseus*	0.09	14
	[8-^{14}C]-10-hydroxynerol	*C. roseus*	0.16	14
	[I-^{3}H$_2$]-citronellol	*C. roseus*	0.0002	14
	[1-^{3}H$_2$]-10-hydroxy-citrollellol	*C. roseus*	0.024	16
	[8-^{14}C]-10-hydroxy-citronellol	*C. roseus*	0.0002	14
	[1-^{3}H$_2$]-linalool	*C. roseus*	0.003	16
	[10-^{3}H]-deoxyloganic acid	*Lonicera japonica*	0.3	60, 61
	[OMe-^{3}H]-deoxyloganin	*C. roseus*	6.4	18
	[3,7,11,-^{14}C]-loganic acid	*C. roseus*	1.1	132
Lysergic acid	$\gamma\gamma$-dimethylallyltryptophan	*C. roseus*	0.78	5
	($\pm$)-[2-^{14}C]-tryptophan	Rye	(+10–37%)	137
	Tryptophan [^{14}CO$_2$H]	Rye	0	138
Lysergic acid amide	$\Delta^{8,\,9}$-lysergic acid	Ergot strains	+	92
	Lysergic acid	Ergot strains	+	92
Lysergic acid α-hydroxy ethyl amide	Lysergylalanine	—	—	101, 102
Mesembrinol	[3,5-^{3}H]-sceletenone	*Sceletium strictum*	2.0	139
	4-0-demethyl [5-^{3}H]-mesembrenone	*S. strictum*	58.6	139
	[4-0-methyl-^{3}H]-mesembrenone	*S. strictum*	1.12	139
	[5-^{3}H]-mesembrenone	*S. strictum*	63.6	139
	L-[3,5-^{3}H$_2$, U-^{14}C]-tyrosine	*S. strictum*	+	140
Mesembrine	[1-^{14}C, 2,6-^{3}H$_2$]-phenylalanine	*S. strictum*	+	141
Methoxy-tabersonine	[5-^{14}C]-tabersonine	*C. roseus*	E, K	31
Minovine	[ar-^{3}H]-tryptophan	*V. minor*	0.08	142
	[3-^{14}C]-tryptophan	*V. minor*	0.24	143
	[ar-^{3}H]-geissoschizine	*V. minor*	0	142
	[ar-^{3}H]-stemmadenine	*V. minor*	0.001	142
	[ar-^{3}H]-16,17-dihydro-secodin-17-ol	*V. minor*	0.001	142, 40, 59, 42

Compound	Precursor	Plant	Incorporation/further information.*	Reference
	[ar-³H]-secodine	*V. minor*	0.001	142, 40, 59
	[ar-³H, ¹⁴CO₂ Me]-secodine	*V. minor*	E, DD	40, 59
	[ar-³H]-tabersonine	*V. minor*	0.002	142
	[ar-³H]-vincaminoreine	*V. minor*	0.3	32
Mitragynine	[3-³H, 6-¹⁴C]-strictosidine	*Mitragyna speciosa*	2.72 PP = 8.40:1	43, 144
	[3-³H, 6-¹⁴C]-vincoside	*M. speciosa*	0.001	144
Mitraphylline	Ajmalicine	*M. parvifolia*	CC	128, 129
	3-Isoajmalicine	*M. parvifolia*	CC	128, 129
Morronoside	[7-³H]-loganin	*Gentiana thumbergii*	0.13	145
	[OMe-¹⁴C]-secologanin	*Coronus officinalis*	0.09	145
Oleuropin	[OMe-¹⁴C]-secologanin	*Olea europea*	0.34	130
Pericalline	[ar-³H]-C-aminomethyl indole	*Aspidosperma pyricollin*	0	146
	[ar-³H]-tryptophan	*A. pyricollin*	0.018	42, 146
	[ar-³H, 2-¹⁴C]-tryptophan	*A. pyricollin*	E, EE	42
	[ar-³H, 3-¹⁴C]-tryptophan	*A. pyricollin*	E, BB	42
	[ar-³H]-stemmadenine	*A. pyricollin*	0.55	146
	[ar-³H]-vallesamine	*A. pyricollin*	0.01	146
	[ar-³H]-16-17-dihydrosecodin-17-ol	*A. pyricollin*	0	31
	[ar-³H]-secodine	*A. pyricollin*	0.01	31
	[ar-³H, ¹⁴CO₂Me]-secodine	*A. pyricollin*	E, BB	31
Perivine	[4-³H, 2-¹⁴C]-4R-mevalonate	*C. roseus*	E, G	12
	[4-³H, 2-¹⁴C]-4S-mevalonate	*C. roseus*	E, I	12
	[4-¹⁴C]-mevalonate	*C. roseus*	E, V	11
	[2-¹⁴C]-geraniol	*C. roseus*	0.04	11
	[1-³H₂, 2-¹⁴C]-geraniol	*C. roseus*	E, G	12
	[2-³H, 2-¹⁴C]-geraniol	*C. roseus*	E, I	12
	[2-³H, 2-¹⁴C]-nerol	*C. roseus*	0.021, G	16
	[1-³H₂]-10-hydroxy-citronellol	*C. roseus*	0.0006	16
	[1-³H₂]-10-oxocitronellol	*C. roseus*	0.0007	16
	[1-³H₂]-linalool	*C. roseus*	0.001	16
	[OMe-³H]-deoxyloganin	*C. roseus*	0.15	18
	[OMe-³H]-loganin	*C. roseus*	0.1	19, 20
	[9-¹⁴C]-loganin	*C. roseus*	0.04	21
	[6,8-³H, OMe-¹⁴C]-loganin	*C. roseus*	0.04	25
	[OMe-³H]-secologanin	*C. roseus*	0.013	26, 27
	[ar-³H]-tryptamine	*C. roseus*	0.053	34, 35
	[OMe-³H]-isovincoside	*C. roseus*	0	24, 35
	[OMe-³H]-vincoside	*C. roseus*	0.05	24, 35

Compound	Precursor	Plant	Incorporation/further information.*	Reference
	[OMe-³H]-vincoside and isovincoside	C. roseus	0.056	24, 35
	[Ar-³H, OMe-³H]-vincoside and isovincoside	C. roseus	0.037	34, 35
	[OMe-³H]-dihydrovincoside	C. roseus	0	24, 35
Quebrachidine	[ar-³H]-5α-carboxy-strictosidine	Rauwolfia vomitoria	0.01	55
	[ar-³H]-5α-carboxyvincoside	R. vomitoria	0.01	55
Quinine	¹⁴CO₂	C. succirubia	E	147, 149
	[2-¹⁴C]-geraniol	C. ledgeriana	0.0001	11
	[3-¹⁴C]-geraniol	C. succirubia	0.015, C	76
	[7-³H]-loganin	C. ledgeriana	0.15, C	76
	[10-¹⁴C]-sweroside	C. succirubia	0.6	122
			0.2	152
	[2-¹⁴C]-tryptophan	C. succirubia	0.7, F	153
	[indole-α-¹⁴C, 1-¹⁵N]-tryptophan	C. succirubia	0.97, BB	151
	[2-¹⁴C, 1-¹⁵N]-tryptophan	C. succirubia	E, BB	77
	[1-³H₂, 1-¹⁴C]-tryptamine	C. ledgeriana	0.33	74
	[ar-³H]-vincoside and isovincoside	C. ledgeriana	0.008	75
	[ar-³H]-corynantheine aldehyde	C. ledgeriana	0	75
	Dihydrocorynantheine	C. succirubia	0	77
	[ar-³H]-corynantheal	C. ledgeriana	0.002	75
	[18-³H]-cinchonidinone	C. ledgeriana	0.002	74
	[ar-³H]-cinchonamine	C. ledgeriana	0.001	74
	[ar-³H]-cinchonamine	C. succirubia	A	77
Reserpiline	[7-³H]-secologanin	Rauwolfia canescens	0.04	43
	[2-¹⁴C]-tryptamine		PP = 0.98:1	
	[3-³H, 6-¹⁴C]-strictosidine	R. canescens	0.56	43
			PP = 0.10:1	
	[3-³H, 6-¹⁴C]-strictosidine	R. canescens	0.34	144
Reserpine	[2-¹⁴C]-acetate	R. serpentina	A, T	50, 51
	[2-¹⁴C]-glycine	R. serpentina	A, F	50, 51
	[¹⁴C]-formate	R. serpentina	0.2, D	51
	[1,3-¹⁴C]-malonate	R. serpentina	0.04, F	47
	[2-¹⁴C]-tyrosine	R. serpentina	0	46
	[2-¹⁴C]-tryptophan	R. serpentina	0.0025	56, 57, 154
Reserpinine	[¹⁴C]-methionine	Vinca major	0.02, M	155
	[1-¹⁴C]-acetate	V. major	0.002, T	155
	[2-¹⁴C]-mevalonate	V. major	0.01	156

Compound	Precursor	Plant	Incorporation/further information.*	Reference
Sarpagine	[ar-³H]-5α-carboxy-strictosidine	*Rauwolfia vomitoria*	0.01	55
	[ar-³H]-5α-carboxyvincoside	*R. vomitoria*	0.01	55
Scandoside	[10-³H]-deoxyloganic acid	*Paederia scandens*	1.6	61, 62
Secologanic acid	[2-³H₂, 2-¹⁴C]-mevalonate	*C. roseus*	0.17, G	132
	[3,7,11-¹⁴C]-loganic acid	*C. roseus*	6.7	132
	[3,7,11-¹⁴C]-loganin	*C. roseus*	2.0	132
Secologanin	[OMe-³H]-loganin	*M. trifoliate*	5.1, G	26, 27
	[7-³H]-loganin	*C. roseus*	approx. 6, C	34, 27
	[3,7,11-¹⁴C]-loganic acid	*C. roseus*	8.8	132
	[2-³H₂, 2-¹⁴C]-mevalonate	*C. roseus*	0.2, G	132
Serpentine	[1,3-¹⁴C]-mevalonate	*Rauwolfia serpentina*	0.16	47
	[2-¹⁴C]-mevalonate	*C. roseus*	0.02	10
	[4,³H, 2-¹⁴C]-4R-mevalonate	*C. roseus*	E, V	11
			E, G	12
	[4-³H, 2-¹⁴C]-4S-mevalonate	*C. roseus*	E, I	12
	[3-¹⁴C]-mevalonate	*C. roseus*	E, V	15
	[4-¹⁴C]-mevalonate	*C. roseus*	E, V	11
	[5-¹⁴C]-mevalonate	*C. roseus*	E, V	11
	[1-³H₂]-geraniol	*C. roseus*	0.14	11
	[1-³H₂, 2-¹⁴C]-geraniol	*C. roseus*	E, G	12
	[2-¹⁴C]-geranyl pyrophosphate	*C. roseus*	0.6	15
	[1-³H]-nerol	*C. roseus*	0.11	11
	[1-³H₂]-10-hydroxygeraniol	*C. roseus*	E, I	12
	[1-³H₂]-10-hydroxycitronellol	*C. roseus*	0.0025	16
	[1-³H₂]-10-oxocitronellol	*C. roseus*	0.0034	16
	[1-³H₂]-linalool	*C. roseus*	0.0007	16
	[OMe-³H]-deoxyloganin	*C. roseus*	0.29	18
	[OMe-³H]-loganin	*C. roseus*	0.45	19, 20
	[9-¹⁴C]-loganin	*C. roseus*	0.6	21
	[7-³H, OMe-³H]-loganin	*C. roseus*	2.0	24
	[OMe-³H]-secologanin	*C. roseus*	0.65	26, 27
	[7-³H]-verbenalin	*C. roseus*	0	19, 20
	[OMe-¹⁴C]-monotropeine methyl ester	*C. roseus*	0	19, 20
	[2-¹⁴C]-tyrosine	*C. roseus*	0	46
	[2-¹⁴C]-tryptophan	*C. roseus*	0.005, F	56, 57, 154
	[ar-³H]-tryptamine	*C. roseus*	2.79	34, 35
	[OMe-³H]-vincoside	*C. roseus*	0.001	36
	[OMe-³H]-vincoside	*C. roseus*	3.9	24, 35
	[OMe-³H]-isovincoside	*C. roseus*	0.51	36

Compound	Precursor	Plant	Incorpora-tion/further informa-tion.*	Refer-ence
	[OMe-³H]-isovincoside	*C. roseus*	0.002	24, 35
	[ar-³H]-vincoside and isovincoside	*C. roseus*	1.6	38
	[³H: ¹⁹C]-strictosidine (7.4:1)	*C. roseus*	1.4 (6.9:1)	36
	[OMe-³H]-vincoside and isovincoside	*C. roseus*	3.91	34, 35
	[OMe-³H]-dihydrovincoside	*C. roseus*	0	24, 35
	[6-¹⁴C]-isovincoside	*C. roseus*	0.70	36
	[ar-³H]-geissoschizine	*C. roseus*	0.58	38
	[ar-³H, OMe-³H]-geissoschizine	*C. roseus*	0.82	38
	[ar-³H]-ajmalicine	*C. roseus*	1.8, J	12
Speciociliatine	[3-³H, 6-¹⁴C]-strictosidine	*Mitragyna speciosa*	0.53 PP=0.1:1	43, 144
	[3-³H, 6-¹⁴C]-vincoside	*M. speciosa*	>0.001	144
Sporidesmin	[3-³H]-tryptophan	*M. speciosa*	+	157
Stemmademine	[2-¹⁴C]-tryptophan	*M. speciosa*	0.001, K	31
	[3-¹⁴C]-tryptophan	*M. speciosa*	0.8, C	42
Strictosidine = (Isovincoside)	[OMe-³H]-loganin	*M. speciosa*	5.2	158
	[7-³H]-loganin	*M. speciosa*	1.5, C	34, 35
	[ar-³H]-tryptophan	*M. speciosa*	1.0	158
Strychnine	[¹⁴CO₂]	*S. nux vomica*	E	159
	[1-¹⁴C]-acetate	*S. nux vomica*	0.004 0.01, FF	160
	[2-¹⁴C]-acetate	*S. nux vomica*	0.021 0.42, GG	160
	[1-¹⁴C]-glycine	*S. nux vomica*	0	161
	[2-¹⁴C]-glycine	*S. nux vomica*	0.18, A, T E, A, T 0.17, A, T	162 161 64
	[2-¹⁴C]-mevalonate	*S. nux vomica*	0.002	160
	[2-¹⁴C]-geraniol	*S. nux vomica*	0.09	160
	[2-¹⁴C]-geranyl pyrophosphate	*S. nux vomica*	0.45	160
	[2-¹⁴C]-tryptophan	*S. nux vomica*	0.15, F	160
	[1-¹⁵N]-tryptophan	*S. nux vomica*	0.25	64
	[CH₂-¹⁴C]-5-hydroxy-tryptophan	*S. nux vomica*	0.0002	64
	[ar-³H]-Weiland–Gumlich aldehyde	*S. nux vomica*	0	160
	[5-¹⁴C]-1-acetyl-Weiland–Gumlich aldehyde	*S. nux vomica*	0	160
	[5-¹⁴C]-1-acetyl-Weiland–Gumlich aldehyde	*S. nux vomica*	+	163

TABLE OF INCORPORATION DATA

Compound	Precursor	Plant	Incorporation/further information.*	Reference
	Geissoschizine	S. nux vomica	+	163
	Diaboline	S. nux vomica	+	163
Sweroside	[2-¹⁴C]-mevalonolactone	Swertia japonica	0.06	117, 118
Swertiamarin	[1-¹⁴C]-acetate	S. japonica	R, D	164
	[2-¹⁴C]-mevalonolactone	S. japonica	0.006	117, 118
Tetrahydro-alstonine	[ar-³H]-strictosidine (isovincoside)	S. japonica	0.1	44
	[³H-¹⁴C]-strictosidine (13:3)	S. japonica	1.2 (13:2)	39
	[³H-¹⁴C]-vincoside (11:06)	S. japonica	—	39
Tetraphyllicine	[ar-³H]-5α-carboxy-strictosidine	Rauwolfia vomitoria	0.01	55
	[ar-³H]-5α-carboxyvincoside	R. vomitoria	0.01	55
Tabersonine	[2-¹⁴C]-tryptophan	C. roseus	0.21, K	51
	[3-¹⁴C]-tryptophan	C. roseus	E, K	69
	[OMe-³H]-stemmadenine	C. roseus	0.27, K	69
	[OMe-³H, 6-¹⁴C]-stemmadenine	C. roseus	0.1, K	69
	[OMe-³H]-catharanthine	C. roseus	0.001, K	69
Uleine	[ar-³H]-3-aminomethyl indole	A. pyricollum	0	146
	[ar-³H]-tryptophan	A. pyricollum	0	59, 46
	[ar-³H₂-2¹⁴C]-tryptophan	A. pyricollum	0	59
	[ar-³H₂-3¹⁴C]-tryptophan	A. pyricollum	0	59
	[ar-³H]-stemmadenine	A. pyricollum	0.0007	146
	[ar-³H]-vallesamine	A. pyricollum	0.003	146
	[ar-³H]-16,17-dihydro-secodin-17-ol	A. pyricollum	0	40, 42, 59
	[ar-³H]-secodine	A. pyricollum	0.001	40, 42, 59
Verbenalin	[¹⁴C]-methionine	Verbena officinalis	E, M	64
	[1-¹⁴C]-acetate	V. officinalis	0.004, B	165
	[2-¹⁴C]-acetate	V. officinalis	0.007, B	165
	[2-¹⁴C]-mevalonate	V. officinalis	0.1, B	165
	[1-¹⁴C]-geraniol	V. officinalis	0.009	165
	[10-³H]-deoxyloganic acid	V. officinalis	11.0	60–62
Vinblastine	[3,6-³H₃-methoxy-¹⁴C]-loganin	C. roseus	+	166
	21-³H-anhydrovinblastine	C. roseus	1.87	167
	[³H-CO₂CH₃]-catharanthine	C. roseus	0.006	168
	[¹⁴C-OCOCH₃]-vindoline	C. roseus	0.009	168
	[ar-³H]-catharanthine	C. roseus	0.005	169

Compound	Precursor	Plant	Incorporation/further information.*	Reference
	[$CO_2C^3H_3$]-catharanthine	C. roseus	0.001	169
	[acetyl^{14}C]-vindoline	C. roseus	0.005	169
Vincadifformine	[2-^{14}C]-tryptophan	V. minor	E	170, 171
	[3-^{14}C]-tryptophan	V. minor	0.24	143
Vincadine	[2-^{14}C]-tryptophan	V. minor	E	170
	[3-^{14}C]-tryptophan	V. minor	0.015	143
	[ar-^3H]-minovine	V. minor	0	143
Vincamedine	[1-^{14}C]-acetate	Vinca difformis	0.0007, T	155
Vincamine	$^{14}CO_2$	V. minor	E	172
	Lysine	V. minor	CC	173
	Tryptophan	V. minor	CC	173
	[ar-^3H]-tryptophan	V. minor	0.09	142
	[ar-^{14}C]-tryptophan	V. minor	E	170, 171
	Tryptamine	V. minor	CC	173
	[ar-^3H]-geissoschizine	V. minor	0.005	142
	[ar-^3H]-16,17-dihydro-secodin-17-ol	V. minor	0.001	142, 40, 59
	[ar-^3H]-secodine	V. minor	0.001	142, 40, 59
	[ar-^3H, $^{14}CO_2Me$]-secodine	V. minor	E, L	40
	[ar-^3H]-stemmadenine	V. minor	+	142
	[ar-^3H]-tabersonine	V. minor	0.07	142
	[ar-^3H]-vincaminoreine	V. minor	0.001	32
Vincaminoreine	[2-^{14}C]-tryptophan	V. minor	E	170
	[3-^{14}C]-tryptophan	V. minor	0.015	143
	[ar-^3H]-minovine	V. minor	0	143
Vincaminorine	[2-^{14}C]-tryptophan	V. minor	E	170, 171
Vincine	Lysine	V. minor	CC	173
	Tryptophan	V. minor	CC	173
	[2-^{14}C]-tryptophan	V. minor	E	170
	Tryptamine	V. minor	CC	173
Vincoside	[7-^3H]-loganin	C. roseus	1.5, C	34, 35
Vindoline	[1-^{14}C]-acetate	C. roseus	0.6, HH A, T	65
			0.01, T	8
			0.9–1.9, Z	29, 66
	[2-^{14}C]-acetate	C. roseus	1.5, 11, A, T	65
			0.38–0.67, Z	29, 66
	[1-^{14}C]-glycine	C. roseus	1.0, U	162
	[2-^{14}C]-glycine	C. roseus	3.4, U, A, T	162

Compound	Precursor	Plant	Incorporation/further information.*	Reference
	Leucine	*C. roseus*	0.007, D	67
	[2-¹³C]-leucine	*C. roseus*	0.037, OO	174
	[1,3-¹⁴C]-malonate	*C. roseus*	0.25, KK A, T	65
	[2-¹⁴C]-malonate	*C. roseus*	0.71, 11	65
	[2-¹⁴C]-glyoxylate	*C. roseus*	11.8, U, A, T	162
	[2-¹⁴C]-pyruvate	*C. roseus*	0.24, JJ	65
	[2-¹⁴C]-mevalonate	*C. roseus*	0.05 0.12 0.02, F	10 156 175
	[4-³H, 2-¹⁴C]-4R-mevalonate	*C. roseus*	E, G	12
	[4-³H, 2-¹⁴C]-4S-mevalonate	*C. roseus*	E, I	12
	[3-¹⁴C]-mevalonate	*C. roseus*	0.6, V	13
	[4-¹⁴C]-mevalonate	*C. roseus*	E, V	11
	[5-¹⁴C]-mevalonate	*C. roseus*	E, V	15
	[6-¹⁴C]-mevalonate	*C. roseus*	E, W	14
	[2-¹⁴C]-mevalonolactone	*C. roseus*	0.6 0.05 0.5, V	65 176 177
	[5-²H₂]-mevalonolactone	*C. roseus*	0.2	178
	[1-²H₂]-geraniol	*C. roseus*	0.8	178
	[2-¹⁴C]-geraniol	*C. roseus*	0.37, N, B 0.1, V E	175 13 178
	[1-³H₂, 2-¹⁴C]-geraniol	*C. roseus*	E	12
	[2-³H, 2-¹⁴C]-geraniol	*C. roseus*	E, I	12
	[3-¹⁴C]-geraniol	*C. roseus*	0.35	68
	[2-¹⁴C]-geranyl pyrophosphate	*C. roseus*	0.2, V	15
	[2-³H, 2-¹⁴C]-nerol	*C. roseus*	E, I	12
	[1-³H₂]-10-hydroxygeraniol	*C. roseus*	0.25, P	16
	[8-¹⁴C]-10-hydroxy-geraniol	*C. roseus*	0.72	14
	[8-¹⁴C]-10-hydroxynerol	*C. roseus*	1.2	14
	[1-³H₂]-citronellol	*C. roseus*	0.0008	14
	[1-³H₂]-10-hydroxy-citronellol	*C. roseus*	0.017	16
	[8-¹⁴C]-hydroxycitronellol	*C. roseus*	0.001	14
	[1-³H₂]-10-oxocitronellol	*C. roseus*	0.011	14
	[1-³H₂]-linalool	*C. roseus*	0.0006	16
	[8-¹⁴C]-10-hydroxylinalool	*C. roseus*	0.003	14
	[7-¹⁴C]-iridodial	*C. roseus*	0.0005	179
	[OMe-³H]-deoxyloganin	*C. roseus*	0.24	18
	[OMe-³H]-loganin	*C. roseus*	0.5	19, 20
	[7-³H]-loganin	*C. roseus*	0.82	23
	[9-¹⁴C]-loganin	*C. roseus*	0.23	21
	[10-¹⁴C]-loganin	*C. roseus*	0.2	22
	[5-³H, OMe-³H]-loganin	*C. roseus*	+	23
	[7-³H, OMe-³H]-loganin	*C. roseus*	0.78	24
	[6,8-³H, OMe-¹⁴C]-loganin	*C. roseus*	0.36	25

Compound	Precursor	Plant	Incorporation/further information.*	Reference
	[OMe-³H]-secologanin	C. roseus	0.12	26, 27
	[10-¹⁴C]-sweroside	C. roseus	11	121, 122
	[7-⁵H]-verbenalin	C. roseus	0	19, 20
	[OMe-¹⁴C]-monotropeine methyl ester	C. roseus	0	19, 20
	[U-¹⁴C]-shikimic acid	C. roseus	3.2, U	28–30
	[3-³H, 6-¹⁴C]-strictosidine	C. roseus	0.19	43
	[6-¹⁴C]-strictosidine	C. roseus	1.80	36
	[O-methyl-³H]-strictosidine	C. roseus	1.90	36
	[2-¹⁴C]-tryptophan	C. roseus	0.32	8
	[3-¹⁴C]-tryptophan	C. roseus	E, K	31
			1.0–4.3	29, 66
	[3-¹⁴C, 1-¹⁵N]-tryptophan	C. roseus	3.2, BB	65
	[ar-³H]-tryptamine	C. roseus	0.0003	32, 33
			0.39	34, 35
	[OMe-³H]-isovincoside	C. roseus	0	24, 35
	[OMe-³H]-vincoside	C. roseus	0.78	24, 35
	[OMe-³H]-vincoside	C. roseus	0.001	36
	[OMe-³H]-vincoside and isovincoside	C. roseus	0.57	34, 35
	[ar-³H]-vincoside and isovincoside	C. roseus	0.11	38
	[ar-³H, OMe-³H]-vincoside and isovincoside	C. roseus	0.31	34, 35
	[³H: ¹⁴C]-strictosidine	C. roseus	1.31 (7.0:1)	36
	[ar-³H]-vincoside	C. roseus	0.96	37
	[ar-³H]-isovincoside	C. roseus	0.01	37
	[OMe-³H]-dihydrovincoside	C. roseus	0	24, 35
	[ar-³H]-geissoschizine	C. roseus	0.13	38
	[OMe-³H]-geissoschizine	C. roseus	0.35	40
	[ar-³H, OMe-³H]-geissoschizine	C. roseus	0.41, LL	38
	[OMe-³H]-corynantheine aldehyde	C. roseus	0.1, K 0.003, J	69
	[ar-³H]-ajmalicine	C. roseus	0.6, K 0.004, K	69 69
	[OMe-³H]-ajmalicine	C. roseus	0.001, K	12
	[ar-³H]-geissoschizine oxindole	C. roseus	0.55	40
	[OMe-³H]-stemmadenine	C. roseus	1.76, K	69
	[OMe-³H, 6-¹⁴C]-stemmadenine	C. roseus	0.95, K	69
	[ar-³H]-16,17-dihydro-secodin-17-ol	C. roseus	0.001	40, 59
	[ar-³H]-secodine	C. roseus	0.02	142, 42, 40
	[ar-³H, ¹⁴CO₂Me]-secodine	C. roseus	0.04 MM E, L, MM	42 40
	[19-³H, ¹⁴CO₂Me]-secodine	C. roseus	0.07, BB	42
	[3,14,15,21-³H]-secodine	C. roseus	0.03, BB	40, 42
	[ar-³H]-strictosidine	C. roseus	0.25	44

Compound	Precursor	Plant	Incorpora-tion/further informa-tion.*	Refer-ence
	[ar-³H]-tabersonine	C. roseus	0.03	32, 33
	[OMe-³H, 6-¹⁴C]-tabersonine	C. roseus	4.8, K	180
	[OMe-³H, 6-¹⁴C]-tabersonine	C. roseus	1.1, K	180
	[ar-³H]-15,20-dihydro-vincadine	C. roseus	0	32, 33
	[ar-³H]-carbomethoxy-cleavamine	C. roseus	9	32, 33
	[1-¹⁴C]-acetate	C. roseus	0.65–0.98, Z	29, 66
	[2-¹⁴C]-acetate	C. roseus	0.97, Z	27, 66
	[3-¹⁴C]-tryptophan	C. roseus	2.6–4.0	29, 66
Vomicine	[1-¹⁴C]-glycine	S. nux vomica	0	160
Alkaloids of V. minor	[2-¹⁴C]-acetate	V. minor	E	172
	Lysine	V. minor	E	173
	Anthranilic acid	V. minor	E	173
	Tyrosine	V. minor	0	181
	[2-¹⁴C]-tryosine	V. minor	0	171
	[2-¹⁴C]-tryptophan	V. minor	E	173, 181, 142
	[2-¹⁴C]-tryptamine	V. minor	E	173
	[U-¹⁴C]-vincamine	V. minor	E	172
	[5-¹⁴C]-vincamine	V. minor	E	181
α-Yohimbine	[3-³H, 6-¹⁴C]-strictosidine	Rauwolfia canescens	1.63 PP = 5.8:1	43
	[3-³H, 6-¹⁴C]-strictosidine	R. canescens	0.7[¹⁴C]	144
	[3-³H, 6-¹⁴C]-vincoside	R. canescens	<0.001	144
Yohimbine	[7-³H]-secologanin	R. canescens	0.14[¹⁴C]	174
	[2-¹⁴C]-tryptamine	R. canescens	PP = 81.5:1	43

Symbols used

A = low incorporation; B = activity in sugar moiety; C = dilution analysis; F = specific; G = 50 per cent ³H retention; H = anomolous ³H:¹⁴C ratio; I = no ³H activity retained; J = plants; K = seedlings; L = ³H activity mostly retained; M = —OMe label; N=3:1 mixture with nerol; P = 3.8:1 mixture with nerol; Q = isolated as ipecoside; R = moderate incorporation; S = 2-year-old plants; T = random incorporation into C_{9-10} unit; U = activity mainly in tryptamine portion; V = theoretical activity; W = approximately 20 per cent activity at C-22; X = N—Me labelled; Y = 50 per cent activity at N—Me; Z = random incorporation; AA = 3-month-old plants; BB = incorporation intact; CC = increased level; DD = some ^{14}C activity lost; EE = incorporation with loss of ^{14}C; FF = 50 per cent activity at C-22; GG = 85 per cent activity at C-32; HH = 50 per cent activity at acetyl group; II = mainly at C—24 of acetyl group; JJ = mainly at C-23 of acetyl group; KK = O-acetyl group labelled; LL = no NIH shift; MM = NIH shift; NN = 60 per cent loss of tritium; OO = incorporation randomized over several carbon atoms; PP = $^{3}H:^{14}C$ ratio; RR = percentage tritium retention; SS = deuterium distribution; TT = little or no change in isotope ratio; UU = percentage tritium retained.

References

1. FEHR, T., ACKLIN, W., and ARIGONI, D. *Chem. Commun.* 801 (1966).
2. GROGER, D., ERGE, D., and FLOSS, H. G. *Z. Naturforsch.* **21b**, 827 (1966).
3. VOIGT, R., BORNSCHEIN, M., and RABITZSCH, G. *Pharmazie* **22**, 326 (1967).
4. FLOSS, H. G., HOMEMANN, V., SCHILLING, N., GROGER, D., and ERGE, D. *J. Am. Chem. Soc.* **90**, 6500 (1968).
5. PLIENINGER, H. and MEYER, E. *Liebigs Ann. Chem.* 813 (1978).
6. PACHLATKO, P., TABACIK, C., ACKLIN, W., and ARIGONI, D. *Chimia* **29**, 526 (1975).
7. FLOSS, H. G., TCHEN-LIN, CHANG, C., NAIDOO, B., BLAIR, G. E., ABOU-CHAAR, C. I., and CASSADY, J. M. *J. Am. Chem. Soc.* **96**, 1898 (1974).
8. LEETE, E., AHMAD, A., and KOMPIS, I. *J. Am. Chem. Soc.* **87**, 4168 (1965).
9. KUTNEY, J. P., BECK, J. F., NELSON, V. R., STUART, K. L., and BOSE, A. K. *J. Am. Chem. Soc.* **92**, 2174 (1970).
10. BATTERSBY, A. R., BROWN, R. T., KAPIL, R. S., PLUNKETT, A. O., and TAYLOR, J. B. *Chem. Commun.* **46**, (1966).
11. BATTERSBY, A. R., BROWN, R. T., KAPIL, R. S., KNIGHT, J. A., MARTIN, J. A., and PLUNKETT, A. O. *Chem. Commun.* 810, 888 (1966).
12. BATTERSBY, A. R., BYRNE, T. C., KAPIL, R. S., MARTIN, J. A., PAYNE, T. G., ARIGONI, D., and LOEW, P. *Chem. Commun.* 951 (1968).
13. LOEW, P., GOEGGEL, H., and ARIGONI, D. *Chem. Commun.* 347 (1966).
14. ESCHER, S., LOEW, P., and ARIGONI, D. *Chem. Commun.* 823 (1970).
15. BATTERSBY, A. R., BROWN, R. T., KNIGHT, J. A., MARTIN, J. A., and PLUNKETT, A. O. *Chem. Commun.* 346 (1966).
16. BATTERSBY, A. R., BROWN, S. H., and PAYNE, T. G. *Chem. Commun.* 827 (1970).
17. LEE, S. L., HIRATA, T., and SCOTT, A. I. *Tetrahedron Lett.* 691 (1979).
18. BATTERSBY, A. R., BURNETT, A. R., and PARSONS, P. G. *Chem. Commun.* 826 (1970).
19. BATTERSBY, A. R., BROWN, R. T., KAPIL, R. S., MARTIN, J. A., and PLUNKETT, A. O. *Chem. Commun.* 812 (1966).
20. *Idem. Ibid.* 890 (1966).
21. BATTERSBY, A. R., KAPIL, R. S., MARTIN, J. A., and MO, L. *Chem. Commun.* 133 (1968).
22. LOEW, P. and ARIGONI, D. *Chem. Commun.* 137 (1968).
23. BATTERSBY, A. R. and GIBSON, K. H. *Chem. Commun.* 902 (1971).
24. BATTERSBY, A. R., BURNETT, A. R., HALL, E. S., and PARSONS, P. G. *Chem. Commun.* 1582 (1968).

25. BATTERSBY, A. R., HUTCHINSON, C. R., and LARSON, R. A. *A.C.S. National Meeting, Boston, Mass Abstracts Organic Section No. 11* 1972.
26. BATTERSBY, A. R., BURNETT, A. R., and PARSONS, P. G. *J. Chem. Soc.(C)* 1187 (1969).
27. BATTERSBY, A. R., BURNETT, A. R., and PARSONS, P. G. *Chem. Commun.* 1280 (1968).
28. STOLLE, K., GROGER, D., and MOTHES, K. *Chem. Ind. (Lond.)* 2065 (1965).
29. STOLLE, K., GROGER, D., and MOTHES, K. *Abh. Dtsch. Akad. Wiss., Berlin* 5, 497 (1966).
30. GROGER, D., STOLLE, K., and MOTHES, K. *Z. Naturforsch.* **B21**, 206 (1966).
31. SCOTT, A. I., REICHARDT, P. B., SLAYTOR, M. B., and SWEENEY, J. G. *Bioorg. Chem.* **1**, 157 (1971).
32. KUTNEY, J. P., CRETNEY, N. J., HADFIELD, J. R., HALL, E. S., NELSON, V. R., and WIGFIELD, D. C. *J. Am. Chem. Soc.* **90**, 3566 (1968).
33. KUTNEY, J. P. *Heterocycles* **4**, 169 (1976).
34. BATTERSBY, A. R., BURNETT, A. R., and PARSONS, P. G. *J. Chem. Soc.*(C) 1282 (1968).
35. BATTERSBY, A. R., BURNETT, A. R., and PARSONS, P. G. *J. Chem. Soc.*(C) 1193 (1969).
36. STOCKIGT, J. and ZENK, M. H. *J. Chem. Soc. Chem. Commun.* 646 (1977).
37. BATTERSBY, A. R., LEWIS, N. G., and TIPPETT, J. M. *Tetrahedron Lett.* 4849 (1978).
38. BATTERSBY, A. R. and HALL, E. S. *Chem. Commun.* 793 (1969).
39. SCOTT, A. I., LEE, S. L., DECAPITE, P., and CULVER, M. G. *Heterocycles* **7**, 979 (1977).
40. SCOTT, A. I. *Acc. Chem. Res.* **3**, 151 (1970).
41. STOCKIGT, J. *J. Chem. Soc. Chem. Commun.* 1097 (1978).
42. KUTNEY, J. P., BECK, J. F., EGGERS, N. J., HANSSEN, H. W., SOOD, R. S., and WESCOTT, N. D. *J. Am. Chem. Soc.* **93**, 7322 (1971).
43. NAGAKURA, N., RUFFER, M., and ZENK, M. H. *J. Chem. Soc. Perkin Trans.* I 2308 (1979).
44. LEETE, E. and MINICH, M. L. *Phytochemistry* **16**, 149 (1977).
45. BARTON, D. H. R., KIRBY, G. W., PRAGER, R. H., and WILSON, E. M. *J. Chem. Soc.* 3990 (1965).
46. LEETE, E., GHOSAL, S., and EDWARDS, P. N. *J. Am Chem. Soc.* **84**, 1068 (1962).
47. LEETE, E. and GHOSAL, S. *Tetrahedron Lett.* 1179 (1962).
48. BATTERSBY, A. R., BINKS, R., LAWRIE, W., PARRY, G. V., and WEBSTER, B. R. *Proc. Chem. Soc.* 369 (1963).
49. GEAR, J. R. and LEETE, E. unpublished results.
50. GARG, A. K. and GEAR, J. R. *Chem. Commun.* 1442 (1969).
51. GARG, A. K. and GEAR, J. R. *Phytochemistry* **11**, 689 (1972).
52. EDWARDS, P. N. and LEETE, E. *Chem. Ind. (Lond.)* 1666 (1961).
53. BATTERSBY, A. R., BINKS, R., LAWRIE, W., PARRY, G. V., and WEBSTER, B. R. *J. Chem. Soc.* 7459 (1965).
54. BATTERSBY, A. R. and PARRY, G. V. *Tetrahedron Lett.* 787 (1964).
55. STOCKIGT, J. *Tetrahedron Lett.* 2615 (1979).
56. LEETE, E. *Chem. Ind. (Lond.)* 692 (1960).
57. LEETE, E., *J. Am. Chem. Soc.* **82**, 6338 (1960).
58. SCOTT, A. I., CHERRY, P. C., and QURESHI, A. A. *J. Am. Chem. Soc.* **91**, 4932 (1969).

59. KUTNEY, J. P., BECK, J. F., EHRET, C., POULTON, G., SOOD, R. S., and WESTCOTT, N. D. *Bio-org. Chem.* **1**, 194 (1971).
60. INOUYE, H., UEDA, S., AOKI, Y., and TAKEDA, Y. *Tetrahedron Lett.* 2351 (1969).
61. INOUYE, H., UEDA, S., AOKI, Y., and TAKEDA, Y. *Chem. Pharm. Bull. (Tokyo)* **20**, 1287 (1972).
62. INOUYE, H., UEDA, S., and TAKEDA, Y. *Tetradedron Lett.* 3351 (1970).
63. INOUYE, H., UEDA, S., and TAKEDA, Y. *Z. Naturforsch.* **B24**, 1666 (1969).
64. HUNI, J. E. S., HILTEBRAND, H., SCHMID, H., GROGER, D., JOHNE, S., and MOTHES, K. *Experientia* **22**, 656 (1966). MAIER, W. and GROGER, D. *Z. Naturforsch.* **B25**, 1192 (1970).
65. GROGER, D., STOLLE, K., and MOTHES, K. *Arch. Pharm. (Weinheim)* **300**, 393 (1967).
66. GROGER, D., STOLLE, K., and MOTHES, K. *Tetrahedron Lett.* 2579 (1964).
67. WIGFIELD, D. C., LEM, B., and SRINIVASAN, V. *Tetrahedron Lett.* 2659 (1972).
68. LEETE, E. and UEDA, S. *Tetrahedron Lett.* 4915 (1966).
69. QURESHI, A. A. and SCOTT, A. I. *Chem. Commun.* 948 (1968).
70. GARG, A. K. and GEAR, J. R. *Tetrahedron Lett.* 4377 (1969).
71. GEAR, J. R. and GARG, A. K. *Tetrahedron Lett.* 141 (1968).
72. BATTERSBY, A. R. and GREGORY, B. *Chem. Commun.* 134 (1968).
73. BATTERSBY, A. R. and PARRY, R. J. *Chem. Commun.* 901 (1971).
74. BATTERSBY, A. R. and PARRY, R. J. *Chem. Commun.* 31 (1971).
75. BATTERSBY, A. R. and PARRY, R. J. *Chem. Commun.* 30 (1971).
76. BATTERSBY, A. R. and HALL, E. S. *Chem. Commun.* 194 (1970).
77. BLACKSTOCK, W. P., BROWN, R. T., CHAPPLE, C. L., and FRASER, S. B. *Chem. Commun.* 1006 (1972).
78. SCOTT, A. I. and WEI, C. C. *J. Am. Chem. Soc.* **94**, 8266 (1972).
79. STEYN, P. S., VLEGGAR, R., FERREIRA, N. P., KIRBY, G. W., and VARLEY, M. J. *J. Chem. Soc. Chem. Commun.* 465 (1975).
80. BATTERSBY, A. R. and BHATNAGAR, A. K. *Chem. Commun.* 193 (1970).
81. FLOSS, H. G. *Chem. Commun.* 804 (1967).
82. MARCHELLI, R., DOSSENA, A., and CASNATI, G. *Chem. Commun.* 779 (1975).
83. SLAYTOR, G. P., MACDONALD, J. C., and NAKASHIMA, R. *Biochemistry* **9** 2886 (1970).
84. ALLEN, C. M. *Biochemistry* **11**, 2154 (1972).
85. ALLEN, C. M. *J. Am. Chem. Soc.* **95**, 2386 (1973).
86. BIRCH, A. J. and FARRAR, K. R. *J. Chem. Soc.* 4277 (1963).
87. BIRCH, A. J., BLANCE, G. E., DAVID, S., and SMITH, H. *J. Chem. Soc.* 3128 (1961).
88. WEYGAND, F. and FLOSS, H. G. *Angew. Chem.* **75**, 783 (1963).
89. BATTERSBY, A. R. *Specialist periodical report, The alkaloids*, Vol. 1, pp. 30–40. Chemical Society, London (1971).
90. NAIDOO, B., CASSADY, J. M., BLAIR, G. E., and FLOSS, H. G. *Chem. Commun.* 471 (1970).
91. CASSADY, J. M., ABOU-CHAAR, C. I., and FLOSS, H. G. *Lloydia* **36**, 390 (1973).
92. ABOU-CHAAR, C. I., GUENTHER, H. F., MANUEL, M. F., ROBBERS, J. E., and FLOSS, H. G. *Lloydia* **35**, 272 (1972).
93. OTSUKA, H., ANDERSON, J. A., and FLOSS, H. G., *J. Chem. Soc. Chem. Commun.* 660 (1979).

94. ACKLIN, W., FEHR, T., and STADLER, P. A. *Helv. Chim. Acta* **58**, 2492 (1975).
95. GROGER, D., MOTHES, K., SIMON, H., FLOSS, H. G., and WEYGAND, F. Z. *Naturforsch.* **16b**, 432 (1961).
96. TAYLOR, E. H. and RAMSTAD, E. *Nature (Lond.)* **188**, 494 (1960).
97. GUNTHER, H. F. and FLOSS, H. G. Unpublished results.
98. PLEININGER, H., IMMEL, H., and VOLKEL, L. *Justus Liebigs Ann. Chem.* **706**, 223 (1967).
99. NELSON, V. and AGURELL, S. *Acta Chem. scand.* **23**, 3393 (1969).
100. FLOSS, H. G., BASMADJIAN, G. P., GROGER, D., and ERGE, D. *Lloydia* **34**, 449 (1971).
101. BASMADJIAN, G., FLOSS, H. G., GROGER, D., and ERGE, D. *Chem. Commun.* 418 (1969).
102. MINGHETII, A. and ARCAMONE, F. *Experientia* **25**, 926 (1969).
103. LIN, C. C. L., BLAIR, G. E., CASSADY, J. M., GROGER, D., MAIER, W., and FLOSS, H. G. *J. org. Chem.* **38**, 2249 (1973).
104. FLOSS, H. G., BASMADJIAN, G., TCHENG, M., SPALLA, C., and MINGHETTI, A. *Lloydia* **34**, 442 (1971).
105. BASSETT, R. A., CHAIN, E. B., and CORBETT, K. *Biochem. J.* **134**, 1 (1973).
106. ROBBERS, J. E., LI-JUNG, C., ANDERSON, J. A., and FLOSS, H. G. *J. Nat. Prod.* **42**, 537 (1979).
107. LOEW, P., VON SZCZEPANSKI, C. H., COSCIA, C. J., and ARIGONI, D. *Chem. Commun.* 1276 (1968).
108. FLOSS, H. G., BASMADJIAN, G. P., TCHENG, M., GROGER, D., and ERGE, D. *Lloydia* **34**, 446 (1971).
109. BULOCK, J. D. and RYLES, A. P. *Chem. Commun.* 1404 (1970).
110. FLOSS, H. G., MOTHES, K., and RETTIO, A. Z. *Naturforsch.* **B19**, 1106 (1964).
111. MAREKOV, N. L., ARNANDOV, M., and POPOV, S. *Dolk. Akad. Nauk.* **23**, 169 (1970); MAREKOV, N. L., MONDESHKI, L., and ARNADOV, M. *ibid.* **23**, 803 (1970).
112. COSCIA, C. J., GUARNACCIA, R., and BOTTA, L. *Biochemistry* **8**, 5036 (1969).
113. COSCIA, C. J. and GUARNACCIA, R. *Chem. Commun.* 138 (1968).
114. COSCIA, C. J. and GUARNACCIA, R. *J. Am. Chem. Soc.* **89**, 1280 (1967).
115. BIRCH, A. J. and SMITH, E. *Aust. J. Chem.* **9**, 234 (1956).
116. GUARNACCIA, R., BOTTA, L., and COSCIA, C. J. *J. Am. Chem. Soc.* **91**, 204 (1969).
117. INOUYE, H., UEDA, S., and NAKAMURA, Y. *Tetrahedron Lett.* 3221 (1967).
118. INOUYE, H., UEDA, S., and NAKAMURA, Y. *Chem. Pharm. Bull. (Tokyo)* **18**, 2043 (1970).
119. GROGER, D. and SIMCHEN, P. Z. *Naturforsch.* **B24**, 356 (1969).
120. SLAYTOR, M. and MCFARLANE, I. J. *Phytochemistry* **7**, 605 (1968).
121. INOUYE, H., UEDA, S., and TAKEDA, Y. *Tetrahedron Lett.* 3453 (1968).
122. INOUYE, H., UEDA, S., and TAKEDA, Y. *Chem. Pharm. Bull. (Tokyo)* **19**, 587 (1971).
123. GROSS, D., LEHMANN, H., and SCHUTTE, H. R. *Tetrahedron Lett.* 4047 (1971).
124. O'DONOVAN, D. and LEETE, E. *J. Am. Chem. Soc.* **85**, 461 (1963).
125. YAMAZAKI, M. and LEETE, E. *Tetrahedron Lett.* 1499 (1964).
126. HORNEMANN, U., HURLEY, L. H., SPEEDLE, M. K., and FLOSS, G. H. *Tetrahedron Lett.* 2255 (1970).
127. HORNEMANN, U., HURLEY, L. H., SPEEDLE, M. K., and FLOSS, H. G. *J. Am. Chem. Soc.* **93**, 3028 (1971).

128. SHELLARD, E. J. and HOUGHTON, P. J. *Planta Med.* **21**, 16 (1972).
129. SHELLARD, E. J., SARPONG, K., and HOUGHTON, P. J. *J. Pharm. Pharmacol.* **23**, 2445.
130. INOUYE, H., UEDA, S., INOUE, K., and TAKEDA, Y. *Tetrahedron Lett.* 4073 (1971).
131. GUARNACCIA, R., BOTTA, L., and COSCIA, C. J. *J. Am. Chem. Soc.* **92**, 6098 (1970).
132. GUARNACCIA, R. and COSCIA, C. J. *J. Am. Chem. Soc.* **93**, 6320 (1971).
133. MADYASTHA, K. M., GUARNACCIA, R., and COSCIA, C. J. *FEBS Lett.* **14**, 175 (1971).
134. BRECHBULER-BADER, S., COSCIA, C. J., LOEW, P., VON SZCZEPANSKI, C. H., and ARIGONI, D. *Chem. Commun.* 136 (1968).
135. BATTERSBY, A. R., KAPIL, R. S., and SOUTHGATE, R. *Chem. Commun.* 131 (1968).
136. BATTERSBY, A. R., HALL, E. S., and SOUTHGATE, R. *J, Chem. Soc. (C)* 721 (1969).
137. HESSE, M. *Indolalkaloide in Tabellen.* Springer-Verlag, Berlin (1964).
138. MANSKE, R. H. F. (ed.) *The alkaloids chemistry and physiology*, Vols. VIII, XI, and XIII. Academic Press, New York (1965, 1968, 1971).
139. JEFFS, P. W. and KARLE, J. M. *J. Chem. Soc. Chem. Commun.* 60 (1977).
140. JEFFS, P. W., JOHNSON, D. B., MARTIN, N. H., and RAUCKMAN, B. S. *J, Chem. Soc. Chem. Commun.* 82 (1976).
141. JEFFS, P. W., CAMPBELL, H. F., FARRIER, D. S., GANGULI, G., MARTIN, N. H., and MOLINA, G. *Phytochemistry* **13**, 933 (1974).
142. KUTNEY, J. P., BECK, J. F., NELSON, V. R., SOOD, R. S. *J. Am. Chem. Soc.* **93**, 255 (1971).
143. KUTNEY, J. P., EHRET, C., NELSON, V. R., and WIGFIELD, D. C. *J. Am. Chem. Soc.* **90**, 5929 (1968).
144. RUEFFER, M., NAGAKURA, N., and ZENK, M. H. *Tetrahedron Lett.* 1593 (1978).
145. INOUYE, H., UEDA, S., and TAKEDA, Y. *Tetrahedron Lett.* 4069 (1971).
146. KUTNEY, J. P., NELSON, V. R., and WIGFIELD, D. C. *J. Am. Chem. Soc.* **91**, 4279 (1969).
147. DE MOERLOOSE, P. and RUYSSEN, R. *J. Pharm. Belg.* **8**, 156 (1953). *Chem. Abst.* **47**, 11354 (1953).
148. DE MOERLOOSE, P. and RUYSSEN, R. *Pharm. Tijdschr. Belg.* **30**, 97 (1953); *Chem. Abst.* **48**, 820 (1954).
149. DE MOERLOOSE, P. *Pharm. Weekblad.* **89**, 541 (1954).
150. LEETE, E. and WEMPLE, J. N. *J. Am. Chem. Soc.* **88**, 4743 (1966).
151. LEETE, E. and WEMPLE, J. N. *J. Am. Chem. Soc.* **91**, 2698 (1969).
152. INOUYE, H., UEDA, S., and TAKEDA, Y. *Tetrahedron Lett.* 407 (1969).
153. KOWANKA, N. and LEETE, E. *J. Am. Chem. Soc.* **84**, 4919 (1962).
154. LEETE, E. *Tetrahedron* **14**, 35 (1961).
155. GOEGGEL, H. and ARIGONI, D. *Experientia* **21**, 369 (1965).
156. GOEGGEL, H. and ARIGONI, D. *Chem. Commun.* 538 (1965).
157. KIRBY, G. W. and VARLEY, M. J. *J. Chem. Soc. Chem. Commun.* 833 (1974).
158. BROWN, R. T., SMITH, G. N., and STAPLEFORD, K. S. J. *Tetrahedron Lett.* 4349 (1968).
159. GROGER, D. and MAIER, W. *4th Int. Symp. Biochem. Physiol. Alkaloids* (ed. K. MOTHES) Akadamie, Berlin (1969); *Chem. Abstr.* **77**, 98828 (1972).

160. SCHLATTER, G., WALDNER, E., GROGER, D., MAIER, W., and SCHMID, H. *Helv. Chim. Acta* **52**, 776 (1969).
161. MAIER, W. and GROGER, D. *Arch. Pharm. (Weinheim)* **304**, 351 (1971).
162. GROGER, D., MAIER, W., and SIMCHEN, P. *Experientia* **26**, 820 (1970).
163. HEIMBERGER, S. I. and SCOTT, A. I. *J. Chem. Soc. Chem. Commun.* 217 (1973).
164. SUGII, M. and HASHIMOTO, Y. *Bull. Inst. Chem. Res. Kyoto Univ.* **36**, 127 (1968); *Chem. Abstr.* **53**, 10395 (1959).
165. HORODYSKY, A. G., WALLER, G. R., and EISENBRAUN, E. J. *J. biol. Chem.* **244**, 3110 (1969).
166. DADDONA, P. E. and HUTCHINSON, C. R. *J. Am. Chem. Soc.* **96**, 6806 (1974).
167. BAXTER, R. L., DORSCHEL, C. A., LEE, S-L., and SCOTT, A. I. *J. Chem. Soc. Chem. Commun.* 257 (1979).
168. HASSAM, S. B. and HUTCHINSON, C. R. *Tetrahedron Lett.* 1681 (1978).
169. SCOTT, A. I., GUERITTE, F., and LEE, S. L. *J. Am. Chem. Soc.* **100**, 6253 (1978).
170. VERGAR-PETRI, G., SZARVAS, T., and VARADI, J. *Gyogyszereszet* **14**, 184 (1970): *Chem. Abstr.* **73**, 127808 (1970).
171. VERGAR-PETRI, G. *4th Int. Symp. Biochem. Physiol. Alkaloids* (ed. K. MOTHES) Akadamie, Berlin (1969). *Chem. Abstr.* **77**, 98808 (1972).
172. VERZAR-PETRI, G. *Acta. Biol. Acad. Sci. Hung.* **22**, 413 (1971); *Chem. Abstr.* **76**, 110357 (1972).
173. BOJTHE-HORVATH, K., VARGA-BALAZS, M., and CLAUDER, O. *Planta Med.* **17**, 328 (1969); *Chem. Abstr.* **71**, 109801 (1969).
174. WIGFIELD, D. C. and WEN, B. P. *Bio-org. Chem.* **6**, 511 (1977).
175. MONEY, T., WRIGHT, I. G., McCAPRA, F., HALL, E. S., and SCOTT, A. I. *J. Am. Chem. Soc.* **90**, 4144 (1968).
176. MONEY, T., WRIGHT, I. G., McCAPRA, F. and SCOTT, A. I. *Proc. Natl. Acad. Sci. U.S.A.* **53**, 901 (1965).
177. McCAPRA, F., MONEY, T., SCOTT, A. I., and WRIGHT, I. G. *Chem. Commun.* 537 (1965).
178. HALL, E. S., McCAPRA, F., MONEY, T., FUKUMOTO, K., HANSON, J. R., MOOTOO, B. S., PHILLIPS, G. T., and SCOTT, A. I. *Chem. Commun.* 348 (1966).
179. BOWMAN, R. M. and LEETE, E. *Phytochemistry* **8**, 1003 (1969).
180. QURESHI, A. A. and SCOTT, A. I. *Chem. Commun.* 948 (1968).
181. VERZAR-PETRI, G. *Bot. Kozlem.* **57**, 121 (1970); *Chem. Abstr.* **74**, 1125 (1971).
182. BATTERSBY, A. R., BURNETT, A. R., KNOWLES, G. D., and PARSONS, P. G. *Chem. Commun.* 1277 (1968).

9

.OCCURRENCE OF INDOLE ALKALOIDS IN PLANTS AND MICRO-ORGANISMS

Plant			Class		
	I	II	III	IV	V
Aceraceae					
Acer rubrum L.				IV	
A. saccharinum L. (=*A. dasycarpum* Ehrd.)				IV	
Alangiaceae					
Alangium lamarckii Thw.					V
Amanitaceae					
Amanita citrina Pers.				IV	
A. *mappa* Batsch.				IV	
A. *muscaria* L.				IV	
A. *pantherina* DC.				IV	
A. *porphyria*				IV	
A. *tomentella*				IV	
Amaranthaceae					
Charpentiera obovata Gaudich.				IV	
Anacardiaceae					
Draoconomelum magniferum Bl.				IV	
Apocynaceae					
Alstonia angustiloba Mig.	I				
A. *congensis* Engl. ∠ *A. Boonei* I	I				
A. *constricta* F. Muell	I				
A. *deplanchei*					V
A. *gilletii* De Wild	I				
A. *lanceolifera*	I				
A. *macrophylla* Wall.	I	II			V
A. *neriifolia* D. Don.	I				
A. *muelleriana* Domin. ∠ *odontophora* I	I			IV	V
A. *quaternata*	I				
A. *scholaris* R. Br.	I				
A. *somersetensis* F. M. Bailey					V
A. *spathulata* Blume	I				
A. *spectabilis* R. Br.	I				
A. *venenata* R. Br.	I	II			
A. *verticillosa* F. Muell.	I				
A. *villosa* Blume					V
A. *vitiensis*	I				
Amsonia angustifolia Michx.	I	II			V
A. *elliptica* Roem; et Schult.					V
A. *tabernaemontana* Walt.	I	II			V

Plant	I	II	Class III	IV	V
Aspidosperma album Vahl. (R. Benth, ex M. Pichon).		II			
A. auriculatum Mgf.	I				
A. australe Muell. Arg.	I				
A. carapanauba M. Pichon	I				
A. chakensis Spegazzini	I	II			
A. compactinervium Kuhlm.	I	II			
A. cuspa		II			
A. cylindirocarpon Muell. Arg.		II			
A. dasycarpon A. DC.	I				
A. desmantham	I	II			
A. discolor A. DC.	I	II			
A. dispersum Muell. Arg.		II			
A. duckei Hub.		II			
A. eburneum Fr. Allem. ex Sald.	I	II			
A. exalalum Monachino		II			
A. excelsum Benth	I				
A. fendleri Woods.		II			
A. formasanum		II			
A. gomezianum A, DC.	I	II			
A. hilarianum Muell. Arg.	I				
A. limae Woods.	I				
A. longipetiolatum Kuhlm.	I				
A. macrocarpon Mart.		II			
A. marcgravianum Woods.	I	II			
A. megalocarpon Muell. Arg.		II			
A. multiflorum A. DC.	I	II			
A. neblinae Monachino.		II			
A. nigricans Handro	I				
A. nitidum Benth. ex Muell. Arg.	I				
A. oblongum A. DC.	I	II			
A. obscurinervium Azambuja		II			
A. olivaceum Muell. Arg.	I	II			
A. parvifolium A. DC.	I	II			
A. peroba F. Allem ex Sald.	I	II			
A. polyneuron Muell. Arg.	I	II		IV	
A. populifolium A. DC.	I	II			
A. pyricollum Muell. Arg.	I	II			
A. pyrifolium Mart.		II			
A. quebrachoblanco Schlocht.	I	II			
A. quirandy Hassl.		II			
A. refractum Mart.		II			
A. rigidum (*A. laxiflorum* Kuhlm) Rusby	I	II			
A. sandwithianum Mgf.		II			
A. sessiliforum F. Allem.		II			
A. spegazzinii Molf. ex. Meyer	I				
A. spruceanum Benth.		II			
A. subincanum Mart. ex A. DC.	I			IV	
A. tomentosum Mart.	I				
A. tritenatum Rojas Acosta.		II			
A. ulei Mgf.	I				
A. veninata	I				
A. verbascifolium Muell. Arg.		II			

} synon.!

ч —

Plant	Class				
	I	II	III	IV	V
Bleekerea vitiensis	I				
Cabucala erythrocarpa	I				
C. fasciculata	I				
C. striolata	I				
C. torulosa	I				
Callichilia (*Hedanthera*) *Barteri* (Hook. f.) Pichon		II		IV	V
C. stenosepala Stapf.		II			V
C. subsessilis Stapf.	I				V
Capuronetta elegans				IV	V
Catharanthus ovalis i C· lanceus, longifolius I, I		II	III		
C. pusillus (Murr.) G. Don (= *Vinca pusilla* Murr., = *Lochnera pusilla* (Murr.) K. Schum).	I	II			
C. roseus (L.) G. Don (see *Vinca rosea* L.)	I	II	III		V
C. trichophyllus (Baker) Pichon.	I	II			
Conopharyngia durissima Stapf (= *Plumeria durissima* ort).	I				
C. johnstonii	I				
C. holstii Stapf G. Don (= *Tabernaemontana holstii*)	I				
C. jollyana Stapf. G. Don. (= *Tabernaemontana jollyana* Pierre ex Stapf)			III		
Craspidospermum verticillatum		II			
Crioceras dipladeniiflorus					V
C. longiflorus		II			
Diplorrhyncus condylocarpon (Muell. Arg.) Pichon ssp. *mossambicensis* (Benth). Duvign.	I				
Ervatamia coronaria Stapf (see *Tabernaemontana coronaria* Willd.)	I	II	III		
E. dichotoma Roxb.			III		
E. divaricata Burkill			III		
E. orientalis				IV	
Excavatia coccinea (T. et B.)	I				
Gabunia eglandulosa Stapf	I		III		V
G. odoratissima Stapf	I		III		
Geissospermum laeve Baill. (= *Geissospermum vellossii* Allem.)	I				
G. argenteum		II			
G. sericeum Benth. et. Hook. f.					V
G. vellosii Allem. (= *Tabernaemontana laevis* Vell.)	I				V
Gonioma kamassi E. Mey.	I	II		IV	
G. malagasy					V
Haplophyton cimicidum A. DC. ∠ Hajunta modesta		II			V
Hunteria eburnea Pichon	I	II			
H. umbellata (K. Schum.) Hall. f. (*Carpodinus umbellatus* K. Schum. *Polyadoa umbellata* Stapf. *Picralima umbellata* Stapf.)	I	II		IV	
Kopsia albiflora Boerl. (*Kopsia flavida* Blume)		II			
K. arborea Blume		II			
K. flavida Blume (see *K. albiflora* Boerl.)		II			
K. fruticosa (Ker.) A. DC. (= *K. pruniformis* Reichb. f. et. Zoll. ex. Bakh f.)		II			
K. longiflora Merrill		II			

Plant	I	II	Class III	IV	V
K. pruniformis Reichb. f. et. Zoll. ex Bakh. f. (see *K. fruticosa* (Ker.) A. DC.)		II			
K. singapurensis Ridley		II			
Lochnera lancea Boj. ex A. DC. (see *Vinca lancea* Boj. (ex A. DC.) K. Schum.)	I	II		V	
L. pusilla (Murr.) K. Schum. (see *Catharanthus pusillus* (Murr.) G. Don).	I	II			
L. rosea Reichb. (see *Vinca rosea* L.)	I	II	III		V
Melodinus australis F. Muell.	I	II			
M. balansae		II			
M. celastroides		II			
M. scandens Forst.	I	II			
Muntafera sessilifolia	I		III		V
Ochrosia balansae	I				
O. elliptica Labill.	I				
O. glomerata Valeton	I				
O.lifauna					V
O. moorei F. Muell.	I				
O. miana	I				V
O. nakaiana	I		IV		
O. oppositifolia K. Schum. (= *Cerbera oppositifolia* (Lam.)	I				
O. poweri Bailey	I				
O. sandwicensis A. DC.	I				
O. vieldardii	I				
Pandaca caduciflia	I			IV	
P. calcarea	I		III		
P. debrayi	I		III		
P. eusepala			III		
P. multiflora		II	III		
P. mocquerysii			III		
P. minutiflora					V
P. ochrasens	I		III		
P. retus		II			
P. retusa			III		
P. mauritiana					V
P. speciosa					V
Peschiera laeta	I				V
P. affinis (Muell. Arg.) Miers (see *Tabernaemontana affinis* Muell. Arg.)	I				
Picralima klaineana Pierre (see *Picralima nitida* (Stapf) Th. et H. Durand).	I				
P. nitida (Stapf) Th. et H. Durand	I				
P. umbellata Stapf (see *Hunteria umbellata* (K. Schum.) Hall. f.)	I	II		IV	
Pleiocarpa flavescens Stapf.		II			
P. mutica Benth.	I	II			V
P. tubicina Stapf.	I	II		IV	
P. talbotii	I				
P. pycnantha (K. Schum.) Stapf. var. *tubicina* (Stapf.) Pichon.	I	II			
Prestonia amazonica (Benth.) Macbride (= *Haemadictyon amazonicum* Benth.)				IV	

Plant	Class I	II	III	IV	V
Rauwolfia affinis Muell. Arg.	I				
R. amsonidefolia A. DC.	I				
R. bahiensis A. DC. ⟵ *R. balansae* I	I				
R. beddomei Hook. f.	I				
R. boliviana Mgf.	I				
R. caffra Sond. (= *R. natalensis* Sond., *R. welwitschii* Stapf)	I				
R. cambodiana Pierre ex Pitard	I				
R. canescens L. (= *tetraphylla*)	I				
R. chinensis Hemsl.	I				
R. confertiflora	I				
R. cubana A. DC.	I				
R. cummunsii Stapf	I				
R. decurva Hook. f.	I				
R. degeneri Sherff.	I				
R. densiflora Benth ex Hook. f.	I				
R. discolor	I		III ? *cf Index Kewensis!*		
R. fruticosa Burck.	I				
R. grandiflora Mart. ex A. DC. ~~heterophylla~~ !	I				
R. heterophylla Roem. et Schult. (= *tetraphylla*)	I				
R. hirsuta Jacq. (= *R. canescens*) ?	I				
R. indecora Woods.	I				
R. inebrians K. Schum. (= *caffra*)	I				
R. javanica Koord. et Val.	I				
R. lamarckii A. DC. (= *R. viridis* Roem. et Schult)	I				
R. ligustrina Willd. ex Roem. et Schult.	I				
R. littoradis Rusby (*R. macrocarpa* Stapf).	I				
R. longeacuminata de Wild. et Th. Dur.	I				
R. longifolia A. DC. (see *Tonduzia longifolia* (A. DC.) Mgf.)	I				
R. macrocarpa Stapf (see *R. littoralis* Rusby)	I				
R. macrophylla Stapf.	I				
R. mannii Stapf	I				
R. mattfeldiana Mgf.	I				
R. mauiensis Sherff.	I				V
R. micrantha Hook. f.	I				
R. micrantha	I				
R. mombasiana Stapf	I				
R. nana E. A. Bruce	I				
R. natalensis Sond. (see *R. caffra* Sond).	I				
R. nitida Jacq.	I				
R. obscura K. Schum.	I				
R. oreogiton	I				
R. paraensis Ducke	I				
R. pentaphylla Ducke	I				
R. perakensis King et Gamble *R. reflexa* I	I				
R. rosea K. Schum.	I				
R. salicifolia Griseb.	I				
R. sandwicensis A. DC.	I				
R. sarapiquensis Woods.	I				
R. schueli Spegazzinii	I				
R. sellowii Muell. Arg.	I				

Plant	I	II	III	IV	V
R. serpentino (L.) Benth. ex Kurz *‹seveneti̇̀*	I				
R. sprucei Muell. Arg. *‹s pathulata*	I				
R. suaveolens	I				
R. stricta	I				
R. sumatrana (Miq.) Jack	I				
R. ternifolia HBK. (= *ligustrina*)	I				
R. tetraphylla L.	I				
R. verticillata (Lour.) Baill.	I				
R. viridis (Muell. Arg.) Guillaumin.	I				
R. volkensii Stapf.	I				
R. vomitoria Afz.	I				
R. welwitschii Stapf (= *R. caffra* Sond).	I				
R. yunnanensis	I				
Rejoua aurantiaca Gaudich.			III		V
Rhazya stricta Decaisne	I	II			
R. orientalis	I				
Schizozygia caffaeoides (Boj.) Baill.		II			
Stemmadenia donnellsmithii (Rose Woods.)		II	III		V
S. galeottiana (A. Rich.) Miers.			III		
S. pubescens Benth. (*S. obovata* K. Schum.)	I	II	III		
S. tomentosa Greenman var. *palmeri*	I	II	III		
Tabernaemontana accedens	I				
T. affinis Muell. Arg. (*Peschiera affinis* (Muell. Ar. g) Miers).	I				
T. alba Mill. or Nickolson (*T. citrifolia* L.)		II	III		
T. apoda			III		
T. armeniaca		II			
T. amygdalifolia Sieber ex A. DC.	I	II			
T. australis Muell. Arg. ‹ brachyantha I			III		V
T. coronaria Willd. (*Ervatamia coronaria* Stapf.)	I		III		
T. cumminsii			III		
T. divaricata		II	III		
T. elegans	I	II			V
T. fuchsiaefolia DC	I				
T. A. DC. heyneana Wall.			III		
T. johnstonii Pichon					V
T. laevis Vell. (see *Geissospermum vellosii* Allem).	I				V
T. laurifolia Blanco			III		
T. mucronata Merrill olivacea			III		
T. oppositifolia Urb.			III		V
T. pachysiphon Stapf var. *cumminsii* H. Huber	I	II	III		
T. pandacaqui Poir.	I	II	III		
T. psychotrifolia H. B. et K.			III		V
T. sphaerocarpa			III		
T. rigida		II			
T. rupicola Benth.			III		
T. riedelii		II			
T. iboga Baill.			III		V
Tonduzia longifolia (A. DC.) Jgf. (*Rauwolfia longifolia* A. DC)	I				
Vallesia dichotoma Ruiz et Pav.	I	II			
V. glabra (Cav.) Link		II			
Vinca difformis Pouvr.	I	II			

Plant	Class				
	I	II	III	IV	V
V. elegantissima	I				
V. erecta Rgl. et Sehmalh	I				
V. herbacea Waldst et Kit. var. *libanotica* (Zucc.) Pichon.	I				
V. lancea Boj. (ex. A. DC) K. Schum. (*Lochnera lancea* K. Schum., *Catharanthus lanceus* Boj. ex A. DC.)	I	II			V
V. libanotica	I	II			
V. major L.	I	II			
V. minor L.	I	II			
V. pubescens Urv. (see *Vinca major* L)	I	II			
V. pusilla	I	II			
V. rosea (L.) Reichb. (*Catharanthus roseus* (L.) G. Don, *Lochnera rosea* Reichb.)	I	II	III		V
V. rosea var. *alba*	I				
Voacanga africana Stapf ex S. Elliot	I	II			V
V. bracteata Stapf			III		
V. chalotiana Pierre ex Stapf	I	II			V
V. dregei E. Mey	I		III		
V. globosa Merrill (*Tabernaemontana globosa*)			III		V
V. grandifolia					V
V. megacarpa Merrill					V
V. papuana (F. Muell.) K. Schum.			III		V
V. schweinfurthii Stapf			III		V
V. thousarsii Roem. et Schult. var. *obtusa* Pichon		II	III	IV	V
Annonaceae					
Menodora teniufolia				IV	
Araceae					
Symplocarpus foetidus Nutt.				IV	
Asclepiadaceae					
Cryptolepis sanguinolenta (Lindl.) Schlechter				IV	
C. triangularis N. E. Br.				IV	
Aizoaceae					
Delosperma sp.				IV	
Bignoniaceae					
Newbouldia laevis Benth. et Hook. f.				IV	
Bromeliaceae					
Ananas sativus Schult.				IV	
Calycanthaceae					
Calycanthus floridus L.				IV	
C. glaucus Wild				IV	
C. occidentalis Hook. et Arn.				IV	
Chimonanthus fragrans Lindle (= *Moretia praecox* Rehd. et Wils)				IV	
Meratia praecox Rehd. et. Wils. (see *Chimonanthus fragrans* Lindle)				IV	

Plant	I	II	Class III	IV	V
Caricaceae					
Carica papaya L				IV	
Chenopodiaceae					
Arthrophytum leptocladum M.Pop.				IV	
A. wackchanica				IV	
Girgensohnia diptera Bunge				IV	
Hammada leptoclada (Popov) ll jin. (=*Anthrophytum leptocladum*)				IV	
Combretaceae					
Aristotelia chilensis	I				
A. serrata	I				
A. peduncularis	I				
Convolvulaceae					
Argyeia barnesii				IV	
A. bizoninsin				IV	
A. capitata				IV	
A. cureata				IV	
A. maingayi				IV	
A. mollis				IV	
A. nervosa				IV	
A. obtusifolia				IV	
A. philippinensis				IV	
A. reticulata				IV	
A. ridleyi				IV	
A. rubicunda				IV	
A. splendens				IV	
Ipomoea hildebrandtii				IV	
I. muelleri				IV	
I. quamoclit				IV	
I. uniflora				IV	
I. violacea				IV	
Rivea corymbosa				IV	
Coprinaceae					
Coprinus micaceus Bull				IV	
Panaeolus acuminatus (Schff. ex Fr.) Quelet				IV	
P. companulatus (Fr.) Quelet				IV	
P. foenesecii Pers.(=*Panaeolina foenesessi* (Pers.) R. Mre.)				IV	
P. fontinalis				IV	
P. gracilis				IV	
P. semiovatus Fr.(=*Anellaria semiovata* (Sow.) Pears. et Denn.)				IV	
P. solidipes				IV	
P. sphinctrinus				IV	
P. subalteatus Berk. et Br.				IV	
P. texensis				IV	
Cucurbitaceae					
Citrullus vulgaris				IV	

Plant	Class				
	I	II	III	IV	V
Cyperaceae					
Carex brevicollis DC.				IV	
Dilleniaceae (Polygonaceae)					
Calligonium alatum				IV	
C. caput-medusae Schrenk				IV	
C. eripodum Bunge				IV	
C. macrocarpum Borszez				IV	
C. minimum Lipski				IV	
Elaeagnaceae					
Elaeagnus angustifolia L.				IV	
E. commutata				IV	
E. densiflorus				IV	
E. hortensis Bieb. (= *E. angustifolia* L.)				IV	
E. orientalis L. (= *E. angustifolia* L.)				IV	
E. spinosa L. (= *E. angustifolia* L.)				IV	
E. umbellata				IV	
Hippophae rhamnoides				IV	
Shepherelia argintia				IV	
Ericaceae					
Vaccinium oxycoccus	I				
Euphorbiaceae					
Alchornea floribunda Muell. Arg.	I				
Aleuritis fordii				IV	
Hippomane mancinella L.				IV	
Gramineae					
Arundinella hirta L.				IV	
Arundo donax L.				IV	
Avena sativa L.				IV	
Hordeum vulgare L				IV	
H. jubatum L.				IV	
H. nodosum L.				IV	
Imperata cylindrica				IV	
Oryza sativa L.				IV	
Phalaris arundinaceae L.				IV	
P. tuberosa				IV	
Phallostachys reticulata				IV	
Triticum vulgare				IV	
Zea mays L.				IV	
Zizania caudiflora				IV	
Juglandaceae					
Juglans regia L.				IV	
Lauraceae					
Nectandra megapotamica				IV	
Persia gratissima Gaertn.				IV	
Persea americana				IV	

Plant	Class				
	I	II	III	IV	V

Lecythidaceae
Couroupita gueanensis   IV

Leguminosae
A. precatorius L.   IV
Acacia acuminata Benth.   IV
A. cardiophylla A. Cunn. ex Benth.   IV
A. confusa Merrill   IV
A. cultiformis A. Cunn. ex G. Don   IV
A. floribunda Willd. (= *longifolia* Willd.)   IV
A. maidenii F. Muell   IV
A. podalyriaefolia A. Cunn.   IV
A. pruinosa A. Cunn. ex Benth.   IV
A. vestita Ker-Gawl.   IV
Allizzia julibrissin Durazz   IV
Aotus subglauca   IV
Burkea africana Hook.   IV
Desmodium gangeticum   IV
D. gyrans   IV
D. pulchellum Benth. ex Baker   IV
D. tiliaefolium G. Don   IV
D. triflorum D.C.   IV
Dicorynia guianensis Amsh.   IV
Dioclea bicolor Benth   IV
D. lasiocarpa Benth   IV
D. macrocarpa Huber   IV
D. reflexa Hook.f.   IV
D. violacea Mart. ex Benth.   IV
Erythrina aborescens   IV
E. abyssinica Lam.   IV
E. acanthocarpa E. Mey   IV
E. americana (= *E. carnea* Ait)   IV
E. berteroana Urb.   IV
E. costaricensis M. Micheli   IV
E. cristagalli L.   IV
E. dominguezii Hassler   IV
E. excelsa Baker   IV
E. falcata Benth   IV
E. flabelliformis Kearn.   IV
E. folkersii Krukoff et Moldenke   IV
E. fusca Lour   IV
E. glauca Willd.   IV
E. grisebachii Urb.   IV
E. herbacea L.   IV
E. hypaphorus Boerl. ex Koord.   IV
E. macrophylla DC   IV
E. orophila Ghesq.   IV
E. pallida Britton et Rose   IV
E. poeppigiana O. F. Cook   IV
E. rubrinerva H. B. et K.   IV
E. sandwicensis Dogner   IV
E. senegalensis DC   IV

Plant	I	II	Class III	IV	V
E. subumbrans Merrill (Hypaphorus subumbrans Hassk.)				IV	
E. tholloniana Hua				IV	
E. variegata L. var. orientalis (= indica Lam)				IV	
E. velutina Willd.				IV	
Griffonia simplicifolia				IV	
Lens esculenta (Moench) Meth.				IV	
Lesspedeza bicolour Turez var. japonica Nakai				IV	
Lupinus albus L.				IV	
L. angustifolius L.				IV	
L. hartwegii Hindl.				IV	
L. hispanicus				IV	
L. polyphyllus Lindl				IV	
L. luteus L.				IV	
Mimosa hostilis Benth.				IV	
M. verrucosa				IV	
Mucuna pruriens DC.				IV	
Petalostylis labicheoides				IV	
Phaseolus vulgaris				IV	
Physostigma cylindrospermum Holmes				IV	
P. venenosum Balf				IV	
Piptadenia colubrina Benth.				IV	
P. excelca Lillo				IV	
P. falcata Benth.				IV	
P. pegrina Benth.				IV	
Pisum sativum L.				IV	
Prosopsis juliflora DC				IV	
P. alba Gris.				IV	
P. nigra				IV	
Samaena saman Merr.				IV	
Icacinaceae					
Cassinopsis ilicifolia					V
Loganiaceae					
Calebassen–Curare	I				
Gardneria multiflora	I				
G. nutans	I				
Gelsmium elegans (Gardn.) Benth.	I				
G. sempervirens Ait.	I				
Mostuea buchholzii Engl.	I				
M. stimulans A. Chev.	I				
Strychnos aculeata	I				
S. angustiflora	I			IV	
S. amazonica Kruk.	I				
S. brachiata	I				
S. brasiliensis	I				
S. camptoneura					V
S. chlorantha Prog.	I				
S. cinnamomifolia Thw.	I				
S. colubrina L.	I				
S. dale	I				
S. diaboli Sandwith	I				

Plant	Class				
	I	II	III	IV	V
S. dinklagei	I				V
S. divaricans Ducke	I				V
S. elaeocarpa	I				V
S. froesii Ducke	I				V
S. gaultheriana Pierre ex C. B. Clarke (= S. malaccensis Benth.)	I				
S. gossweileri	I				
S. henningsii Gilg.	I				
S. holsii. Gilg ex Engl. var. reticulata f. condensata	I				
S. icaja Baill.	I				
S. ignatii Berg.	I				
S. jobertiana	I				
S. kipapa Gilg.	I				
S. KL, 1929	I				
S. lanceolaris Miq.	I				
S. ligustrina Blume	I				
S. lucida R. Br.	I				
S. macrophylla Barb. Rodr.	I				
S. malaccensis Benth. (see S. gaultheriana Pierre ex C. B. Clarke).	I				
S. madeda	I				
S. malacoclados	I				V
S. melinoniana Baill.	I			IV	
S. mitscherlichii R. Schomb. (S. smilacina Benth.)	I				V
S. nux-vomica	I				
S. paramensis	I				
S. potatorum	I				
S. psilosperma F. Muell.	I				
S. quaqua Gilg.	I				
S. rheedei C. H. Clarke	I				
S. romeu belenii	I				
S. rubiginosa A. DC.	I				
S. smilacina Benth. (see S. mitscherlichii R. Schomb.)	I				
S. sobrensis	I				V
S. solimoesana Kruk.	I				V
S. splendens Gilg.	I				
S. subcordata Spruce	I				
S. tabascana	I				
S. tieute Lesch.	I				
S. tomentosa Benth.	I				V
S. toxifera R. Schomb.	I				V
S. toinervis (Vell.) Mart.	I				V
S. usambarensis				IV	
S. variablis	I				
S. wallichiana	I				
Malphighiaceae					
Banisteria caapi Spruce				IV	
Banisteriopsis argentea Spring ex Juss.				IV	
B. inebrians Morton Cabi paraensis Ducke				IV	
B. rusbyana				IV	

Plant	Class				
	I	II	III	IV	V
Malvaceae					
Abelmoschus esculentus (Moench) Menth				IV	
Banisteriopsis caapi				IV	
Gossypium hirsutum L. (= *herbaceum* L.)				IV	
Meliaceae					
Hedranthera barteri	I				V
Musaceae					
Musa paradisiaca L. (see *M. sapientium* L.)				IV	
M. sapientium L.				IV	
Myristicaceae					
Virola sp.				IV	
Gymnacranthera panticulata (A. D. C) Warb				IV	
Naucleaceae					
Cephalanthus occidentalis	I				
Mitragyna pavrifolia				IV	
Neonauclea schlechteri	I				
Ochnaceae					
Testulae gabonensis				IV	
Passifloraceae					
Passiflora actinea Hook.				IV	
P. alata Ait.				IV	
P. alba Link et Otto				IV	
P. bryonioides H.B.K.				IV	
P. capsularis L.				IV	
P. coerulae				IV	
P. decaisneana				IV	
P. edulis Sims.				IV	
P. eichleriana Mast.				IV	
P. foetida L.				IV	
P. incarnata L.				IV	
P. quadrangularis L.				IV	
P. ruberosa				IV	
P. subpeltata				IV	
P. warmingii				IV	
Polygalaceae					
Polygala tenuifolia Willd.				IV	
Ranunculaceae					
Ranunculus sceleratus L.				IV	
Rosaceae					
Prunus demostica L.				IV	
Rhamnaceae					
Zizyphus amphibia				IV	

Plant			Class		
	I	II	III	IV	V
Rubiaceae					
Adina cordifolia Hook.	I				
A. rubescens	I				
A. rubrostipulata K. Schum. (see *Mitragyna* Havil.)	I				
Anthocephalus cadamba	I				
Antirrhea putaminosa (F. v. Muell.) Baill.	I				
Cinchona calisaya Wedd.	I				
C. caloptera Miq.	I				
C. carabayensis Wedd.	I				
C. condaminea Humb. et Bonpl. (= *C. officinalis* L.)	I				
C. cordifolia Mutis (*C. pubescens* Vehl.)	I				
C. corymbosa Karst. (*C. pitayensis* Wedd.)	I				
C. erythrantha Pav	I				
C. erythroderma Wedd.	I				
C. hasskarliana Miq.	I				
C. lanceolata Ruiz et Pav. (= *C. officinalis* L.)	I				
C. lancifolia Mutis	I				
C. ledgeriana Moens	I				
C. lucumaefolia Pav. (= *C. macrocalyx* Pav.)	I				
C. micrantha Ruiz et Pav.	I				
C. nitida Ruiz et. Pav.	I				
C. oblongifolia Mutis (= *Cascarilla oblongifolia* Wedd).	I				
C. officinalis L.	I				
C. ovata Ruiz et Pav.	I				
C. pahundiana Howard	I				
C. pelletieriana Wedd. (= *C. pubescens* Vahl.)	I				
C. pitayensis Wedd. (see *Cinchona corymbosa* Karst.)	I				
C. pubescens Vahl. (see *C. pelletieriana* Wedd.)	I				
C. robusta Howard	I				
C. rosulenta Howard	I				
C. scrobiculata Humb. et Bonpl.	I				
C. succirubra Pav.	I				
C. tucujensis Karst.	I				
Corynanthe macroceras (K. Schum. Pierre (*pausinystalia macroceras* Pierre ex Beille)	I				
C. paniculata Welv.	I				
C. yohimbe K. Schum. (= *Pausinystalia yohimba* (K. Schum.) Pierre).	I				
Coutarea latiflora Sesse et Moe, ex DC	I				
Guettarda eximia	I				
Hodgkinsonia frutescens F. Muell.				IV	
Leptactina densiflora Hook. f.				IV	
Mitragyna africana Korth	I				
M. ciliata aubrev et Pellegr	I				
M. hirsuta Havil.	I				
M. inermis O. Kuntze	I				
M. javanica (Koord.) Korth. (= *Stephagine parvifolia*).	I			IV	
M. macrophylla Heirn (*M. stipulosa* O. Kuntze)	I				
M. parvifolia Korth	I				

Plant	Class				
	I	II	III	IV	V
M. rotundifolia (Roxb.) O. Kuntze (*M. diversifolia* Hook. f.)	I				
M. rubrostipulacea Havil. (= *Adina rubrostipulata* K. Schum.)	I				
M. speciosa Korth	I				
M. stipulosa O. Kuntze (see *M. macrophylla* Hiern).	I				
Nauclea latifolia	I			IV	
Oldenlandia affinis D.C.				IV	
Ourouparia africana Baill.	I				
O. formosana Matsumura et Hayata (*Nauclea formosana* Matsumura).	I				
Nauclea diderrichii	I				
N. latifolia				IV	
N. parva				IV	
Pauridiantha lyalli				IV	
Psychotria poychotriaefolia (Seem) Stand.				IV	
P. viridis R&P				IV	
P. alpina	I				
Rutaceae					
Citrus aurantium L.				IV	
C. sinensis Osbeck				IV	
Clausena anisata				IV	
C. heptaphylla				IV	
Dictyoloma incanescens D.C				IV	
Vepris ampody H. Perr.				IV	
Euxylophora paraensis				IV	
Evodia alata F. Muell.				IV	
E. rutaecarpa Hook. f. et Thoms.				IV	
Hortia arborea Engl.				IV	
H. braziliana Vel.				IV	
Murraya exotica				IV	
M. koenigii				IV	
Pentaceras australis Hook. f.				IV	
Zanthoxylum (*Xanthoxylum*) *budrunga* Wall. (see *Z. rhetsa* A. DC)				IV	
Z. flavum				IV	
Z. dinklagei				IV	
Z. oxyphyllum Edgew.				IV	
Z. rhetsa A. DC. (*Z. budrunga* Wall).				IV	
Z. suberosum C. T. White				IV	
Sapotaceae					
Poutaria sp.		I			
Simarubaceae					
Ailanthus altissima				IV	
Dictyolama incanescens DC. (= *Dictyolama vandellianum* A. Juss.)				IV	
Picrasma ailanthoides Sieb et Zucc.				IV	
P. crenata (Vill.) Engl.				IV	
P. excelsa				IV	

Plant			Class		
	I	II	III	IV	V
Pausinystalia trillesii Beille	I				
Pavetta panceopata				IV	
Pseudocinchona africana A. Chev.	I				
P. mayumbensis R. Hamet	I				
Pogonopus tubulosus (DC) Schum.					V
Remijia pedunculata Flueck.	I				
R. purdieana Wedd.	I				
Riccardea sinuata				IV	
Samadera indica				IV	
Sickingia rubra K. Schum. (*Arariba rubra* Mart).				IV	
Uncaria attenuata	I				
U. bernaysii F. Muell.	I				
U. buluanensis	I				
U. callophylla	I				
U. canescens	I				
U. elliptica	I				
U. ferrea DC	I				
U. gambier Roxb. (= *Orouparia gambir* Baill.)	I				
U. guaianensis	I				
U. kawakamii Hayata	I				
U. longiflora	I				
U. pteropoda Miq.	I				
U. rhynchophylla miq. (= *Orouparia rhynchophylla* Matsumura)	I				
U. tomentosa DC. (= *Orouparia guianesis* Aubl).	I				
U. macrophylla	I				
U. orientalis	I				
U. perottetlii	I				
Picrasma javanica				IV	
Solanaceae					
Lycopersicon esculentum Mill.				IV	
Nicotiana tabacum L.				IV	
Solanum melongena L.				IV	
S. nigrum L.				IV	
Strophariaceae					
Psilocybe atrobrunnea				IV	
P. aztecorum Heim				IV	
P. baeocystis Singer et Smith				IV	
P. coerulescens Murr. var. *mazatecorum* Heim.				IV	
P. caevulipes				IV	
P. cyanescens				IV	
P. mexicana Heim				IV	
P. semperviva Heim et Cailleux				IV	
P. stricticeps				IV	
P. zapotecorum Heim				IV	
Stropharia cubensis Earle				IV	
Strychnaceae					
Gardneria nutans ← Loganiaceae!	I				

Plant			Class		
	I	II	III	IV	V
Thymelaceae					
Nectandra megapotamica				IV	
Tovariaceae					
Tovaria pendula Ruiz & Pavon				IV	
Urticaceae					
Girardinia heterophylla (Decne)				IV	
Laportea moroides				IV	
Urtica dioica L.				IV	
U. pilulifera L.				IV	
U. parviflora Roxb.				IV	
Verbenaceae					
Clodendron trichotomum					V
Zygophyllaceae					
Nitraria schoberi				IV	
Peganum harmala L.				IV	
Zygophyllum fabago L.				IV	
Z. elephantiasis				IV	

Micro-organism			Class		
	I	II	III	IV	V
Absidia ramosa				IV	
Aspergillus amstelodami				IV	
A. caespitosus				IV	
A. chevalieri				IV	
A. clavatus				IV	
A. echinulatus				IV	
A. fumigatus				IV	
A. glaucas				IV	
A. niger				IV	
A. ruber				IV	
A. ustus				IV	
Boletus edulis				IV	
Chaetomium cochlioides				IV	
C. globusum				IV	
Chromobacterium violaceum				IV	
Claviceps anisata				IV	
C. fusiformis				IV	
C. purpurea				IV	
C. paspali				IV	
C. penniseti				IV	
Convolvulus major				IV	
Escherichia coli				IV	
Elymus sp.				IV	
Gliocladium fimbriatum				IV	
Pellicularia filamentosa				IV	

Micro-organism			Class		
	I	II	III	IV	V
Penicillium aurantio-virens				IV	
P. cinarescens				IV	
P. concavoregulosum				IV	
P. cyclopium				IV	
P. islandicum				IV	
P. italicum				IV	
P. jenseni				IV	
P. notatum				IV	
P. ochraceum				IV	
P. oxalicum				IV	
P. paxilli				IV	
P. regulosum				IV	
P. roquefortii				IV	
P. terlikowski				IV	
P. verrucolosum				IV	
Psilocybe aztecorum				IV	
P. caerulescens				IV	
P. mexicana				IV	
P. sempervirens				IV	
P. zapoticorum				IV	
Rhizopus nigricans				IV	
R. suinus				IV	
Schizophyllum commune				IV	
Stamptomyces griseus				IV	
Stropharia cubensis				IV	
Thermoactinomycetes sp.				IV	
Trichoderma viride				IV	

INDEX